Soil Biology

Volume 14

Series Editor
Ajit Varma, Amity Institute of Microbial Sciences,
Noida, UP, India

Volumes published in the series

Applied Bioremediation and Phytoremediation (Vol. 1)
A. Singh, O.P. Ward (Eds.)

Biodegradation and Bioremediation (Vol. 2)
A. Singh, O.P. Ward (Eds.)

Microorganisms in Soils: Roles in Genesis and Functions (Vol. 3)
F. Buscot, A. Varma (Eds.)

In Vitro Culture of Mycorrhizas (Vol. 4)
S. Declerck, D.-G. Strullu, J.A. Fortin (Eds.)

Manual for Soil Analysis – Monitoring and Assessing Soil Bioremediation (Vol. 5)
R. Margesin, F. Schinner (Eds.)

Intestinal Microorganisms of Termites and Other Invertebrates (Vol. 6)
H. König, A. Varma (Eds.)

Microbial Activity in the Rhizosphere (Vol. 7)
K.G. Mukerji, C. Manoharachary, J. Singh (Eds.)

Nucleic Acids and Proteins in Soil (Vol. 8)
P. Nannipieri, K. Smalla (Eds.)

Microbial Root Endophytes (Vol. 9)
B.J.E. Schulz, C.J.C. Boyle, T.N. Sieber (Eds.)

Nutrient Cycling in Terrestrial Ecosystems (Vol. 10)
P. Marschner, Z. Rengel (Eds.)

Advanced Techniques in Soil Microbiology (Vol. 11)
A. Varma, R. Oelmüller (Eds.)

Microbial Siderophores (Vol. 12)
A. Varma, S. Chincholkar (Eds.)

Microbiology of Extreme Soils (Vol. 13)
P. Dion, C.S. Nautiyal (Eds.)

Petr Karlovsky
Editor

Secondary Metabolites in Soil Ecology

Springer

Professor Dr. Petr Karlovsky
University of Göttingen
Molecular Phytopathology and Mycotoxin Research
Grisebachstraße 6
37077 Göttingen
Germany
E-mail: pkarlov@gwdg.de

ISBN 978-3-540-74542-6 e-ISBN 978-3-540-74543-3

Soil Biology ISSN: 1613–3382

Library of Congress Control Number: 2007934831

Cover design: WMXDesign GmbH, Heidelberg, Germany

Printed on acid-free paper 5 4 3 2 1 0

springer.com

Preface

We have know since decades that the structural variation, richness and inventiveness of natural product chemistry revealed by secondary metabolites by far exceed those of primary metabolites, but we are just beginning to understand their relevance for the functioning of ecosystems. The goal of this volume is to corroborate the role of secondary metabolites in organismic interactions in soil and inspire ecologists to pay more attention to chemical phenomena beyond the concept of food webs.

Two chapters describe new approaches relevant for the analysis of secondary metabolites in soil: Linda Thomashow describes how antibiotics can be detected in situ and Sanjay Swarup's team gives an account of the application of metabolomics techniques to rhizosphere research. Most of the chapters, however, focus on functional aspects, limiting the descriptive approach to a necessary minimum. In an attempt to achieve a balanced coverage, we devoted two to three chapters to bacterial, fungal and plant metabolites each. Topics covered bacterial metabolites involve quorum sensing (Venturi's group), the application of fluorescent Pseudomonas in biological control of plant pathogens (Velusamy Palaniyandi and S.S. Gnanamanickam) and chemical interactions between Streptomyces, fungi and plants (Mika Tarkka and Rüdiger Hampp). The role of secondary metabolites in biological control is also the subject of the first of chapter on fungal metabolites (Mathivanan's group), while the remaining two chapters summarize knowledge and speculations pertinent to the role of truffle metabolites in "burnt" phenomenon (Richard Splivallo) and put mycotoxins in soil into an ecological perspective (Susanne Elmholt). Two chapters are devoted to plant metabolites: Franz Hadacek' review deal with the biological effects of constitutive plant metabolites and Jorge Vivanco's group remind us that root exudates, which is the group of plant metabolites largely neglected by phytochemists, play a crucial role in controlling the constitution or rhizosphere microflora. The final two chapters summarize the effects of volatiles on soil invertebrates (Ron Wheatley) and focus on selected model compounds in detail, revealing the importance of natural conditions for toxicity testing (Neal Sorokin and Jeanette Whitaker). I am indebted to thank all authors for making this volume a comprehensive source on the wide range of biological activities and functions exerted by secondary metabolites in soil. Complete treatment of the topic obviously cannot be achieved in a single volume, but

I feel that the chapters well convey the major message and complement chemically oriented, descriptive treatments that have been available on the topic so far.

I am grateful to Prof. Ajit Varma for inviting me to assemble the volume, many instructive talks and emails and his diligent editorial support. I would also like to express my thanks to Springer with its Editorial Director Dr. Dieter Czeschlik and Dr. Jutta Lindenborn, who is responsible for "Soil Biology" series, for making this endeavor possible. I am particularly thankful to Dr. Lindenborn for her support and patience.

Current research on secondary metabolite advances from a descriptive to functional approach, necessitating a conceptual shift from chemistry to biology. We need to join the expertise and technologies of all relevant disciplines in focusing on the biological role of secondary metabolite synthesis. This volume proves that this process has already started.

Göttingen
July 2007

Petr Karlovsky

Contents

Contributors

Harsh P. Bais
Department of Plant and Soil Science, Delaware Biotechnology Institute, University of Delaware, Newark, DE 19711, USA

V.S. Bhinu
Agriculture and Agri-Food Canada, 107 Science Place, Saskatoon, Saskatchewan, Canada S7N0X2

Robert F. Bonsall
Department of Plant Pathology, Washington State University, Pullman, WA, USA

Corey D. Broeckling
Department of Horticulture and Landscape Architecture, Colorado State University, Fort Collins, CO 80523-1173, USA and Center for Rhizosphere Biology, Colorado State University, Fort Collins, CO 80523-1173, USA

Susanne Elmholt
Department of Agroecology, Danish Institute of Agricultural Sciences, Research Centre Foulum, DK-8830 Tjele, Denmark, susanne.elmholt@agrsci.dk

Sara Ferluga
Bacteriology Group, International Centre for Genetic Engineering & Biotechnology Area, Padriciano 99, 34012 Trieste, Italy

S.S. Gnanamanickam
Centre for Advanced Studies in Botany, University of Madras, Guindy Campus, Chennai 600 025, India

Franz Hadacek
Department of Chemical Ecology and Ecosystem Research, Faculty of Life Sciences, University of Vienna, Althanstrasse 14, A-1090 Vienna, Austria, franz.hadacek@univie.ac.at

Rüdiger Hampp
Botanical Institute, Physiological Ecology of Plants, University of Tübingen, Auf der Morgenstelle 1, D-72076 Tübingen, Germany, ruediger.hampp@uni-tuebingen.de

Petr Karlovsky
Molecular Phytopathology and Mycotoxin Research Unit, University of Goettingen, Germany, pkarlov@gwdg.de

N. Mathivanan
Biocontrol and Microbial Metabolites Lab, Centre for Advanced Studies in Botany, University of Madras, Guindy Campus, Chennai 600 025, India

V.R. Prabavathy
Biocontrol and Microbial Metabolites Lab, Centre for Advanced Studies in Botany, University of Madras, Guindy Campus, Chennai 600 025, India, prabhamathi@yahoo.com.

Sheela Reuben
Small Molecule Biology Laboratory, Department of Biological Sciences, National University of Singapore, Singapore 117 543

Neal Sorokin
Reckitt Benckiser Healthcare (UK) Ltd., Dansom Lane Hull, HU8 7D9, neal.sorokin@Reckittbenckiser.com

Richard Splivallo
Department of Plant Biology, University of Torino, IPP-CNR, 25 Viale Mattioli, 10125 Torino, Italy

Laura Steindler
Bacteriology Group, International Centre for Genetic Engineering & Biotechnology Area, Padriciano 99, 34012 Trieste, Italy

Sanjay Swarup
Small Molecule Biology Laboratory, Department of Biological Sciences, National University of Singapore, Singapore 117 543, dbsss@nus.edu.sg

Mika Tarkka
UFZ, Helmholtz-Centre for Environmental Research, Department of Soil Ecology, Theodor-Lieser-Strasse 4, D-06120 Halle, Germany

Linda S. Thomashow
USDA-ARS, Root Disease and Biological Control Research Unit, Washington State University, Pullman, WA, USA, thomashow@wsu.edu

P. Velusamy
Centre for Advanced Studies in Botany, University of Madras, Guindy Campus, Chennai 600 025, India, velusamy22973@yahoo.com

Vittorio Venturi
Bacteriology Group, International Centre for Genetic Engineering & Biotechnology Area, Padriciano 99, 34012 Trieste, Italy and Plant Bacteriology Group, International Centre for Genetic Engineering & Biotechnology, Biosafety Outstation, Via Piovega 23, 31050 Ca' Tron di Roncade, Treviso, Italy, venturi@icgeb.org

V.R. Vijayanandraj
Biocontrol and Microbial Metabolites Lab, Centre for Advanced Studies in Botany, University of Madras, Guindy Campus, Chennai 600 025, India

Jorge M. Vivanco
Department of Horticulture and Landscape Architecture, Colorado State University, Fort Collins, CO 80523-1173, USA and Center for Rhizosphere Biology, Colorado State University, Fort Collins, CO 80523-1173, USA, j.vivanco@colostate.edu

David M. Weller
USDA-ARS, Root Disease and Biological Control Research Unit, Washington State University, Pullman, WA, USA, wellerd@wsu.edu

Ron E. Wheatley
Environment–Plant Interactions Programme, Scottish Crop Research Institute, Invergowrie, Dundee, DD2 5DA, UK, ron.wheatley@scri.ac.uk

Jeanette Whitaker
Centre for Ecology & Hydrology, Lancaster Environment Centre, Library Avenue, Bailrigg, Lancaster LA1 4AP, UK, jhart@ceh.ac.uk

Chapter 1
Secondary Metabolites in Soil Ecology

Petr Karlovsky

1.1 Introduction: Chemical Interactions in Soil

Interactions among organisms are central to understanding any ecosystem, perhaps with the exception of a short period when a newly created niche is colonized by its first inhabitants. Soil environment is not an exception, but biotic interactions dominating soil biology differ from those in other systems because of the dominating role of sessile organisms and the lack of autotrophy in soil (chemolithoautotrophs being an interesting but not significant exception). When chemical processes in soil are discussed, the traditional concept of food webs comes first to mind as a framework for the exchange of organic substances and flow of energy. Feeding, predation, degradation of macromolecular substrates and absorption of nutrients have dominated thinking about biogenic chemical processes in soil. The food web approach proved extremely fruitful in generating hypotheses and inspiring experimental approaches concerning the bulk transformation of organic matter, but it did not address phenomena related to chemical interactions which are more specific both on the chemical and on the taxonomical level and which cannot be adequately described in terms of energy flow and biomass transformation. These interactions involve compounds named secondary metabolites, which are not strictly needed for the survival and reproduction of their producers. Secondary metabolites are structurally highly diverse and each of them is produced only by a small number of species. They exert various biological effects, often at very low concentrations, and can be regarded as carriers of chemical communication among soil inhabitants.

The high complexity and heterogeneity of soil makes this matrix recalcitrant to chemical analysis. Methods for the determination of pesticides, polychlorinated biphenyls and other xenobiotics in soil have existed for a long time to monitor pollution of the environment, but it is only recently that dedicated analytical methods for natural metabolites in soil have been available (Mortensen et al. 2003). Apart

Petr Karlovsky
Molecular Phytopathology and Mycotoxin Research Unit, University of Goettingen, Germany
e-mail: pkarlov@gwdg.de

P. Karlovsky (ed.), *Secondary Metabolites in Soil Ecology. Soil Biology 14.*

from the complexity of soil matrix, analytical methods for secondary metabolites in the soil have to cope with the enormous diversity of the analyte itself. The resolution of current metabolomics approaches is far from adequate even for the metabolome of a single organism, let alone for systems orders of magnitude more complex. Adsorption phenomena, large differences in concentrations among metabolites and their heterogeneous distribution further complicate profiling of secondary metabolites in soil by current metabolomics techniques. We may need to focus on dominant metabolites and major effects first, gradually zooming into the system as the progress of analytical techniques allows us.

Most secondary metabolites produced by soil microbes appear to be secreted, an observation which corroborates their role if controlling biotic interactions. The research field addressing the role of secreted metabolites in an ecosystem is ecological chemistry. Concerning soil microorganisms, the antibiotics paradigm has dominated experimental approaches to the ecological role of secondary metabolites so far, followed by pathogenic interactions between microorganisms and plants. Other roles of secondary metabolites, such as facilitating symbiosis with insects, plants and higher animals, are documented but have rarely been addressed (Demain and Fang 2000; Sect. 1.6.3). For instance, it has been known since ancient times that fungal products may poison animals, but the idea that microbes produce toxins to protect their substrates from ingestion by animals did not surface until Janzen's pioneering paper was published in *The American Naturalist* (Janzen 1977). Even then, attempts to test this hypothesis experimentally have rarely been reported. The role of secondary metabolites in interactions among soil microorganisms or between a microorganism and a plant might appear to be easier to address, but rigorous testing of a working hypothesis in this area is tricky (see Sects. 1.6.2, 1.6.3, 1.7). Without the capability of manipulating secondary metabolite synthesis or their targets genetically, conclusive results are difficult if not impossible to obtain.

1.2 Should the Term "Secondary Metabolites" Be Abandoned?

More than a century ago, Kossel (1891) defined secondary metabolites by exclusion (compounds that do not belong to primary metabolites), provoking criticism which has never ceased. The current, generally accepted concept in line with Kossel's view is that primary metabolites are chemical components of living organisms that are vital for their normal functioning, while secondary metabolites are compounds which are dispensable. A distinguishing feature of secondary metabolites is that their production is limited to a group of species or genera and is rarely conserved over a wide taxonomical range, while primary metabolism is conserved among phyla and across kingdoms.

The specificity of secondary metabolism encouraged botanists and mycologists to use secondary metabolite production as a taxonomical characteristic in plants (Smith 1976) and fungi (Frisvad et al. 1998). Chemotaxonomy harbors risks, because on the one hand a single-point mutation might block a whole biosynthetic pathway, and on the other hand there are indications that some gene clusters involved in secondary metabolite biosynthesis have been transmitted among species by a horizontal gene transfer. The use of chemotaxonomy for elucidating phylogenetic relationships was therefore limited, and it became obsolete with ready access to DNA sequences. However, chemotaxonomy has not lost its appeal as a rapid and inexpensive support for taxonomical classification of microbial isolates.

Many scientists studying secondary metabolites dislike the term, because it appears to imply an auxiliary importance of secondary metabolites compared with the importance of primary metabolites. Numerous attempts to replace "secondary metabolites" by other labels were undertaken without gaining wide acceptance. Several initiatives emphasized the biological role of these compounds. For example, the designation "ecological metabolites" stresses the role of secreted metabolites in interactions of their producers with other organisms. Similarly, Frisvad's creation of "extrolites" (an outwardly producers directed chemically differentiated product of a living organism) is based on the notion that the function of many secondary metabolites is to control or modulate interactions with the environment. In the meantime the author has been using his term as a synonym for all secondary metabolites (Frisvad et al. 2004). The problem is that not all secondary metabolites fit his definition of extrolites, and for the majority of secondary metabolites we do not know whether they are "outwardly directed" or not. I suppose this is the reason why the term "extrolites" has not been embraced by the scientific community. In their recent review of fungal metabolomics, Frisvad's colleagues abandoned the term "extrolites" completely, consistently using "secondary metabolites" (Smedsgaard and Nielsen 2005). Substitutes like the term "extrolites" will unlikely replace the established term "secondary metabolites" because their definitions do not cover the full range of natural products known as secondary metabolites, and because their applicability relies on information which is seldom available. Let us look at a couple of examples. Leaf-movement factors in nyctinastic plants are clearly secondary metabolites, but one would not call them ecological metabolites or extrolites. Sometimes both secondary and primary metabolites serve the same purpose, defeating any classification based on function. For instance, pyochelin is secreted by *Pseudomonas* sp. and citric acid is secreted by plant roots, both facilitating the uptake of mineral nutrients by their producers. A functional classification would blur the distinction between secondary metabolites (pyochelin as a nonribosomal peptide) and primary metabolites (citric acid as a member of the Krebs cycle), the preservation of which is desirable. From a practical point of view, the main problem with functional classifications is that for most newly described natural products we do not know anything beyond their structure and taxonomical affiliation of the producer, the latter information often being limited to a genus.

The traditional distinction between primary and secondary metabolism is straightforward and knowledge of the structure is usually sufficient for the assignment of a compound to primary or secondary metabolism. As useful as some of the suggested substitutes are in emphasizing functional aspects, terms like "extrolites," "special metabolites" (Gottlieb 1990), "idiolites" (Demain 1986), "ecological metabolites" (Sirenko et al. 1979) and so on will unlikely replace the term "secondary metabolites." We do not need to search for a substitute as long as we do not associate "secondary" with unimportant or uninteresting.

1.3 Overcoming the Phytochemist's Approach to Secondary Metabolites

Secondary metabolites are the study object of natural product chemistry. The amazing structural variability of these compounds has attracted the curiosity of chemists and the biological activities possessed by natural products have inspired the pharmaceutical industry to search for lead structures in microbial cultures and plant extracts. This strategy proved highly successful: until the advent of molecular genetics, natural product chemistry was the main source of innovation in drug development. An impressive number of compounds have been purified and their structures elucidated in the past four decades. Neither computer-aided drug design nor combinatorial chemistry has surpassed nature as a source of structural variability.

Paradoxically, the success of natural product chemistry in applied research and product development steered the field towards a dead end in basic research. While commercial interests generated pressure to purify and run though bioassays more and more compounds each year, little effort has been devoted to questions of primary scientific interest—namely, for what reasons plants and microbes make them and what happens to them in nature. The vast majority of publications on secondary metabolites have been limited to structure elucidation, at best accompanied by arbitrarily selected bioassays. Any randomly selected issue of the *Journal of Natural Products* will illustrate this practice. This situation is reminiscent of old-time entomology, when scholars were collecting and meticulously describing insects but devoted little effort to the physiology, genetics, ecology or ethology of their subjects.

Apart from searching for new structures and commercially exploitable biological activities, a natural product chemistry field progressing well in the past few decades was the elucidation of biosynthetic pathways. Feeding isotopically labeled precursors proved an efficient strategy to this end even before the implementation of spectroscopic techniques, when stepwise chemical degradation and elementary analysis dominated the tedious process of structure elucidation. Labeling with heavy isotopes remained a major tool of pathway elucidation after the coupling of nuclear magnetic resonance with mass spectrometry became the workhorse of natural product chemistry because both techniques can distinguish isotopes. A practical reason for the interest in biosynthetic pathways was that feeding different precursors provides access to new derivatives with potentially improved properties. The elucidation of biosynthetic pathways by natural product chemists was limited to establishing sequences of intermediates, and it usually failed short of experimentally addressing the enzymatic reactions involved. Enzymes were a domain of biochemistry, which was well isolated from organic chemistry at traditional universities, being affiliated with the faculty of biology rather than that of chemistry. Biochemists was still busy investigating the intricacies of primary metabolism, while natural product chemists were publishing hundreds of weird and beautiful structures each year as on the assembly line.

Chemical ecology has formally existed for more than a century (Mitchell-Olds et al. 1998), but compared with the proliferation of natural product chemistry its achievements have been modest. It is difficult to understand why so few people seriously addressed the question why those fancy structures published by phytochemists each year actually existed in nature. It seems that the voluminous literature on natural products remained largely unnoticed by biologists, and those who were aware of the growing need for a scientific inquiry did not possess the expertise and tools needed. For natural product chemists, describing new structures was what describing new species was for a traditional taxonomist. Bioassays were used to assess the potential commercial value of new metabolites rather than a means of addressing their function in nature. Describing and cataloging items is a necessary first step towards understanding, but it is not more than a first step. Resources available for research are limited and it is my view that rather than following a convenient routine purify–elucidate–publish–abandon (and purify–elucidate–patent–license in rare lucky cases), natural product chemistry needs to attach more meaning to its results. For example, chemists occasionally experimented with growing conditions in order to maximize the yield or to generate new products. The inventor of the one strain, many compounds (OSMAC) concept, A. Zeeck, explicitly suggested that varying cultivation parameters could provide insights into the role of secondary metabolites in microbial communities (Bode et al. 2002), but the approach has never been used systematically to this end. Another rarely used option is to select bioassays applied to new metabolites according to the natural environment of the producer. The narrow traditional concept of natural product chemistry and its isolation from microbiology and biochemistry contributed to the discrepancy between the volume of descriptive work and the scarcity of functional approaches.

Only in the 1990s did research on secondary metabolites began to overcome its limits. On one hand, biologists installed gas chromatography and high-performance liquid chromatography systems in their laboratories and learned how to purify secondary metabolites from plant extracts and microbial cultures. On the other hand, chemists learned that apart from growing producing strains in fermenters, they can genetically manipulate biosynthetic pathways and use cell-free extracts or purified enzymes to perform biosynthetic reactions in test tubes. The transition was all but smooth because questions arising in biology traditionally caused little excitement in chemistry. Natural product chemists retiring these days remember how difficult it was at the beginning of their careers to compete for chemistry grants with projects proposals on natural products. As it took time for them to establish the same reputation as physical and synthetic chemists had, concepts like metabolomics face difficulties now to be accepted within the realm of chemistry. But the new paradigm has been set. Not only research on natural products became interdisciplinary, involving fields as diverse as molecular genetics and entomology, but the boundary between disciplines has started to dissolve as laboratory members are compelled to learn techniques adequate for their research subjects, rather than picking topics amenable to techniques which they have mastered for years. My laboratory in the Department of Crop Sciences uses mass spectrometry to elucidate biochemical transformations of secondary metabolites and my colleagues in the Institute of Botany study biosynthetic pathways. Our colleagues in the Faculty of Chemistry investigate the biophysics of biological membranes and perform transposon mutagenesis in *Actinomyces*. This development was a necessary prerequisite for natural product chemistry to overcome its descriptive tradition.

1.4 Chemical Ecology of Microorgansims Has Been Neglected

Ecological chemistry of soil is dominated by microbes. Most research activities labeled as chemical ecology worldwide have so far been concerned with interaction between insects and plants. The selection of papers published in the *Journal of Chemical Ecology* provide a good example. According to its mission statement, the journal is devoted to "promoting an ecological understanding of the origin, function, and significance of natural chemicals that mediate interactions within and between organisms," but the majority of its articles deal with insect–plant interactions. This is just another manifestation of a phenomenon known from systematic biology: the smaller the dimensions of members of a taxonomical group are, the more species the group possesses and the fewer the taxonomists that deal with it. While whole institutes are devoted to ecological studies of insect–plant interactions, only a handful of laboratories seriously investigate chemical communication among microbes in nature. Three systems with a high potential for practical applications are prominent exceptions: quorum sensing in bacteria, biological control of plant diseases, and interaction of plant pathogens with their hosts. A review of advances in ecological chemistry written by the late Jeffrey B. Harborne (1999), one of the most influential doyens of phytochemistry, nicely documents this bias. The review is divided into four sections according to interacting organisms: animal–animal, plant–animal, plant–plant and plant–microbe. A section on microbe–microbe interactions, which would arguably be concerned with chemical interactions more substantial for the survival of their participants than any of the four combinations listed above, just did not occur. Similarly Bell (2001) claims in his review on ecological biochemistry to have selected "examples … of different types of biochemical relationships," but he presents merely the following sections (apart from the introduction and conclusions): beetles and seeds, caterpillars and leaves, biochemical polymorphism in plants, biochemical polymorphism in herbivores and, finally, induced response to herbivory. Sections on microbes such as "bacteria and plants" or "induced response to fungi" are missing, though the title of the review "Ecological biochemistry and its development" did not indicate that it is limited to plant–insect interactions. Overcoming a bias towards creatures that can be seen by the naked eye and collected by hand is the first prerequisite for maintaining progress in chemical ecology in a broader sense.

1.5 The Origin of Chemical Diversity in Soil

Secondary metabolism continues to be a rich source of new and often surprising structures. The number of secondary metabolites discovered so far, which is estimated to be at most 50,000 (Demain and Fang 2000), appears to represent only a fraction of the chemical diversity possessed by extant plants and cultivable fungi, bacteria and protists. Even worse is the fact that the vast majority of microbes inhabiting natural biota cannot be cultivated under laboratory conditions. The metagenome approach pioneered by Diversa Corporation is unlikely to recover intact and functional biosynthetic pathways involving several enzymes, nonubiquitous cofactors or specific precursors. The consequence is that most of the chemical diversity on Earth is not accessible for humans and it is likely to remain out of our reach in the foreseeable future.

An intuitive concept that the force driving the diversification of secondary metabolites produced by soil-borne or soil-inhabiting microorganisms is competition is widespread. In terms of interference competition, an organism which acquires the ability to produce a new antibiotic will experience a gain in fitness. The efficiency of the antibiotic declines as resistance mechanisms arise and spread, in analogy to the race between the pharmaceutical industry and human-pathogenic bacteria. Intuitively, this situation appears to favor diversity in antimicrobial metabolites. This view has recently been corroborated by the outstanding work by Czaran et al. (2002). The authors simulated an evolutionary arms race which takes place in a spatially structured environment. The basic idea was that the production of a secondary metabolite which blocks competitors either increases or decreases the net fitness of the producer, depending on the presence of the competitor and its resistance towards this particular toxin. The crucial point that led to the generation of diversity was the introduction of costs of resistance. In a spatially segmented, two-dimensional substrate, several strains survived at a stable total density but with periodically fluctuating abundance at local regions. The final version of the model consisted of 14 systems, each containing an immune producer, a resistant nonproducer and a sensitive nonproducer. It is significant that this groundbreaking result was achieved by a computer simulation. Because of the enormous complexity of soil ecosystems and the inherent limits of our experimental tools, numerical simulations are likely to play an important role in research into chemical interactions in soil in the future.

The most valuable outcome of computer modeling is an experimentally testable hypothesis. Davelos et al. (2004) recently documented spatial fragmentation of interference competition in soil experimentally. The authors showed that in *Streptomyces* from prairie soil, antibiotic production is highly variable in space, implying that thc fitness benefit resulting from antibiotic production varies among locations. Resistance patterns were consistent across locations, indicating that the costs of resistance were low. This contradicts the results of Czaran et al. (2002), because selection against resistance was a crucial factor promoting chemical diversity in their model. The apparent discrepancy shows that we are still at the beginning of understanding chemical diversity in ecosystems. In addition to variation in space, variation in time needs to be addressed experimentally. Maintenance of chemical diversity by selection in a fragmented environment is one of the most promising areas of current secondary metabolite research.

A factor not considered in the model of Czaran et al. (2002) is that secondary metabolites may act additively, synergistically or antagonistically. Challis and Hopwood (2003), again focusing on *Streptomyces*, investigated antibiotic effects regarding synergy and contingency, which they defined as the production of several metabolites targeting the same competitor. Their work took advantage of rich data on the production of antibiotics by *Actinomyces* and the complete genome sequence of two *Streptomyces* species. The coproduction of clavulanic acid and cephamycin C, the common regulation of both pathways (both are controlled by *ccaR* protein) and the location of the gene clusters in the genome, as well as the comparison of clavulanic acid and cephamycin C production by different strains, supported a view that clavulanic acid synthesis developed as a response to the acquisition of β-lactamase by one of the organisms targeted by cephamycin C. Similar arguments are presented for siderophores (iron chelators), streptogramins and further secondary metabolites, showing that the synergistic and contingent effect of secondary metabolites against the same competitor was one of the reasons for the development of multiple pathways for antimicrobial secondary metabolites.

How do microorganisms generate and maintain chemical diversity on a biochemical level? Firn and Jones (2000, 2003) suggested that a small set of enzymes with relaxed specificities may generate a large set of different but structurally related metabolites. Only some among these products exert effects which enhance the fitness of their producer under current conditions. The other metabolites serve merely as a supply of diversity for future needs. Apart from postulating how relaxed enzyme specificities generate structural diversity, which can easily be accommodated by the current framework of evolutionary theory, a novel and controversial aspect of their metabolic grid concept is the notion that evolution optimized retention of chemical diversity at minimum metabolic cost, including the production of metabolites which do not exert any beneficial effect on their producers. If such "useless" metabolites exist, one might suggest an alternative explanation by considering them to be side products of biosynthetic pathways which have not been optimized yet for specificity. Structurally related metabolites usually exert similar effects, while the efficiencies of individual metabolites differ. This is well known not only for antibiotics, but also for all groups of mycotoxins (e.g., fumonisins, trichothecenes, aflatoxins, enniatins and zearalenone derivatives). Apart from the hard-to-swallow idea of evolution maintaining chemical diversity for future needs, a problem with the hypothesis is that it is impossible to prove for any secondary metabolite that it does not enhance the fitness of its producer under certain conditions. The concept was derived from the so-called screening hypothesis, which sought to reconcile the diversity of natural products with the observation that the majority of these compounds are not active in bioassays used in screening programs developed by the pharmaceutical industry. Even if the assertion that "potent biological activity is a rare property for any one molecule to possess" is true, it may not be relevant for ecosystems with complex interorganismal interactions, because activity does not need to be strong in order to positively affect the fitness of its producer. Moreover, even potent activity may remain unnoticed in bioassays unless adequate target organisms are used. Because most natural targets of metabolites secreted in soil are unknown and possibly uncultivable, the value of in vitro bioassays for explaining the biological role of secondary metabolites in soil is inherently limited.

1.6 Secondary Metabolites and Fitness: Evolution Meets Ecology

1.6.1 Chemical Interactions and Coevolution of Soil Species

Metabolites involved in interorganismal interactions affect the relative fitness of interacting partners in a distinctive way. The simplest scenario is that the biological activity of an ecological metabolite has been optimized by evolution to affect a target organism in a way benefiting the producer. This idea is the basis of many concepts of metabolite-mediated interaction, including interference competition among fungi, attraction of pray by carnivorous plants and protection of plants from herbivores by repellant volatiles and antifeedants. These ideas are straightforward and as long as the production of the metabolite in question is amenable to control by genetic engineering or by induction/suppression of its synthesis, it is relatively easy to design experiments for testing working hypotheses in natural environments. Elementary evolutionary considerations require us to assume that the selection pressure exerted by a secreted secondary metabolite on the population of the target organism will affect allele frequencies, speed up the elimination of genotypes responding in unfavorable ways and facilitate fixation of mutations enhancing the fitness of the target under the effect of the metabolite. Eventually, an evolutionary change will occur which will overcome the fitness depression of the target organism and eliminated fitness gain, benefiting the metabolite producer. In reality, both interacting partners are subjected to selection pressures at the same time, leading to reciprocal adaptation in a process called coevolution.

Coevolution became the basic explanatory framework in research on plant–insect interactions, which is a field in which ecological chemistry has been developed most extensively. In spite of relentless criticism by Jermy (1988, 1998), the coevolutionary theory proliferated and ramified into its most recent incarnation known as geographic mosaic theory of coevolution (Thompson 2005). Unfortunately, this development has little benefited ecological chemistry of soil. Belowground research has always played a poor cousin's role in ecology, possibly because field trips, insects and flowering plants are more attractive for most students than soil microcosms, complex instrumentation and methods requiring considerable training time. But even when we compare applications of the same technique to aboveground and belowground space, soil biology gets the short end of the stick. Studies of volatiles provide a revealing example. Volatile compounds in soil are likely to be more important for the orientation of invertebrates than in aboveground environments because visual orientation in soil is impaired. Furthermore, concentration gradients of volatiles in soil air are more stable than gradients in aboveground space because of limited air convection. In spite of this, students of plant volatiles rarely turn their headspace gas chromatography (Tholl et al. 2006) and insect-antenna-derived sensors (Weissbecker et al. 2004) to rhizosphere air. Although experimental data are largely lacking, volatile-mediated relationships similar to those known from aboveground ecosystems (Harrewijn et al. 2005) are likely to have been established by the coevolution of herbivorous invertebrates and plants in soil. Volatiles generated by soil microorganisms, plant roots and germinated seeds are well known to affect soil fungi and stimulate plant growth (Schenck and Stotzky 1975; Ryu et al. 2003; Kai et al. 2007). Coevolutionary relationships based on chemical communication via nonvolatile components of soil solutions, including olfactory cues evaluated by soil invertebrates, are likely to play an even more significant role, but available experimental data are equally scarce.

1.6.2 Cost of Biosynthesis

Let us look at the metabolic costs of secondary metabolite synthesis, which can be easily investigated in simple systems. In plant–insect interactions this issue has been extensively addressed (Gershenzon 1994). Determining the cost of biosynthesis of a particular metabolite by a microorganism appears to be a straightforward issue, providing suitable mutants are available. Wilkinson et al. (2004) recently determined the effect of a stepwise deactivation of the sterigmatocystin biosynthesis pathway in *Aspergillus nidulans* on the fitness of the fungus. Their result was surprising: the number of conidia produced in axenic cultures increased with the progression of sterigmatocystin synthesis. The lowest number of conidia was found in cultures of a mutant in which the complete pathway had been shut off via a regulatory gene *afl*R; the highest number of conidia was found in the wild-type strain. Because the strains were isogenic, hidden effects of additional mutations can be excluded. The authors showed that the effect cannot be explained by protection against light.

The result of Wilkinson et al. (2004) is counterintuitive: the synthesis of sterigmatocystin is thought to provide ecological benefits to its producer called indirect effects (Strauss et al. 2002), but the direct effect of the biosynthesis on the fitness of its producer is expected to be negative because it consumes energy and metabolic precursors, which could otherwise be used to build up biomass and reproductive structures. Because the experiments were performed in axenic cultures, observed positive effects of sterigmatocystin synthesis on conidia formation did not involve interactions of *A. nidulans* with other organisms. Sterigmatocystin is known as a carcinogenic mycotoxin (it serves as a precursor of aflatoxin synthesis in other *Aspergillus* species) and although its ecological role is not known, it is a common belief that its function is to inhibit organisms which compete for resources with sterigmatocystin producers. An alternative explanation to direct benefits to the fungus as postulated by the authors is that the observed effect could have resulted from regulatory phenomena. This hypothesis is corroborated by the fact that both conidia development and sterigmatocystin synthesis are derepressed by a common activator FluG, which counteracts the affect of the repressor SfgA (Seo et al. 2006).

The work of Wilkinson et al. (2004) was the first one addressing the effect of a stepwise deactivation of a biosynthetic of a secondary metabolite on fungal fitness, but the observation of a negative rather than a positive effect of the loss of a dispensable pathway on fitness under axenic conditions is not unique. For example, Gaffoor et al. (2005) disrupted all polyketide synthase (PKS) genes of *Fusarium graminearum* and observed inhibition of mycelial growth in mutants that lost two out of 15 PKS genes. Similarly, Zhou et al. (2000) observed growth inhibition in *A. parasiticus* after disruption of PKS FLUP. The mechanisms of these effects are unknown. Regulatory phenomena may be responsible for apparent benefits caused by the synthesis of these metabolites in axenic cultures. To test this hypothesis, one would need to isolate regulatory mutants which reverse the effect of the disruption of the biosynthesis on fitness. In axenic cultures, the fitness of double mutants should be even higher than the fitness of the wild-type, nondisrupted strain.

The work of Wilkinson et al. (2004) makes clear that the effect of the synthesis of a secondary metabolite presumed to have ecological roles in the fitness of its producer needs to be assessed experimentally on a case-to-case basis. Knockout mutants are now available for many secondary metabolite pathways in fungi, but most of them are not ideal for experiments involving fitness estimation because they contain genes conferring resistance against hygromycin, phleomycin or other antibiotics used for selection of transformants. These resistance genes are expressed constitutively and are likely to have a negative impact on fitness. The best strategy for experiments involving fitness estimation appears to be the use of clean gene deletions, which can be achieved with the help of site-specific excision by recombinases such as Cre or ϕC31. However, this procedure is much more laborious than gene disruption. Alternatively, ectopic insertions can be used as controls instead of wild-type strains. Because the insertion of the resistance cassettes into the genome may cause unpredictable effects, several independently generated ectopic transformants have to be used.

Well-designed experiments with carefully engineered strains in axenic and mixed culture will allow us to assess the affect of selected metabolites on the fitness of their producer and on other organisms in the system. The interpretation of the results may be complicated by regulatory effects (see later), synergy or contingency effects (Challis and Hopwood 2003) or detoxification (Karlovsky 1999). In spite of this complexity, carefully engineered mutants in systems imitating natural conditions open the only window currently available for unbiased direct observation of biological functions of secondary metabolites in soil.

1.6.3 Complexity of Chemical Interactions in Soil

Microbial populations in soil are complex and their total population density is high. One-to-one correspondence between a producer of a metabolite and its target, as known from insect–plant interactions, will rarely be encountered. In interactions among soil microorganisms, all partners are producers and many if not all are targets of ecological metabolites. In terms of fitness, the outcome of chemical interactions of a particular microorganism will be determined by how well the blend of its own secondary metabolites is adapted to the current environment and how efficiently its countermeasures (resistance, detoxification, export, etc.) prevent the harmful effects of metabolites produced by other inhabitants of the niche. As already mentioned in the context of fitness, this inherent complexity needs to be taken into account when studying effects of perturbations of chemical interaction (e.g., by gene knockouts) in natural systems.

Growth inhibition or toxicity in general are not the only effects exerted by metabolites involved in chemical warfare in soil. Microorganisms may avoid harmful effects of antimicrobial compounds produced by their competitors by suppressing their synthesis. The interpretation of such effects from an ecological point of view is straightforward. For example, fusaric acid is a mycotoxin and presumably a virulence factor of *F. oxysporum*. Plant infection by *F. oxysporum* can be suppressed by certain strains of *Pseudomonas fluorescens* which produce the antifungal metabolite 2,4-diacetylphloroglucinol (see Chap. 5). Notz et al. (2002) showed that fusaric acid suppresses the production of 2,4-diacetylphloroglucinol by *P. fluorescens*. Importantly, this effect was demonstrated not only in vitro, but strains carrying reporter fusions for 2,4-diacetylphloroglucinol synthesis were investigated in the rhizosphere and the effects of *F. oxysporum* strains producing different amounts of fusaric acids were compared.

Secondary metabolites involved in antagonistic interaction may affect other functions and activities of competitors to benefit their producers. For instance, mycotoxin deoxynivalenol produced by *F. graminearum* appears to inhibit the expression of a chitinase gene in *Trichoderma atroviride* (Lutz et al. 2003). Because chitinase activity is a decisive factor determining the efficiency of the biocontrol agent *T. atroviride* against *F. graminearum*, the repression of chitinase production by deoxynivalenol may be regarded as a defense mechanism. This results revealed a new ecological role for mycotoxin deoxynivalenol, which was known to act as a virulence factor of *F. graminearum* in wheat. Deoxynivalenol obviously plays at least two different and unrelated ecological roles. (Because of the induction of vomiting and food refusal by deoxynivalenol in mammals, the mycotoxin might also be involved in interference competition between *Fusarium* and grain- or seed-consuming animals.)

Detoxification is a widespread mechanism of defense of target organisms against harmful secondary metabolites (Karlovsky 1999). Antimicrobial plant metabolites are often detoxified by a phytopathogenic microorganism (Pedras and Suchy 2005; Pedras and Hossain 2006; Morrissey and Osbourn 1999; Glenn et al. 2003). These processes have been studied with plant metabolites extracted from leaves and stems, but plant phytoalexins and phytoanticipins also reach soil with root exudates (see Chap. 11) and with plant debris (see Chap. 10). Detoxification of plant defense chemicals is therefore as important in the rhizosphere as it is in aboveground plant organs.

The effects of secondary metabolites on the biology of soil inhabitants are too numerous to list here exhaustively. Metabolites of plant origin induce germination of fungal spores and microsclerotia, attract and repel nematodes, mediate allelopathy among plants and induce chemotaxis in zoospores and protozoans. Strigolactones (Humphrey and Beale 2006) belong to the most interesting compounds not discussed in this volume. These plant secondary metabolites, which are secreted by roots in extremely low quantities challenging our most sensitive analytical techniques, stimulate the germination of parasitic weeds and mycorrhiza fungi. Siderophores are another group of secreted metabolites involved in complex interactions. They are synthesized to facilitate the uptake of iron by their producers, but many microorganisms hijack foreign siderophores to lower their costs of iron extraction, or even use them decadently as a cheap nutrient. Similarly as in marine ecosystems (Engel et al. 2002), nontoxic concentrations of antimicrobial compounds involved in interference competition may effect microbial behavior, corroborating the view that chemical communication is the primary factor controlling interorganismal interactions in soil.

1.6.4 Regulation of Biosynthesis as a Key to Function

Producers of metabolites with ecological roles need to adapt to changing environments by controlling their biosynthesis pathways because the mobility of microbes is limited and the production of any ecological metabolite incurs metabolic costs. Therefore, the regulation of the production of secondary metabolites, regarding both their qualitative spectrum and their quantities, appears to be a crucial factor affecting the success of a microbe in a biotope. Apart from commercially relevant antibiotics, the most thoroughly investigated regulation of secondary metabolite synthesis includes mycotoxins. In line with the prediction that a well-tuned regulation is an important factor maximizing fitness, the regulation of the synthesis of mycotoxins by *Aspergillus* spp., *Penicillium* spp. and *Fusarium* spp. appears to be very complex. The effects of many environmental factors on the synthesis of a number of mycotoxins have been experimentally determined and regulatory elements involved in the control of mycotoxin synthesis have been identified and cloned. Unfortunately, we have not been able to extract much biological meaning out of these data so far. For example, we know how nitrogen, phosphorus and starch affect fumonisin synthesis in *F. verticillioides*, that a very high sugar concentration is needed for zearalenone synthesis and that deoxynivalenol is produced in media with a high amount of yeast extract. The effects of water activity, temperature and substrate on mycotoxin production have been mapped in detail in Naresh Magan's group. We know that the highest amounts of fumonisins and zearalenone accumulate when their producers are grown on rice, which is not their natural substrate. What does it all mean? We do not know yet, but it is reasonable to assume that mycotoxin synthesis is regulated in order to limit metabolic costs and/or self-poisoning. Deciphering regulatory patterns of mycotoxin biosynthesis should therefore provide us with clues about their biological function. In general, it appears that we do not have data from relevant conditions yet, or we were unable to look at the data in the right way.

In the course of the characterization of PKS genes in *F. graminearum*, Gaffoor et al. (2005) investigated the expression of all 15 PKS genes of this fungus under 18 culture conditions and discerned seven expression patterns, some of which can be interpreted in ecological terms (e.g., plant infection-specific expression and grain-specific expression). On a different note, their work documents an immense gap in our understanding of fungal secondary metabolites: although *F. graminearum* is the most thoroughly studied *Fusarium* species, its whole genome has been sequenced and disruptions of all its 15 PKS genes are available, the chemical products of nine of its PKS genes are still unknown!

The induction of the synthesis of metabolites putatively involved in interference competition by cultivation of their producers in the presence of competitors provides information which may be more valuable than the results of bioassays. This strategy was used successfully for *Heterobasidion annosum* (Sonnenbichler et al. 1989) at times when gene disruptions in fungi were not readily available. Apart from corroborating the role of certain secondary metabolites with antifungal activity in the interaction of this tree pathogen with antagonistic fungi, these experiments revealed that the antifungal metabolites produced by *H. annosum* can be detoxified by putative target organisms (Sonnenbichler et al. 1993).

1.7 Pitfalls in Search for Function

Interference competition dominated thinking about chemical interactions in soil, inspired by the potent effects of antibiotics isolated from soil *Actinomyces*. Competition among coprophilous fungi in dung was a popular experimental system for these studies because of easy experimental access and a well-described, predictable sequel of colonizing organisms. However, most of the investigations were performed on isolated organisms. For example, Gloer and Truckenbrod (1988) began their report by stating "Isoepoxydon has been established as the major causative agent of interference competition between *Poronia punctata…*," while, in fact, only in vitro effects have been established. The bioassay used by Gloer and Truckenbrod was based on a species competing with the producer of isoepoxydon, but too often the role of a secondary metabolite in interference competition is postulated on the basis of bioassays with human pathogens or other ecologically inappropriate organisms.

On the other hand, antibacterial or antifungal effects may be overlooked when a metabolite is well known in a different context. For instance, a strong toxic effect of mycotoxin zearalenone on filamentous fungi remained unnoticed for decades (Utermark and Karlovsky 2007). Zearalenone is known as a potent estrogen and the ingestion of contaminated food and feeds poses a health risk to humans and farm animals. This prominent biological activity and the label "mycotoxin" apparently prevented people working with zearalenone from subjecting it to a standard antifungal assay.

Zearalenone provides an instructive example of a wrong assignment of function too. The estrogenic activity of the metabolite inspired speculations about its role as a sex hormone and regulator of reproduction in *Gibberella zeae* (Nelson 1971). The hypothesis was seemingly corroborated by observations that zearalenone added to *G. zeae* cultures increased perithecia production (Wolf and Mirocha 1973) and that dichlorvos, an inhibitor of zearalenone biosynthesis, reduced perithecia production (Wolf et al. 1972). In spite of the facts that many chemicals, including commercial fungicides in sublethal doses, stimulate perithecia formation, that dichlorvos unspecifically inhibits many PKSs, and that *F. culmorum*, which does not possess a sexual stadium, produces large amounts of zearalenone, the sex hormone hypothesis survived for over three decades. Nelson's idea was so appealing that it persisted even after the exposure of zearalenone–perithecia correlation as a fallacy (Windels et al. 1989).

Neither isoepoxydon nor zearalenone has been shown to enhance the fitness of their producers in the presence of competing fungi in natural environments so far, but the role of zearalenone in interference competition is strongly supported by finding that mycoparasite *Gliocladium roseum*, which preys on *Fusarium* spp., developed an enzymatic detoxification mechanism for zearalenone (El-Sharkawy and Abul-Hajj 1988). *G. roseum* is resistant to zearalenone and the inactivation of its detoxification activity renders it susceptible (Utermark and Karlovsky 2007).

Research on secondary metabolites involved in interaction of microbial pathogens with plants suffered from serious setbacks. Gäumann (1954) and his disciples postulated half a century ago that phytotoxins are causally involved in all plant diseases. A generation of phytopathologists generated phytotoxicity data to support their hypothesis, but a convincing proof did not surface even for a single toxin at that time because of the lack of appropriate experimental tools. Referring to this era, Robert Scheffer and Steve Briggs once wrote: "The literature on toxins affecting plants is vast, but much of it is meaningless." Their harsh judgment was embraced by the next generation of phytopathologists, who went to the other extreme and abandoned research into secondary metabolites acting as virulence factors for nearly three decades. (Host-specific toxins were a noticeable exception.) As a consequence, opportunities to design novel resistance mechanisms for crops based on detoxification of fungal toxins were considerably delayed and our understanding of pathogen–plant relationships was deprived of one of its principal facets. A renewed interest of phytopathologists in non-host-specific toxins, as we experience it now, will likely benefit not only plant protection but also basic research on secondary metabolites in general.

1.8 Future of Secondary Metabolite Research

Thousands of secondary metabolite structures have been published, but educated guesses about biological function are possible only for a negligibly small fraction of them. Besides, they are seldom more than guesses: when a bioassay demonstrates toxic effects upon a competitor, we still do not know whether the substance is produced under relevant conditions in nature, whether its local concentration is sufficient to exert the effects observed in vitro and how adsorption, degradation and interaction with other metabolites modulates its toxicity in situ. It is not possible to determine or control all these factors. The only reliable way to address the biological role of a particular metabolite is to manipulate its biosynthesis or degradation by genetically engineering interacting organisms and investigating the consequences of the perturbation under natural conditions. This strategy has been used extensively and successfully in interactions between plant pathogens and their hosts. In a few cases, the role of secondary metabolites in biological control of plant pathogens has also been studied with the help of genetically engineered microbes. It is time now to extend the concept to chemical ecology of soil in a broad sense.

How is secondary metabolite research advancing beyond its traditionally descriptive approach? Natural product chemistry is extending its scope and embracing techniques and concepts originating from biochemistry and genetics, while ecologists and environmental microbiologists recognize that chemical interactions mediated by secondary metabolites are crucial for our understanding of soil ecosystems. Empirical screening of natural products for biological activities, as well as high-throughput purification and structure elucidation of natural products from arbitrarily selected sources, should be left to the responsibility of the pharmaceutical industry and service laboratories, releasing capacity in academia and basic research to address fundamental questions. The following emerging approaches and technologies are likely to play a role in this transition:

- Application of genetic engineering in systematically controlling the production and/or degradation of secondary metabolites, followed by monitoring how these perturbations affect the system, allows us to assess the effect of secondary metabolites on the fitness of soil organisms.
- Analytical techniques for the quantification of many metabolites in matrices as complex as soil are needed to follow the dynamics of secondary metabolite production, transformation and degradation in soil. In situ detection and nondestructive analysis are needed in order to take into account the heterogeneous structure of soil ecosystems.
- Routine techniques available for monitoring microbial populations in soil are differential gradient gel electrophoresis (DGGE) of amplified ribosomal RNA genes or reverse-transcribed ribosomal RNA, terminal restriction fragment length polymorphism (T-RFLP) of ribosomal RNA genes and in situ hybridization of taxon-specific oligonucleotides labeled by fluorescent dyes (FISH). In future these techniques they will be extended by large-scale metagenome sequencing (Eisen 2007; Rusch et al. 2007).
- Modeling chemical interactions in microbial ecosystems and their evolutionary consequences will be increasingly important. The interplay of factors such as metabolic costs, competition, spatial heterogeneity, synergy of antibiotic effects of many metabolites, adsorption and detoxification can be investigated by computer modeling, while it is difficult to address more than one factor experimentally.

Soil is arguably the most complex and difficult system to chose for the study of ecological functions of secondary metabolites. However, soil is also the ecosystem in which chemical interactions play the most substantial role, and from where major insights into the evolution of chemical diversity are expected to come.

References

Bell EA (2001) Ecological biochemistry and its development. Phytochemistry 56:223–227

Bode HB, Bethe B, Höfs R, Zeeck A (2002) Big effects from small changes: possible ways to explore nature's chemical diversity. Chembiochem 3:619–627

Challis GL, Hopwood DA (2003) Synergy and contingency as driving forces for the evolution of multiple secondary metabolite production by *Streptomyces* species. Proc Natl Acad Sci USA 100:14555–14561

Czaran T, Hoekstra RF, Pagie L (2002) Chemical warfare between microbes promotes biodiversity. Proc Natl Acad Sci USA 99:786–790

Davelos AL, Kinkel LL, Samac DA (2004) Spatial variation in frequency and intensity of antibiotic interactions among *Streptomycetes* from prairie soil. Appl Environ Microbiol 70:1051–1058

Demain AL (1986) Regulation of secondary metabolism in fungi. Pure Appl Chem 58:219–226

Demain AL, Fang A (2000) The natural functions of secondary metabolites. In: Scheper T (ed) Advances in biochemical engineering/biotechnology, vol. 69. Springer Berlin, pp 1–39

Eisen JA (2007) Environmental shotgun sequencing: its potential and challenges for studying the hidden world of microbes. PLoS Biol 5:e82

El-Sharkawy S, Abul-Hajj YJ (1988) Microbial cleavage of zearalenone. Xenobiotica 18:365–371

Engel S, Jensen PR, Fenical W (2002) Chemical ecology of marine microbial defense. J Chem Ecol 28:1971–1985

Firn RD, Jones CG (2000) The evolution of secondary metabolism—a unifying model. Mol Microbiol 37:989–994

Firn RD, Jones CG (2003) Natural products—a simple model to explain chemical diversity. Nat Prod Rep 20:382–391

Frisvad JC, Thrane U, Filtenborg O (1998) In: Frisvad J (ed) Chemical fungal taxonomy. CRC, London, pp 289–321

Frisvad JC, Smedsgaard J, Larsen TO (2004) Mycotoxins, drugs and other extrolites produced by species in Penicillium subgenus Penicillium. Stud Mycol 49:201–241

Gaffoor I, Brown DW, Plattner R, Proctor RH, Qi W, Trail F (2005) Functional analysis of the polyketide synthase genes in the filamentous fungus *Gibberella zeae* (anamorph *Fusarium graminearum*). Eukaryot Cell 4:1926–1933

Gäumann E (1954) Toxins and plant diseases. Endeavour 13:198–204

Gershenzon J (1994) The cost of plant chemical defense against herbivory: a biochemical perspective. In: Bernays EA (ed) Insect–plant interactions, vol 5. CRC, Boca Raton, pp 105–173

Glenn AE, Meredith FI, Morrison WH 3rd, Bacon CW (2003) Identification of intermediate and branch metabolites resulting from biotransformation of 2-benzoxazolinone by *Fusarium verticillioides*. Appl Environ Microbiol 69:3165–3169

Gloer JB, Truckenbrod SM (1988) Interference competition among coprophilous fungi: production of (+)isoepoxydon by *Poronia punctata*. Appl Environ Microbiol 54:861–864

Gottlieb OR (1990) Phytochemicals: differentiation and function. Phytochemistry 29:1715–1724

Harborne JB (1999) Recent advances in chemical ecology. Nat Prod Rep 16:509–523

Harrewijn P, Minks AK, Mollema C (2005) Evolution of plant volatile production in insect-plant relationships. Chemoecology 5:55–73

Humphrey AJ, Beale MH (2006) Strigol: biogenesis and physiological activity. Phytochemistry 67:636–640

Janzen DH (1977) Why fruits rot, seeds mold, and meat spoils. Am Nat 980:691–713

Jermy T (1988) Can predation lead to narrow food specialization in phytophagous insects? Ecology 69:902–904

Jermy T (1998) The major transitions in evolution: what has driven them? Trends Ecol Evol 13:199–200

Kai M, Effmert U, Berg G, Piechulla B (2007) Volatiles of bacterial antagonists inhibit mycelial growth of the plant pathogen *Rhizoctonia solani*. Arch Microbiol 187:351–360

Karlovsky P (1999) Biological detoxification of fungal toxins and its use in plant breeding, feed and food production. Nat Toxins 7:1–23

Kossel A (1891) Ueber die Chemische Zusammensetzung der Zelle. Arch Physiol 181–186

Lutz MP, Feichtinger G, Défago G, Duffy B (2003) Mycotoxigenic *Fusarium* and deoxynivalenol production repress chitinase gene expression in the biocontrol agent *Trichoderma atroviride* P1. Appl Environ Microbiol 69:3077–3084

Mitchell-Olds T, Gershenzon J, Baldwin I, Boland W (1998) Chemical ecology in the molecular era. Trends Plant Sci 3:362–365

Morrissey JP, Osbourn AE (1999) Fungal resistance to plant antibiotics as a mechanism of pathogenesis. Microbiol Mol Biol Rev 63:708–724

Mortensen GK, Strobel BW, Hansen HC (2003) Determination of zearalenone and ochratoxin A in soil. Anal Bioanal Chem 376:98–101

Nelson RR (1971) Hormonal involvement in sexual reproduction in the fungi, with special reference to F-2, a fungal estrogen. In: In Akai S, Ouchi S (eds) Morphological and biochemical events in plant-parasite interaction. Phytopathological Society of Japan, Tokyo, pp 181–200

Notz R, Maurhofer M, Dubach H, Haas D, Défago G (2002) Fusaric acid-producing strains of *Fusarium oxysporum* alter 2,4-diacetylphloroglucinol biosynthetic gene expression in *Pseudomonas fluorescens* cha0 in vitro and in the rhizosphere of wheat. Appl Environ Microbiol 68:2229–2235

Pedras MS, Hossain M (2006) Metabolism of crucifer phytoalexins in *Sclerotinia sclerotiorum*: detoxification of strongly antifungal compounds involves glucosylation. Org Biomol Chem 4:2581–2590

Pedras MS, Suchy M (2005) Detoxification pathways of the phytoalexins brassilexin and sinalexin in *Leptosphaeria maculans*: isolation and synthesis of the elusive intermediate 3-formylindolyl-2-sulfonic acid. Org Biomol Chem 3:2002–2007

Ryu CM, Farag MA, Hu CH, Reddy MS, Wei HX, Paré PW, Kloepper JW (2003) Bacterial volatiles promote growth in *Arabidopsis*. Proc Natl Acad Sci USA 100:4927–4932

Rusch DB, Halpern AL, Sutton G, Heidelberg KB, Williamson S, Yooseph S, Wu D, Eisen JA, Hoffman JM, Remington K, Beeson K, Tran B, Smith H, Baden-Tillson H, Stewart C, Thorpe J, Freeman J, Andrews-Pfannkoch C, Venter JE, Li K, Kravitz S, Heidelberg JF, Utterback T, Rogers YH, Falcón LI, Souza V, Bonilla-Rosso G, Eguiarte LE, Karl DM, Sathyendranath S, Platt T, Bermingham E, Gallardo V, Tamayo-Castillo G, Ferrari MR, Strausberg RL, Nealson K, Friedman R, Frazier M, Venter JC (2007) The Sorcerer II global ocean sampling expedition: Northwest Atlantic through Eastern Tropical Pacific. PLoS Biol 5:e77

Schenck S, Stotzky G (1975) Effect on microorganisms of volatile compounds released from germinating seeds. Can J Microbiol 21:1622–1634

Seo JA, Guan Y, Yu JH (2006) FluG-dependent asexual development in *Aspergillus nidulans* occurs via derepression. Genetics 172:1535–1544

Sirenko LA, Malyarevskaya AY, Birger T, Kirpenko YA (1979) Ecological metabolites. In: Krasnov EV (ed) Vzaimodeistne vodoi zhiuym veshchestuom, vol 2. Izdania, Moscow, pp 48–56

Smedsgaard J, Nielsen J (2005) Metabolite profiling of fungi and yeast: from phenotype to metabolome by MS and informatics. J Exp Bot 56:273–286

Smith PM (1976) The chemotaxonomy of plants. Arnold, London

Sonnenbichler J, Bliestle IM, Peipp H, Holdenrieder O (1989) Secondary fungal metabolites and their biological activities, I. Isolation of antibiotic compounds from cultures of *Heterobasidion annosum* synthesized in the presence of antagonistic fungi or host plant cells. Biol Chem Hoppe Seyler 370:1295–1303

Sonnenbichler J, Peipp H, Dietrich J (1993) Secondary fungal metabolites and their biological activities, III. Further metabolites from dual cultures of the antagonistic basidiomycetes *Heterobasidion annosum* and *Gloeophyllum abietinum*. Biol Chem Hoppe Seyler 374:467–473

Strauss SY, Rudgers JA, Lau JA, Irwin RE (2002) Direct and ecological costs of resistance to herbivory. Trends Ecol Evol 17:278–285

Tholl D, Boland W, Hansel A, Loreto F, Röse US, Schnitzler JP (2006) Practical approaches to plant volatile analysis. Plant J 45:540–560

Thompson JN (2005) The geographic mosaic of coevolution. University of Chicago Press, Chicago

Utermark J, Karlovsky P (2007) Role of zearalenone lactonase in protection of *Gliocladium roseum* from fungitoxic effects of the mycotoxin zearalenone. Appl Environ Microbiol 73:637–642

Weissbecker B, Holighaus G, Schutz S (2004) Gas chromatography with mass spectrometric and electroantennographic detection: analysis of wood odorants by direct coupling of insect olfaction and mass spectrometry. J Chromatogr A 1056:209–216

Windels CE, Mirocha CJ, Abbas HK, Xie W (1989) Perithecium production in Fusarium graminearum populations and lack of correlation with zearalenone production. Mycology 81:272–277

Wolf JC, Mirocha CJ (1973) Regulation of sexual reproduction in *Gibberella zeae* (*Fusarium roseum* “graminearum”) by F-2 (zearalenone). Can J Microbiol 19:725–734

Wolf JC, Lieberman JR, Mirocha CJ (1972) Inhibition of F-2 (zearalenone) biosynthesis and perithecium production in *Fusarium roseum* ‘Graminearum’. Phytopathology 62:937–939

Wilkinson H, Ramaswamy A, Sim SC, Keller NP (2004) Increased conidiation associated with progression along the sterigmatocystin biosynthetic pathway. Mycologia 96:1190–1198

Zhou R, Rasooly R, Linz JE (2000) Isolation and analysis of fluP, a gene associated with hyphal growth and sporulation in *Aspergillus parasiticus*. Mol Gen Genet 264:514–520

Part I
Detection and Analysis

Chapter 2
Detection of Antibiotics Produced by Soil and Rhizosphere Microbes In Situ

Linda S. Thomashow (✉), Robert F. Bonsall, and David M. Weller

2.1 Introduction

It has long been known that certain antibiotic-producing soil microorganisms are inhibitory to plant pathogens, both in the laboratory and in the field (Stallings 1954). The exploitation of these natural antagonistic interactions has been a driving force in research on the biological control of plant pathogens over the past century, but only in recent decades has pathogen control by antibiotics produced at biologically relevant levels in the environment been demonstrated conclusively. This progress, resulting from conceptual and technological advances made initially in the laboratory and then extended to the field, has set new standards for biocontrol research involving antibiotics. More generally, the approaches used in these studies may be useful in exploring the significance of other bioactive metabolites produced by microorganisms in their native habitats.

Among the conceptual advances underpinning progress towards understanding the role of antibiotics in the environment has been recognition that individual strains often are capable of producing more than one inhibitory compound. Detection methods based on the biochemical properties of a particular antibiotic therefore must be specific enough to distinguish among the repertoire of possible products (the number and optimal conditions for production of which usually must be determined empirically). In addition, assays based on biological activity must eliminate or compensate for the effects of metabolites other than the one of interest. This element of specificity generally was lacking in traditional studies in which activity against a target pathogen or indicator organism in vitro was taken as evidence of activity against that organism in situ. Perhaps more difficult is the need to

Linda S. Thomashow
USDA, ARS Washington State University, Pullman, WA, USA
e-mail: thomashow@wsu.edu

P. Karlovsky (ed.), *Secondary Metabolites in Soil Ecology. Soil Biology 14.*

compensate adequately for biological effects of metabolites other than the one under investigation, in part because the full range of products an organism is capable of producing seldom is known. A case in point is the well-known biocontrol strain *Pseudomonas fluorescens* Pf-5, in which DNA sequence analysis recently resulted in the discovery of a previously unknown group of bioactive cyclic lipopeptides (Gross et al. 2007).

Antibiotic detection in microbial habitats also has been facilitated by knowledge of the regulatory mechanisms by which microorganisms integrate antibiotic biosynthesis with growth and other metabolic processes, all of which require adequate nutrient supplies. Carbon and energy resources are scarce in bulk soil and, consequently, metabolic levels in microorganisms are low. Microbiological activity is more intensive in the spermosphere and rhizosphere of plants and in or around plant debris and fungal propagules where resources are comparatively abundant, and it is in these habitats that antibiotics are likely to be detected. The availability of metabolizable substrates places spatial constraints on the ability of microbial populations to produce antibiotics, and this influences the choice of method and sensitivity of detection of antibiotics, as well as the sample sizes that must be analyzed.

Technological advances in molecular biology and biochemistry, developed initially to dissect the genetics and regulation of antibiotic synthesis in vitro, have been indispensable in achieving the sensitivity and specificity needed to detect and assess the activity of antibiotics in situ. Biochemical approaches involving antibiotic extraction, fractionation, and characterization, usually on the basis of chemical and physical properties, provide direct and incontrovertible evidence of antibiotic production. However, direct approaches have limited sensitivity that depend not only on the physicochemical and biological properties of individual antibiotics, but also on the efficiency of recovery during extraction and the detection limit attainable with the instrumentation available. Amounts of antibiotic recovered usually are expressed relative to the size of the initial sample, but it must be remembered that these are average values, and localized antibiotic concentrations may be considerably higher in sites where microbial activity occurs. Thus, when biological aspects of antibiotic production are of interest, molecular approaches based on the detection of antibiotic activity or gene expression may be preferable to direct bioanalytical methods, assuming that the antibiotic biosynthesis genes themselves already have been identified. Molecular techniques enable the construction of mutant strains defective only in synthesis of the compound of interest, providing the specificity needed to assess the impact of particular antibiotics on other organisms in the soil environment. Alternatively, when the impact of physiological or edaphic conditions on antibiotic production is of concern, reporter strains can be constructed in which a readily monitored gene product, rather than the antibiotic itself, is assayed. Molecular approaches are fundamentally indirect and subject to limitations discussed more extensively in the following text, but they can provide a degree of sensitivity not achieved by direct analysis.

2.2 Direct Analysis: Sample Preparation and Chromatography

2.2.1 Collection and Storage

The detection of antibiotics in complex environmental matrices is influenced by the physical properties of the sample as well as the procedures used to process it prior to and during extraction. For rhizosphere samples, it is important to note whether specific portions of the root system have been harvested, how much plant tissue and how much soil are present, and how the tissue and soil have been separated. Our rhizosphere samples typically include the entire root systems of young seedlings as well as the soil particles that adhere to the roots after gentle shaking. Rhizosphere samples that cannot be extracted immediately are frozen and stored in the dark to prevent losses resulting from microbial degradation and sensitivity to heat or light. Bulk soil samples usually are collected to a known depth and broken up or milled, sieved, and stored frozen or dried prior to extraction, depending on the stability of the antibiotic.

The extraction efficiency and the sensitivity of detection are major factors in determining the sample size. Soil samples in the range 1–10 g (Jacobsen et al. 2004; Kim and Carlson 2006; Stoob et al. 2006; Thiele-Bruhn et al. 2004) and root systems of 50–200 seedlings or 25–30 g of roots with adhering soil (Bonsall et al 1997; Raaijmakers et al 1999; Thomashow et al. 2002) are representative. For quantitative determinations, the efficiency of recovery can be estimated from a standard curve in which the antibiotic has been spiked into control samples in amounts spanning the range expected in test samples. In addition, test samples can be amended with an internal standard having properties similar to those of the compound of interest, but which does not occur naturally in the sample matrix or interfere with subsequent analyses.

2.2.2 Sorption and Sample Preparation

The distribution of antibiotics between soil solids and water has gained scientific attention in recent years owing to concern over the environmental fate and consequences of the large amounts of veterinary pharmaceuticals used in animal husbandry. While focused on veterinary antibiotics, the information gained from these studies is largely consistent with earlier work describing the behavior of organic compounds in soil. In general, antibiotics adsorb rapidly to the surfaces of soil particles and dissolved organic matter, and recovery declines continuously over time (Blum et al. 1994; Chiou 1989; Weber and Miller 1989). Thus, recoveries of sulfonamide antibiotics spiked into aged agricultural soils were significantly reduced after 6–17 days compared with a contact time of just 90 min (Stoob et al. 2006). Adsorption does not necessarily inactivate antibiotics, however, as tetracycline and tylosin, two widely used growth promoters in food animal production, remained biologically active even when tightly adsorbed to clay particles (Chander et al. 2005).

Sorption is a complex phenomenon influenced not only by the physicochemical properties of an antibiotic but also by the composition and structure of solid matrices. Most antibiotics are moderately soluble in water and many have pK_a values within the range of pH values found in soils, indicating that their ionic form, solubility, and sorptive properties will be strongly influenced by the pH and ionic composition of the soil (ter Laak et al. 2006a; Tolls 2001). Below pH 6.5, nonionic forms of organic acids and phenolic compounds are readily sorbed by soil organic matter (Chiou 1989), and at pH values above the pK_a, charge interactions occur with inorganic soil constituents. Veterinary pharmaceuticals were associated with dissolved organic matter and soil particles much more strongly than predicted simply on the basis of hydrophobic interactions, perhaps because of cation exchange, cation bridging at clay particle surfaces, surface complexation, and hydrogen bonding (Tolls 2001). Cation bridging is thought to account for association with clay minerals, with sorption strongly related to particle size and, hence, to surface area. Soils rich in aluminum and iron oxyhydroxides have a high sorptive capacity for carboxyl and phenolic hydroxyl groups, and some soils rich in Mn^{2+} have a high capacity for organic acids (Dalton et al. 1989; Lehmann et al. 1987). Considering the variability in composition among soils, it is not surprising that sorption coefficients (the ratio of the concentrations of a compound in the sorbent and aqueous phases at equilibrium) can differ by several orders of magnitude from one soil to another, and efforts to develop models predictive of sorption coefficients based on soil properties have had limited success. Up to 78% of the variation in sorption coefficients of three veterinary compounds among 11 soils could be explained when six soil properties (pH, organic carbon content, clay content, cation-exchange capacity, aluminum oxyhydroxide content, and iron oxyhydroxide content) were integrated, but not when they were considered separately. The remaining variability was related to concentration effects associated with pH-dependent antibiotic ionization (ter Laak et al. 2006b).

In an efficient extraction process, the distribution of an antibiotic in the soil or rhizosphere is shifted from the sorbed form to a solvent. This process is facilitated by stirring, shaking, sonication, or, more recently, pressurized liquid extraction. The last of these requires specialized equipment, but is thought to increase sample wetting, solvent penetration, and diffusion rates (Ramos et al. 2002) and to reduce solvent consumption and extraction time (Stoob et al. 2006). The composition of the liquid phase is determined largely by the solubility and charge properties of the antibiotic, and to a lesser extent by the need to minimize the coextraction of soil organic compounds likely to interfere with subsequent purification and analysis. The liquid phase for soil and rhizosphere antibiotics typically is a mixture consisting of a polar organic solvent in water, adjusted to a pH below the pK_a of the antibiotic to facilitate partitioning into the solvent. The extractant solution may also contain agents such as Na_2EDTA, citric acid, NaCl, and McIlvine buffer to improve the recovery of antibiotics that form strong complexes with divalent and trivalent metal ions present in the soil (Petrović et al. 2005).

Procedures suitable for the extraction of the most frequently identified antibiotics produced in the rhizosphere by *Pseudomonas* spp. have been published (Bonsall et al. 1997; Raaijmakers et al. 1999) and can be adapted for other substances by adjusting the amount of sample required and selecting appropriate solvents. This method can recover phenazine-l-carboxylic acid (PCA), its hydroxyphenazine derivatives, pyrrolnitrin, pyoluteorin, and 2,4-diacetylphloroglucinol (DAPG) from field samples with recoveries of 60% or better. Briefly, roots with adhering rhizosphere soil are shaken in 80% acetone acidified to pH 2.0 with 10% trifluoroacetic acid. The solids are removed by settling or centrifugation and the filtrate is collected and concentrated after passage through a solvent-compatible filter.

Additional concentration and fractionation steps are required prior to chromatographic analysis in order to remove organic matter typically present in large amounts in soil extracts. Dark-colored organic contaminants can be removed by centrifuging solutions of some antibiotics frozen at −20 °C in acidified 35% acetonitrile (Bonsall et al. 1997), but other antibiotics may not remain soluble under these conditions. Antibiotics with ionizable residues can be separated from many contaminants by exploiting the pH-dependent differential solubility of the neutral and charged forms in organic and aqueous solvents. Thus, isolation procedures in the past routinely included at least one liquid–liquid extraction step to partition antibiotics away from salt residues and impurities, and into organic solvents from which they could be concentrated readily (Thomashow et al. 1990). More recently, however, solid-phase cartridges that exploit pH-dependent ionic speciation and polarity differences among antibiotics have largely replaced liquid–liquid extraction procedures. Solid-phase extraction (SPE) offers many advantages over liquid–liquid partitioning: less solvent waste and reduced operator exposure to solvents, and more rapid and efficient isolation and concentration of analytes. Early applications of SPE from soil samples include the preparative enrichment of DAPG on octadecyl silica (Shanahan et al. 1992) and the trapping of macrocyclic xanthobaccin compounds produced by *Stenotrophomonas* sp. SB-K-88 in the rhizosphere of sugar beet by growing the seedlings in a mixture of sand and Amberlite XAD-2 (Nakayama et al. 1999). A variety of SPE sorbents have been evaluated for antibiotic cleanup and recovery in studies addressing the environmental impact of human and veterinary pharmaceuticals. Among the most frequently employed are polymeric Oasis HLB (lipophilic divinylbenzene plus hydrophilic *N*-vinyl pyrrolidone) cartridges because they tolerate a broad range of pH, have greater capacity than alkyl-bonded silicas, and enable good recovery of both polar and nonpolar compounds (Díaz-Cruz and Barceló 2005; Petrović et al. 2005), and Oasis MCX cartridges containing a mixed-mode sorbent with cation-exchange and reversed-phase characteristics effective for polar to medium-polar compounds (Petrović et al. 2005). Tandem SPE employing sequential strong anion exchange and HLB cartridges has been used successfully to simultaneously reduce interfering organic matter and extract veterinary antibiotics (Jacobsen et al. 2004; Renew and Huang 2004).

2.2.3 Chromatography and Detection

Among the chromatographic methods available to fractionate and detect antibiotics from soil, the simplest but most limited in analytical capability is thin layer chromatography (TLC). TLC does not require sophisticated instrumentation, nor do samples generally require extensive cleanup prior to analysis. Compounds can be separated with good resolution and methods are readily adaptable for applications ranging from high throughput to preparative-scale work. Both normal and reversed-phase adsorbents have been used with a variety of mobile-phase solvent systems (Thomashow et al. 2002). Substances are visualized by UV absorption, chromogenic reaction with spray reagents, or bioautography, in which indicator organisms suspended in agar or broth are overlaid on chromatograms to detect bioactive spots. Antibiotics are identified on the basis of the appearance, distance traveled relative to the solvent front (R_f value), and cochromatography with authentic standards in more than one adsorbent or solvent system. Quantities are estimated from spot size and intensity, or size of the inhibition zone for bioautography, at various dilutions relative to known amounts of standards run on the same plate. The need for standards cannot be overemphasized, as variation in preparative methods and the sources and specifications of adsorbents and support media can result in significant differences between observed and published R_f values. Because many of the antibiotics produced by soil microorganisms and rhizosphere microorganisms are not commercially available, well-characterized strains capable of producing these substances in vitro often are the most convenient source of standards.

High performance liquid chromatography (HPLC), coupled with UV spectroscopy or various forms of mass spectrometry (MS), now is used routinely to fractionate, detect, and identify antibiotics produced in the rhizosphere (Bakker et al. 2002; Chin-A-Woeng et al. 1998; Glandorf et al. 2001; Huang et al. 2004; Nielsen and Sørensen 2003; Raaijmakers et al. 1999; Thomashow et al. 2002). These techniques are adaptable to a wide variety of analytes and offer a high degree of reproducibility, resolving capability, sensitivity, and quantitative accuracy. Considerations in developing and optimizing a chromatographic system include selection of the column, the mobile phase, the elution profile, and the mode of detection. Reversed-phase columns with octadecyl (C_{18}) or octyl (C_8) bonded silica packing, and gradient elution with acidified acetonitrile–water or methanol–water are commonly used. For high-throughput applications, isocratic elution avoids time- and solvent-consuming column reequilibration between samples if satisfactory resolution can be achieved. Shorter columns also may speed up analysis, albeit with the risk of reduced separation. Retention time and peak shape typically are optimized via the solvent composition and elution profile.

Detection commonly is by UV absorption, and because photodiode-array detectors concurrently monitor a range of wavelengths, they offer important advantages over fixed-wavelength detectors. Individual components within a mixture can be monitored simultaneously, each at its own absorption maximum, and subsequent spectral analyses can provide insight into peak purity and identity. Alternatively, amperometric detection may provide greater sensitivity and selectivity for some phenolics, and fluorometric detection may offer similar advantages for compounds such as indoles and some phenazines. Detection and quantification are further improved by coupling HPLC with MS, which enables confirmation of the identity of compounds on the basis of their molecular structure. Because samples from complex matrices such as soil typically include organic material that reduces detection sensitivity and interferes with quantification, soil extracts typically are analyzed by MS/MS, time-of-flight (TOF) MS, or triple-quadrupole TOF (Q-TOF) MS. The technical differences and relative merits among these mass analyzers, which increase detection sensitivity by providing an additional degree of chemical separation of analytes from interfering compounds in the matrix, have been considered in several recent reviews (Díaz-Cruz and Barceló 2005; Kim and Carlson 2005; Petrović et al. 2005). Q-TOF-MS, in particular, has become an important analytical tool because it can provide high mass-accuracy data and full MS/MS spectra, enabling both screening and confirmation of analytes in a single run. In our hands, the detection limits of DAPG and phenazine-1-carboxylic acid, produced in the rhizosphere of wheat, by Q-TOF-MS are 15 ng and 800 pg, amounts about 20-fold and 500-fold greater, respectively, than can be detected by photodiode-array spectroscopy. DAPG, produced by indigenous populations of *P. fluorescens* on the roots of field-grown wheat, was present at about 20 ng per gram of root fresh weight (Raaijmakers et al. 1999) and a wide variety of other antibiotics produced in situ have been reported at levels ranging from 5 to 5,000 ng per seed or gram of dry soil or root fresh weight (Thomashow et al. 1997).

2.3 Indirect Evidence of Antibiotic Production

2.3.1 Detection of Antibiotic Biosynthesis Genes

Because the detection of antibiotics by direct methods requires knowledge of their biophysical properties and can be laborious and costly, it often is more convenient to monitor the potential for, or consequences of, antibiotic production in the environment. The detection of antibiotic biosynthesis genes, whether or not expressed, provides insight into the distribution of antibiotic producers in nature and serves as a first indication that antibiotics may be present at biologically relevant levels. For biological phenomena such as the suppression of plant pathogens that are mediated by antibiotic production in situ, the detection of biosynthesis genes is a convenient means of monitoring the frequency, diversity, and dynamics of introduced or indigenous antibiotic-producing populations.

Molecular methods have been published (De La Fuente et al. 2006; Delaney et al. 2001; Mavrodi et al. 2001, 2007; McSpadden-Gardener et al. 2001; Thomashow et al. 2002) for the detection of key biosynthesis genes of the most frequently studied antibiotics produced in the rhizosphere, as well as for some antibiotics (Gross et al. 2007) with emerging roles in microbe–plant interactions. For example, a 745-bp internal fragment from the highly conserved *phlD* gene of the DAPG biosynthesis pathway has been used to enumerate DAPG producers from the roots of wheat (Raaijmakers et al. 1997, 1999) and maize (Picard et al. 2002) by colony hybridization and PCR, and a rapid PCR-based dilution-end-point assay for DAPG producers with a detection limit of approximately 10^3 colony-forming units per rhizosphere also has been developed (McSpadden et al. 2001).

Stringency is a critical determinant of sensitivity and specificity in all hybridization- and PCR-based detection strategies and is modulated by the selection of appropriate probe or primer sequences and by rigorous control of experimental conditions. Primers used to screen for antibiotic genes in environmental isolates or to detect the total population capable of producing a particular antibiotic must be *nonspecific* enough to accommodate templates with minor sequence heterogeneity due to sequence polymorphisms or codon degeneracy. On the other hand, it may be necessary to quantify specific subfractions (genotypes) of an antibiotic-producing population, as is the case for DAPG producers in which *phlD* polymorphisms are indicative of the affinity of a strain for particular host crops (De La Fuente et al. 2006). The typical approach to the design of such primers is to first align the DNA sequences from several homologues of the target gene. The alignment will reveal suitably spaced blocks of conserved or unique sequences from which primer pairs can be selected and PCR conditions optimized to meet the required specificity criteria. An alternative approach, which has been applied to detect genes involved in the synthesis of polyketide antibiotics, involves back-translating the amino acid sequence of a conserved region from related strains according to the preferred codon usage of the target species, and then synthesizing degenerate primers (Metsä-Ketalä et al. 1999; Seow et al. 1997).

Whereas colony hybridization and PCR require culturing of environmental isolates prior to gene detection, real-time PCR provides a culture-independent means of detecting antibiotic biosynthesis genes in DNA isolated directly from environmental sources. Eliminating the requirement for bacterial growth shortens assay turnaround time and avoids questions about the suitability of the culture conditions employed, whether isolates capable of antibiotic production are present in a viable-but-nonculturable condition, and if the populations detected after enrichment are skewed owing to inhibitory interactions among strains during growth (Validov et al. 2005). A culture-independent quantitative real-time PCR method for detection of the *phlD* gene has been developed that has a detection limit comparable to those of culture-based approaches, can detect both introduced and indigenous populations, and is capable of distinguishing among strain genotypes (Mavrodi et al. 2007).

Real-time PCR shares the same principles governing sensitivity, specificity, and primer design as standard PCR, but data collection and analysis in real-time PCR occur as the reaction proceeds in the instrument, making the technique much faster and less prone to contamination than standard PCR. Amplification is detected as an increase in fluorescence emitted by a dye after it has been incorporated into a double-stranded DNA product, the specificity of which is evaluated by melting temperature and melting curve analysis after each reaction. Fluorescence is monitored at each PCR cycle, and because the cycle in which the first significant increase in fluorescence above the background is correlated with the initial amount of target template, the method is inherently quantitative. For measurements to be meaningful, however, reactions must be highly optimized with regard to amplification conditions, amplification efficiency, and primer concentration and specificity. Standard curves must be developed over a range of DNA concentrations and in order to relate template DNA concentration to bacterial population size, the size of the genome and the copy number of the template gene must be known. Procedures for recovering DNA from environmental samples also must be optimized, and because recoveries may differ among matrices differing in their physicochemical properties, recovery values should be determined separately for each sample matrix. Whereas real-time PCR efficiencies for DNA extracted from DAPG-producing bacteria applied to the roots of wheat ranged from 80 to 98%, only about 10% of the DNA present in those bacteria actually was recovered (Mavrodi et al. 2007), suggesting that the sensitivity with which antibiotic producers are detected by real-time PCR will improve as better DNA extraction techniques are developed.

2.3.2 Antibiotic Gene Expression In Situ

Transcriptional analyses of antibiotic gene expression provide a sensitive and convenient alternative to direct antibiotic isolation, especially when biosynthesis over time or in response to environmental conditions is of interest. Such studies typically employ strains in which a reporter gene, the product of which is easily monitored and not naturally present in the environment, is placed under the transcriptional control of a promoter regulating the expression of the antibiotic biosynthesis genes. The speed and sensitivity with which reporters such as the green fluorescent protein gene *gfp* or the ice nucleation gene *inaZ* can be assayed facilitate the use of samples as small as single seeds or seedlings, allowing sufficient replication that significant differences among treatments can be detected despite the sample-to-sample variation inherent in such studies (Loper and Lindow 2002; Thomashow et al. 2002).

Reporter gene expression provides evidence that antibiotic synthesis can occur under prevailing environmental conditions, but the expression level need not be indicative of the actual amount of antibiotic present in a biologically active state. This is partially because transcriptional activity is measured relative to the total population size even though that population is physiologically heterogeneous, having been recovered from a variety of different microhabitats on the roots. A further confounding factor arises from the differential turnover rates of antibiotics and reporter gene products. Antibiotics in soil can lose activity over time owing to adsorption (Chander et al. 2005) or degradation, either by the producer strain itself (Bottiglieri and Keel 2006) or by the indigenous microflora. On the other hand, some reporters, and especially green fluorescent protein, are relatively stable and may more accurately reflect cumulative gene expression than instantaneous transcription rates. Finally, the complex regulatory circuitry involved in antibiotic synthesis (Haas and Keel 2003; Haas et al. 2000) and the structure of the reporter construct itself (Pessi et al. 2002) can influence the relationship between reporter gene expression and the amount of antibiotic actually produced.

2.3.3 Biological Activity In Situ: The Value of Mutants

Because antibiotics in the environment can reach biologically significant concentrations in localized sites while remaining at subthreshold or undetectable levels overall, their presence often is inferred from effects on other organisms that act as indicators of antibiotic activity. Such indirect estimates of antibiotic activity are of particular value when the measured effect can be attributed with certainty to the antibiotic of interest, and are greatly facilitated if wild-type strains can be compared directly with genetically defined, antibiotic-deficient mutants. The use of mutants is a preferred option when the activity of a producer strain is of interest over a range of environmental conditions that may impact on both the synthesis of an antibiotic and its biological availability (Ownley et al. 2003).

Antibiotic-deficient mutants may arise spontaneously or be induced chemically, by UV irradiation, or with molecular genetic techniques. The latter are used almost exclusively nowadays because the site of mutation can be localized, providing insight into the biochemical basis and specificity of the mutant phenotype. Mutants should be complemented with wild-type DNA to restore antibiotic synthesis and to help rule out the involvement of undetected second-site mutations. Thorough phenotypic characterization also is essential, especially when genes other than those involved directly in synthesis of the target antibiotic have been mutated. Because many bacteria produce more than one antibiotic, indirect assays based on inhibitory activity in vitro must eliminate or compensate for the effects of metabolites other than the one of interest.

2.3.4 Nontarget Effects of Antibiotic Production In Situ

Although most studies of antibiotic activity in situ have focused on the suppression of plant pathogens, the broader environmental consequences of antibiotic production in the environment also have received attention, particularly in relation to the introduction of recombinant strains of *P. fluorescens* engineered to produce multiple antibiotics (Bakker et al. 2002; Blouin-Bankhead et al. 2004; De Leij et al. 1995; Glandorf et al. 2001; Iavicoli et al. 2003; Leeflang et al. 2002; Moënne-Loccoz et al. 2001. Natsch et al. 1998; Thirip et al. 2001; Timms-Wilson et al. 2004;

Viebahn et al. 2003, 2005a, b; Winding et al. 2004). These studies have addressed effects on soil enzyme activities and available nutrients as well as impacts on the abundance and community structure of microorganisms that are closely related or unrelated to the introduced bacteria, and on protozoa, nematodes, and plants. Especially noteworthy is a 6-year field study of the effects on the soil microbial community of *P. putida* WCS358r, an antibiotic nonproducer modified to produce either phenazine-1-carboxylic acid or DAPG (Bakker et al. 2002; Glandorf et al. 2001; Leeflang et al. 2002; Viebahn et al. 2003, 2005a, b). The study revealed that the wild-type and recombinant strains both had transient effects on the composition of the fungal and bacterial rhizosphere microflora of wheat, with the effects of the recombinant strains sometimes lasting longer. The impact of the introduced strains differed from year to year, revealing no consistent pattern. The results are typical of those of other studies conducted under a variety of controlled or field conditions, and are consistent with the fact that populations of introduced rhizobacteria generally establish large populations immediately after inoculation onto the seed or into soil and then the densities decline over time and distance from the inoculum source. Collectively, the data indicate that the presence of antibiotic-producing bacteria affects rhizosphere microbial community structure, but the effects vary from study to study, often are less than those associated with routine agronomic practices, and are transient.

References

Bakker PA, Glandorf DC, Viebahn M, Ouwens TW, Smit E, Leeflang P, Wernars K, Thomashow LS, Thomas-Oates JE, van Loon LC (2002) Effects of *Pseudomonas putida* modified to produce phenazine-1-carboxylic acid and 2,4-diacetylphloroglucinol on the microflora of field grown wheat. Antonie Van Leeuwenhoek 81:617–624

Blouin-Bankhead S, Landa B, Lutton E, Weller DM, McSpadden Gardener B (2004) Minimal changes in rhizobacterial population structure following root colonization by wild type and transgenic biocontrol strains. FEMS Microbiol Ecol 49:307–318

Blum U, Worsham AD, King LD, Gerig TM (1994) Use of water and EDTA extractions to estimate available (free and reversibly bound) phenolic acids in Cecil soils. J Chem Ecol 20:341–359

Bonsall RF, Weller DM Thomashow LS (1997) Quantification of 2,4-diacetylphloroglucinol produced by fluorescent *Pseudomonas* spp. in vitro and in the rhizosphere of wheat. Appl Environ Microbiol 63:951–955

Bottiglieri M, Keel C (2006) Characterization of PhlG, a hydrolase that specifically degrades the antifungal compound 2,4-diacetylphloroglucinol. Appl Environ Microbiol 72:418–427

Chander Y, Kumar K, Goyal SM, Gupta SC (2005) Antibacterial activity of soil-bound antibiotics. J Environ Qual 34:1952–1957

Chin-A-Woeng TFC, Bloemberg GV, van der Bij AJ, van der Drift KMGM, Schripsema J, Kroon B, Scheffer RJ, Keel C, Bakker PAHM, Tichy J-V, de Bruijn FJ, Thomas-Oates JE, Lugtenberg BJJ (1998) Biocontrol by phenazine-1-carboxamide-producing *Pseudomonas chlororaphis* PCL 1391 of tomato root rot caused by *Fusarium oxysporum* f. sp. radicis-lycopersici. Mol Plant Microbe Interact 11:1069–1077

Chiou CT (1989) Theoretical considerations of the partition uptake of nonionic organic compounds by soil organic matter. In: Sawhney BL, Brown K (eds) Reactions and movement of organic chemicals in soils. SSSA special publication 22. Soil Science Society of America, Madison, pp 1–29

Dalton BR, Weed SB, Blum U (1989) Differential sorption of exogenously applied ferulic, *p*-coumaric, *p*-hydroxybenzoic, and vanillic acids in soil. Soil Sci Soc Am J 51:1515–1521

De La Fuente L, Mavrodi DV, Landa BB, Thomashow LS, Weller DM (2006) *phlD*-based genetic diversity and detection of genotypes of 2,4-diacetylphloroglucinol-producing *Pseudomonas fluorescens*. FEMS Microbiol Ecol 56:64–78

De Leij FAAM, Sutton EJ, Whipps JM, Fenton JS, Lynch JM (1995) Impact of a field release of genetically modified *Pseudomonas fluorescens* on indigenous microbial populations of wheat. Appl Environ Microbiol 61:3443–3453

Delaney SM, Mavrodi DV, Bonsall RF, Thomashow LS (2001) *phzO*, a gene for biosynthesis of 2-hydroxylated phenazine compounds in *Pseudomonas aureofaciens* 30-84. J Bacteriol 183:318–327

Díaz-Cruz MS, Barceló, D (2005) LC-MS^2 trace analysis of antimicrobials in water, sediment and soil. Trends Anal Chem 24:645–657

Glandorf DC, Verheggen P, Jansen T, Jorritsma JW, Smit E, Leeflang P, Wernars K, Thomashow LS, Laureijs E, Thomas-Oates JE, Bakker PA, van Loon LC (2001) Effect of genetically modified *Pseudomonas putida* WCS358r on the fungal rhizosphere microflora of field-grown wheat. Appl Environ Microbiol 67:3371–3378

Gross H, Stockwell VO, Henkels MD, Nowak-Thompson B, Loper JE, Gerwick, WH (2007) The genomisotopic approach: A systematic method to isolate products of orphan biosynthetic gene clusters. Chem Biol 14:53–63

Haas D, Keel C (2003) Regulation of antibiotic production in root-colonizing *Pseudomonas* spp. and relevance for biological control of plant disease. Annu Rev Phytopathol 41:117–153

Haas D, Blumer C, Keel C (2000) Biocontrol ability of fluorescent pseudomonads genetically dissected: importance of positive feedback regulation. Curr Opin Biotechnol 11:290–297

Huang Z, Bonsall RF, Mavrodi DV, Weller DM, Thomashow LS (2004) Transformation of *Pseudomonas fluorescens* with genes for biosynthesis of phenazine-1-carboxylic acid improves biocontrol of Rhizoctonia root rot and in situ antibiotic production. FEMS Microbiol Ecol 49:243–251

Iavicoli A, Boutet E, Buchala A, Métraux JP (2003) Induced systemic resistance in *Arabidopsis thaliana* in response to root inoculation with *Pseudomonas fluorescens* CHA0. Mol Plant Microbe Interact 16:851–858

Jacobsen AM, Halling-Sørensen B, Ingerslev F, Hansen SH (2004) Simultaneous extraction of tetracycline, macrolide and sulfonamide antibiotics from agricultural soils using pressurised liquid extraction, followed by solid-phase extraction and liquid chromatography-tandem mass spectrometry. J Chromatogr A 1038:157–170

Kim S-C, Carlson K (2005) LS-MS^2 for quantifying trace amounts of pharmaceutical compounds in soil and sediment matrices. Trends Anal Chem 24:635–644

Kim SC, Carlson K (2006) Occurrence of ionophore antibiotics in water and sediments of a mixed-landscape watershed. Water Res 40:2549–2560

Leeflang P, Smit E, Glandorf DCM, van Hannen EJ, Wernars K (2002) Effects of *Pseudomonas putida* WCS358r and its genetically modified phenazine producing derivative on the *Fusarium* population in a field experiment, as determined by 18S rDNA analysis. Soil Biol Biochem 34:1021–1025

Loper JE, Lindow SE (2002) Reporter gene systems useful in evaluating in situ gene expression by soil- and plant-associated bacteria. In: Hurst CJ, Crawford RL, Knudsen GR, McInerney MJ, Stetzenbach LD (eds) Manual of environmental microbiology, 2nd edn. American Society of Microbiology, Washington, pp 627–637

Lehmann RG, Cheng HH, Harsh JB (1987) Oxidation of phenolic acids by soil iron and manganese oxides. Soil Sci Soc Am J 51:352–356

Mavrodi OV, Mavrodi DV, Thomashow LS, Weller DM (2007) Quantification of 2,4-diacetylphloroglucinol-producing *Pseudomonas fluorescens* in the plant rhizosphere by real-time PCR. Appl Environ Microbiol 73:5531–5538

Mavrodi OV, McSpadden-Gardener BB, Mavrodi DV, Bonsall RF, Weller DM, Thomashow LS (2001). Genetic diversity of *phlD* from 2,4-diacetylphloroglucinol-producing fluorescent *Pseudomonas* spp. Phytopathology 91:35–43

McSpadden-Gardener BB, Mavrodi DV, Thomashow LS, Weller DM (2001) A rapid polymerase chain reaction-based assay characterizing rhizosphere populations of 2,4-diacetylphloroglucinol-producing bacteria. Phytopathology 91:44–54

Metsä-Ketalä M, Salo V, Halo L, Hautala A, Hakala J, Mäntsälä P, Ylihonko K (1999) An efficient approach for screening minimal PKS genes from *Streptomyces*. FEMS Microbiol Lett 180:1–6

Moënne-Loccoz Y, Tichy HV, O'Donnell A, Simon R, O'Gara F (2001) Impact of 2,4-diacetylphloroglucinol-producing biocontrol strain *Pseudomonas fluorescens* F113 on intraspecific diversity of resident culturable fluorescent pseudomonads associated with the roots of field-grown sugar beet seedlings. Appl Environ Microbiol 67:3418–3425

Nakayama T, Homma Y, Hashidoko Y, Mizutani J, Tahara S (1999) Possible role of xanthobaccins produced by *Stenotrophomonas* sp. strain SB-K88 in suppression of sugar beet damping-off disease. Appl Environ Microbiol 65:4334–4339

Natsch A, Keel C, Hebecker N, Laasik E, Défago G (1998) Impact of *Pseudomonas fluorescens* strain CHA0 and a derivative with improved biocontrol activity on the culturable resident bacterial community on cucumber roots. FEMS Microbiol Ecol 27:365–380

Nielsen TH, Sørensen J (2003) Production of cyclic lipopeptides by *Pseudomonas fluorescens* strains in bulk soil and in the sugar beet rhizosphere. Appl Environ Microbiol 69:861–868

Ownley BH, Duffy BK, Weller DM (2003) Identification and manipulation of soil properties to improve the biological control performance of phenazine-producing *Pseudomonas fluorescens*. Appl Environ Microbiol 69:3333–3343

Pessi C, Blumer C, Haas D (2002) *lacZ* fusions report gene expression, don't they? Microbiology 147:1993–1995

Petrović M, Hernando MD, Díaz-Cruz MS, Barceló D (2005) Liquid chromatography-tandem mass spectrometry for the analysis of pharmaceutical residues in environmental samples: a review. J Chromatogr A 1067:1–14

Picard C, di Cello F, Ventura M, Fani R, Gluckert A (2000) Frequency and biodiversity of 2, 4-diacetylphloroglucinol-producing bacteria isolated from the maize rhizosphere at different stages of plant growth. Appl Environ Microbiol 66:948–955

Raaijmakers JM, Weller DM, Thomashow LS (1997) Frequency of antibiotic-producing *Pseudomonas* spp. in natural environments. Appl Environ Microbiol 63:881–887

Raaijmakers JM, Bonsall RF, Weller DM (1999) Effect of population density of *Pseudomonas fluorescens* on production of 2,4-diacetylphloroglucinol in the rhizosphere of wheat. Phytopathology 89:470–475

Ramos L, Kristenson EM, Brinkman UAT (2002) Current use of pressurised liquid extraction and subcritical water extraction in environmental analysis. J Chromatogr A 975:3–29

Renew JE, Huang CH (2004) Simultaneous determination of fluoroquinolone, sulfonamide, and trimethoprim antibiotics in wastewater using tandem solid phase extraction and liquid chromatography-electrospray mass spectrometry. J Chromatogr A 1042:113–121

Seow KT, Meurer G, Gerlitz M, Wendt-Pienkowski E, Hutchinson CR, Davies J (1997) A study of iterative type II polyketide synthases, using bacterial genes cloned from soil DNA: a means to access and use genes from uncultured microorganisms. J Bacteriol 179:7360–7368

Shanahan P, Borro A, O'Gara F, Glennon JD (1992) Isolation, trace enrichment and liquid chromatographic analysis of diacetylphloroglucinol in culture and soil samples using UV and amperometric detection. J Chromatogr 606:171–177

Stallings JH (1954) Soil produced antibiotics-plant disease and insect control. Bacteriol Rev 18:131–146

Stoob K, Singer HP, Stettler S, Hartmann N, Mueller S, Stamm CH (2006) Exhaustive extraction of sulfonamide antibiotics from aged agricultural soils using pressurized liquid extraction. J Chromatogr A 1128:1–9 ter Laak TL, Gebbink WA, Tolls J (2006a) The effect of pH and ionic strength on the sorption of sulfachloropyridazine, tylosin, and oxytetracycline to soil. Environ Toxicol Chem 25:904–911

ter Laak TL, Gebbink WA, Tolls J (2006b) Estimation of soil sorption coefficients of veterinary pharmaceuticals from soil properties. Environ Toxicol Chem 25:933–941

Thiele-Bruhn S, Seibicke T, Schulten HR, Leinweber P (2004) Sorption of sulfonamide pharmaceutical antibiotics on whole soils and particle-size fractions. J Environ Qual 33:1331–1342

Thirup L, Johnsen K, Winding A (2001) Succession of indigenous *Pseudomonas* spp. and actinomycetes on barley roots affected by the antagonistic strain *Pseudomonas fluorescens* DR54 and the fungicide imazalil. Appl Environ Microbiol 67:1147–1153

Thomashow LS, Weller DM, Bonsall RF, Pierson LS (1990) Production of the antibiotic phenazine-1-carboxylic acid by fluorescent *Pseudomonas* species in the rhizosphere of wheat. Appl Environ Microbiol 56:908–912

Thomashow LS, Bonsall RF, Weller DM (1997) Antibiotic production by soil and rhizosphere microbes in situ. In: Hurst CJ, Knudsen GR, McInerney MJ, Stetzenbach LD, Walter MV (eds) Manual of environmental microbiology, 1st edn. American Society of Microbiology, Washington, pp 493–499

Thomashow LS, Bonsall RF, Weller DM (2002) Antibiotic production by soil and rhizosphere microbes in situ. In: Hurst CJ, Crawford RL, Knudsen, GR, McInerney MJ, Stetzenbach LD (eds) Manual of environmental microbiology, 2nd edn. American Society of Microbiology, Washington, DC, pp 636–647

Timms-Wilson TM, Kilshaw K, Bailey MJ (2004) Risk assessment for engineered bacteria used in biocontrol of fungal disease in agricultural crops. Plant Soil 266:57–67

Tolls J (2001) Sorption of veterinary pharmaceuticals in soils: a review. Environ Sci Technol 35:3397–3406

Validov S, Mavrodi O, De La Fuente, L, Boronin A, Weller D, Thomashow L, Mavrodi D (2005) Antagonistic activity among 2,4-diacetylphloroglucinol-producing fluorescent *Pseudomonas* spp. FEMS Microbiol Lett 242:249–256

Viebahn M, Glandorf DCM, Ouwens TWM, Smit E, Leeflang P, Wernars K, Thomashow L, van Loon LC, Bakker PAHM (2003) Repeated introduction of genetically modified *Pseudomonas putida* WCS358r without intensified effects on the indigenous microflora of field-grown wheat. Appl Environ Microbiol 69:3110–3118

Viebahn M, Veenman C, Wernars K, van Loon LC, Smit E, Bakker PAHM (2005a) Assessment of differences in ascomycete communities in the rhizosphere of field-grown wheat and potato. FEMS Microbiol Ecol 53:245–253

Viebahn M, Doornbos R, Wernars K, van Loon LC, Smit E, Bakker PAHM (2005b) Ascomycete communities in the rhizosphere of field-grown wheat are not affected by introductions of genetically modified *Pseudomonas putida* WCS358r. Environ Microbiol 7:1775–1785

Weber JB, Miller CT (1989) Organic chemical movement over and through soil. In: Sawhney BL, Brown K (eds) Reactions and movement of organic chemicals in soils. SSSA special publication 22. Soil Science Society of America, Madison, pp 305–354

Winding A, Binnerup SJ, Pritchard H (2004) Non-target effects of bacterial biological control agents suppressing root pathogenic fungi. FEMS Microbiol Ecol 47:129–141

Chapter 3
Rhizosphere Metabolomics: Methods and Applications

Sheela Reuben, V.S. Bhinu, and Sanjay Swarup(✉)

3.1 Introduction

The emerging field of rhizosphere metabolomics involves analysis of entire metabolite complement (metabolome), in an unbiased way to understand complex physiological, pathological, symbiotic and other relationships among the inhabitants of the rhizosphere. Metabolomic studies of the rhizosphere are quite challenging since the rhizosphere is a complex as well as a dynamic microenvironment. Metabolite composition in the rhizosphere is primarily governed by the nature of root exudates, secretions from rhizobacteria, fungi and other soil organisms. Conversely, the nature of these root exudates also directly or indirectly affects microbial growth in the rhizosphere. While some compounds enhance growth, others have antimicrobial activities. Apart from the diverse roles of compounds present, the complexity of the rhizosphere also stems from competition among rhizosphere microbes. Some of them are growth-promoting, while others are pathogenic. These effects are not only confined to the microbes but also extend to the plants growing in the rhizosphere. Hence, gaining knowledge of these rhizosphere metabolites as well as the effect of the biota will help us better understand this ecological niche.

The field of metabolomics utilizes analytical techniques such as chromatography, mass spectrometry (MS), nuclear magnetic resonance (NMR) and spectroscopy to profile, identify and estimate the relative abundance of metabolites at a given time. Various methods involving gene expression studies, enzymatic studies and biochemical techniques have been used to understand the events that occur in the rhizosphere. However, it is often observed that the metabolite levels do not coincide with the activity of the biosynthetic enzymes and their end products. Effects of the metabolites on the system, therefore, cannot be easily studied using solely RNA or enzyme-based techniques. Metabolomics helps to overcome

Sanjay Swarup
Small Molecule Biology Laboratory, Department of Biological Sciences,
National University of Singapore, Singapore 117543
e-mail: sanjay@nus.edu.sg

these pitfalls and provides a comprehensive approach to provide a biochemical status report.

This chapter focuses on metabolomics in relation to the rhizosphere. A brief overview of the components of the rhizosphere, interaction between the different species and competition among rhizobacteria is provided. Many technologies used for metabolic profiling and their role in rhizosphere metabolomics are discussed in some detail. The sole of bioinformatics and data visualization methods are summarized. Finally, this chapter ends with a brief view of the applications of selected metabolomic studies.

3.2 The Rhizosphere—Chemical and Biological Components

The rhizosphere is a complex, dynamic and highly interactive microenvironment. It has an ecological advantage for those organisms that are exclusively associated with the roots of plants. Here, we provide a brief description of the rhizosphere components especially in relation to the origin and nature of metabolites and their effects on the rhizosphere biota and microbiota.

3.2.1 Composition of the Rhizosphere

3.2.1.1 Root Exudates

Root exudates play a key role in the rhizosphere, as their composition and abundance affect the growth and characteristics of the organisms thriving in the rhizosphere. Excellent reviews and books are available on this topic where details of exudate composition are provided (Bais et al. 2006; Mukerji et al. 2006; Pinton et al. 2000). The list of low molecular weight compounds identified in the rhizosphere is very long and broadly consists of amino acids, organic acids, sugars, phenolics and various other secondary metabolites. Exudates vary with respect to signals of biotic or abiotic origins. Allelochemicals in the root exudates govern the type of organisms that grow in the region. Allelochemicals are secondary metabolites that influence the growth of other organisms. Allelopathy, a phenomenon that refers to the role of allelochemicals, is exploited in the control of insects and weed plants. Some of the allelochemicals include tannins, cyanogenic glycosides, benzoquinones, flavonoids and phenolic acids. The biological and physiological mechanisms of allelochemicals have been reviewed (Weir et al. 2004; Inderjit and Duke 2003). Benzoxazinones are an important class of allelochemicals whose sample extraction and separation methods have been reviewed by Eljarrat and Barcelo (2001) and Bonnington et al. (2003), respectively. These compounds are easily hydrolyzed and hence care needs to be taken during their sample preparation.

3.2.1.2 Rhizobacteria

Rhizobacteria form an integral part of the rhizosphere. They include the microorganisms that are both beneficial as well as pathogenic. The term "rhizoengineering" refers to the engineering of rhizobacterial populations in order to improve the interactions and outcomes within the soil environment. Several beneficial microorganisms are known to cause breakdown of natural products or even degrade them to simple sugars that are recycled for other anabolic reactions in the rhizosphere biota. Many of these plant products are terminal metabolites of biosynthetic events in plants and it is not uncommon to find rhizobacteria that utilize these end products for energy generation. In the case of the phenylpropanoid compounds, they are exuded in the rhizosphere and some microbes degrade these compounds through specific metabolic pathways. Two examples are the phenylpropanoid catabolic pathway in plant growth promoting rhizobacterial strains of *Pseudomonas putida* (Pillai and Swarup 2002) and the fluorophenol degradation pathway in different species of *Rhodococcus* (Boersma et al. 2001).

3.2.1.3 Soil Fungi

These organisms could be pathogenic or symbiotic, such as in the case of the mycorrhizae and contribute to the complex biotic interactions in the rhizosphere. Fungal development is often stimulated in the presence of roots especially owing to the nitrogen released by the roots. The presence of some rhizobacteria may cause the inhibition of mycorrhizal growth. For example, the growth of ectomycorrhizae is inhibited by the presence of selected isolates of *Pseudomonas* and *Serratia* in the early infection stage of the fungi (Bending et al. 2002). Such growth inhibition is mostly mediated by the secretion of antibiotics or antimicrobial compounds in the rhizosphere such as phenazines or selected flavonoids.

3.2.1.4 Soil Nematodes

Soil nematodes play an important part in the rhizosphere. Nematodes influence the nature of root exudation, which affects the physiological functioning of microorganisms in the rhizosphere. These exudates may serve as signal molecules for nematode antagonists and parasites (Kerry 2000). Matrix assisted laser desorption ionization (MALDI) time-of-flight (TOF) analysis has been used in analyzing the metabolites from plant nematodes (Perera et al. 2005). These nematodes are often difficult to identify and the technique provides a good opportunity for a rapid and simple identification of plant parasitic nematodes. More details regarding MALDI TOF analysis are provided in Sect. 3.3.4.5.

3.3 Rhizosphere Metabolomics

3.3.1 Metabolomics: an Overview

"Metabolome" refers to the sum total of all the nonproteinaceous small molecules (metabolites) present in an organism. Metabolites are the small molecules that are the end products of enzymatic reactions. "Metabolomics" is unbiased identification and quantification of all metabolites present in a sample from an organism grown under defined conditions. Another term, "metabonomics," is used frequently in the biomedical (toxicology) literature and for methods involving NMR spectroscopy. However, "metabolomics" is a preferred term for unbiased metabolite analyses (Bhalla et al. 2005). Metabolomics is not restricted to *metabolic profiling*; it also encompasses a much broader study including identification of metabolites (to understand the range of metabolites produced by the organism), their quantitation (to detect the abundance of metabolites), comparisons (to understand the differences arising from perturbations in metabolic pathways), data analysis (chemoinformatics) and development of metabolic models. "Metabolic profiling" refers to obtaining a listing of the entire range of the metabolites present in the organism. Such complete profiles are not unattainable in practice, because one single method of extraction or analysis covers only a partial spectrum of the metabolome, as mentioned earlier. Various analytical methods alone, or in conjunction with others, have therefore been used for comprehensive metabolic profiling. Some of these techniques are discussed in Sect. 3.3.4.

Metabolomics has many applications, including but not restricted to (1) understanding the enzyme fluxes, (2) uncovering novel metabolic pathways, (3) unraveling cryptic pathways, (4) identifying biomarkers and (5) metabolic engineering of novel products that are industrially and biomedically relevant. Metabolomics, in conjunction with other "omics" such as functional genomics, proteomics and transcriptomics, has helped in better understanding the biological systems. Integration of data from the various fields has helped in painting a holistic picture of the biological system using the systems approach. Many excellent reviews are available for metabolomic studies (Sumner et al. 2003; Ryals et al. 2004; Villas-Boas et al. 2005; Bhalla et al. 2005; Dunn and Ellis 2005).

The rhizosphere is a constantly changing microenvironment, where there is a flux of energy, nutrients and molecular signals between the plant roots and microbes that affects their mutual interactions. Metabolites exuded from plants as well as the metabolites released or secreted by the microbiota present in the rhizosphere have a considerable effect on this microenvironment. Hence, metabolic profiling constitutes a powerful technique to understand the underlying phenomenon of such exudations and the effects of metabolites on soil ecological relationships, plant–microbe interactions and other soil organisms. The use of improved analytical techniques has helped in the characterization of microorganisms from soil. One example is the characterization of *Bacillus* and *Brevibacillus* strains using UV resonance Raman spectroscopy (López-Díez and Goodacre 2004). Other applications include understanding biotic interactions. For example, gas chromatography (GC)/MS has been used to study symbiotic nitrogen fixation in legume roots and in understanding plant–microbe interactions (Desbrosses et al. 2005). Several techniques have been used in structural elucidation of metabolites, like NMR, IR spectroscopy, Fourier transform (FT) MS and so on. Tandem MS has also been used in such studies especially for studying metabolites from roots or root exudates. It is usually helpful in providing an initial partial structure that can be fully elucidated by NMR spectroscopy. For example, structural elucidation of montecristin, a key metabolite in biogenesis of acetogenins from the roots of *Annona muricata*, was performed using tandem MS and NMR (Gleye et al. 1997). Another application is the identification of allelochemicals (Eljarrat and Barcelo 2001).

3.3.2 Root Exudates Profiling

Root exudates form a major component of the rhizosphere. While numerous reports are available on identification of selected classes of root exudates as mentioned in Sect. 3.2.1.1, we briefly describe here some studies that have employed unbiased analytical methods. High-performance liquid chromatography (HPLC) and NMR spectroscopy have led to the identification and quantification of a number of metabolites in the root exudates of *Arabidopsis thaliana* (Walker et al. 2003). Nearly 289 possible secondary metabolites were quantified and chemical structures of ten compounds were elucidated. The authors conducted a time-course study of root exudates from plants treated with salicylic acid, jasmonic acid, chitosan and two fungal cell wall elicitors. Plants treated with salicylic acid had the maximum number of compounds in their exudates, while elicitation with jasmonic acid had the least effect on exudates. This method of root exudates profiling could identify differences in root exudation with respect to plant stress. Such types of studies can provide indirect reference to the metabolic pathways during the different stress conditions. Some of the compounds reported in the study had previously not been reported from *Arabidopsis*. The authors also tested the antibacterial as well as the antifungal activities of several of these compounds. This study also highlights the constant change in the exudation, which directly affects the microbial populations.

Root exudates profiling in graminaceous plants has been used to understand the acquisition of metal ions from soil. Root exudates profiling in graminaceous plants was conducted using multinuclear and two-dimensional NMR with GC/MS and coupled with high-resolution MS for metabolite identification (Fan et al. 2001). The root profiling method was used to examine the role of exudate metal ion ligands (MILs) in the acquisition of Cd and transition metals in barley and wheat. The change in the root exudate profile was studied in wheat, barley and rice grown on Fe- and Cd-deficient soils. MILs such as 3-epihydroxymugineic acid, mugineic acid, 2′deoxymugineic acid and malate in barley were elevated in Fe-deficient conditions, which in turn increased the Fe-mobilizing substances. The results suggest that enhanced exudation of murigenic acids and malate may be involved in acquisition of transition metals but not of Cd, and also that the mechanisms of acquisition for essential and toxic metal ions may be different.

3.3.3 Current Limitations of Rhizosphere Metabolomics

While metabolomics is highly powerful, the field as such has certain limitations. Four general limitations are described here. One bottleneck at present is that it is technically challenging and expensive to detect a wide range of metabolites often at low concentrations; hence, combinations of techniques need to be used. This not only makes it an expensive approach but also demands high levels of technical expertise in analytical chemistry as well as "chemometrics" (analysis of chemical data), which are described in some detail in Sect. 3.3.4. Each technique has its own set of limitations, owing to which the entire profile is often not obtained. For example, with an ionization-based mass spectrometer such as an electrospray ionization (ESI) mass spectrometer the metabolites detected are often the highly ionizable ones under the particular buffer system and conditions used in specific experiments. Ones that do not ionize in such a buffer system are often not detected. The second challenge lies is the choice of sample extraction procedure. There is no universal extraction method that is suitable for all types of compounds. Often one needs to use different sample extraction techniques and buffer systems depending on the type of target compounds. Thirdly, metabolomics is limited by the requirement of an accurate and well-curated database for the spectra of compounds. Such databases are required for comparisons of established signatures or fingerprints of various metabolites. As more and more laboratories are beginning to use metabolomics, the future of a comprehensive database looks bright. At present, many laboratories use standards for various compounds to establish the identity of a metabolite. Some of the compounds are not commercially available and these need to be isolated from natural sources. Hence, a large proportion of the compounds detected remain unidentified. A major development in solving this problem is the increasing adoption of GC/MS by many metabolomics researchers. As GC/MS generates spectra for fully ionized compounds, such spectra are highly reproducible between laboratories and can be used as references and exchanged between researchers. The fourth bottleneck in this field is the limited availability of chemoinformatics tools that expedite analysis of a large amount of data and convert it to an interpretable form with respect to the biological characteristics of the various systems. There has been considerable progress in the computational biology field in recent years and more tools are being developed. Some of the bioinformatics tools are discussed in Sect. 3.3.5.1.

In addition to the current limitations in metabolomics, the field of rhizosphere metabolomics in particular faces two additional challenges. One is due to the different collection procedures required for root exudates. These are collected by growing plants hydroponically, aeroponically or even in soil; hence, specialized techniques and care are required for collection purposes. Secondly, it is difficult to separate the exudations from various interacting biotic agents in the rhizosphere such as plants and their associated microbes. Further advances in these areas are required to bridge the gap in the knowledge base of gnotobiotic and field-based systems. Spectroscopic methods used for the identification of soil bacteria are becoming increasingly popular as they provide information on nonculturable microbes as well as on the relative abundance of various microbial species. For example, Raman microscopy analysis gives the spectral profile from a single cell, which helps in bacterial species differentiation (Huang et al. 2004).

3.3.4 Rhizosphere Metabolomics Methods

As exemplified by the root exudate profiling studies already described, they rely largely on a highly sophisticated suite of analytical techniques. We briefly describe the major groups of techniques used in metabolomic studies. Some of these include the chromatography techniques, MS, NMR and spectroscopy of various types. The utility of analytical techniques in metabolomics has been more extensively reviewed by Dunn et al. (2005).

3.3.4.1 Chromatography Techniques

Chromatography techniques help in separation and analysis of the metabolites. Different types of such techniques are available for metabolite analyses (Table 3.1). Here, we give a few examples that represent how these techniques can help in understanding the nature of root exudates or interactions in the rhizosphere:

- Thin layer chromatography. This technique involves the separation of metabolites on the basis of differential partitioning between the components of a mixture and the stationary solid phase. This is a very simple and inexpensive analytical method. Reverse-phase thin layer chromatography (TLC), along with some other techniques, has been useful in understanding fungal-bacterial interactions in the rhizosphere. The rhizobacteria *Pseudomonas chlororaphis* PCL1391 produces an antifungal metabolite phenazine-L-carboxamide, which is a crucial trait in its competition with the phytopathogenic fungus *Fusarium oxysporum* f. sp. *radicis-lycopersici* in the rhizosphere (Chin-A-Woeng et al. 2005). In this study, TLC was used to identify autoinducer compounds that were released during the expression of sigma factor psrA in different quorum-sensing gene mutants. In another application, TLC was used in studying the nodulation signaling metabolites that are secreted into the growth medium produced due to the *nod ABC* genes of *Rhizobium* and *Bradyrhizobium* strains (Spaink et al. 1992). TLC can be used to separate polar metabolites and fatty acids as well as to test the purity of compounds.
- Reverse-phase HPLC. In this technique the metabolites are separated on the basis of their hydrophobicity and they can be identified by comparing the retention

Table 3.1 Summary of the different available analytical techniques that have proven valuable in soil biology studies

Techniques	Special features	Examples from soil biology studies
Chromatography		
Thin layer chromatography	Simple and inexpensive	Study of nod metabolites (Spaink et al. 1992)
Anion-exchange chromatography	Effective separation	Study of soluble carbohydrates (Cataldi et al. 2000)
Mass spectrometry		
Matrix-assisted laser desorption ionization	Useful in identification of metabolites	Study of aconitum alkaloids from aconite roots (Sun et al. 1998)
Proton transfer reaction mass spectrometry	Rapid and real-time analysis	Study of rhizosphere volatile organic compounds and their induction by biotic stresses (Steeghs et al. 2004)
Spectroscopy		
Nuclear magnetic resonance	Less sensitive but powerful for chemical structure identification	Building flux maps and metabolic network modeling. (Ratcliffe and Shachar-Hill 2005)
Tandem analysis methods		
Liquid chromatography/electrospray ionization mass spectrometry	Analysis of even labile compounds as there is no need to derivatize	Identification of isoflavone conjugates from roots of lupine species (Kachlicki et al. 2005)
Gas chromatography/mass spectrometry	Low-mass volatile compound can be identified	Identification of signaling molecules during ectomycorrhizae formation (Menotta et al. 2004)
Gas chromatography/combustion/isotope-ratio mass spectrometry	Helps to study dynamic nature of metabolites	To study the nature and dynamics of plant sugars in the rhizosphere (Derrien et al. 2003)

times with those of standard compounds. This method has been used in comparing the root exudates from different cultivars. For example, root exudates from seven accessions were evaluated using HPLC (Czarnota et al. 2003). Another application includes the use of HPLC in quantifying the amount of sorgoleone, a photosynthetic inhibitor in the rhizosphere of sorghum plants (Weidenhamer 2005). Polydimethylsiloxane (PDMS) was used for the study. The amounts of sorgoleone retained on the PDMS increased with time, which could be shown using HPLC methods. The use of HPLC in root exudates profiling has already been mentioned in Sect. 3.3.2 (Walker et al. 2003).

- Anion-exchange chromatography. Another form of chromatography, this technique is based on charge-to-charge interactions between the target compounds and the charges immobilized on the column resin. In anion-exchange chromatography the binding ions are negative and the immobilized functional group is positive. It has been used to determine the composition of soluble carbohydrates in plant tissues such as olive roots (Cataldi et al. 2000). The authors have used the technique for efficient separation of carbohydrates. Such studies can be extended to understand the movement of sugars and the types of sugars that are available in the soil for the rhizobacteria.

Chromatography techniques are powerful tools when used in conjunction with other techniques such as MS. Liquid chromatography (LC) and GC techniques have been used with different types of mass spectrometers as described in the context of rhizosphere metabolomics here.

3.3.4.2 Mass Spectrometry

MS analysis has come a long way since 1954, when John Beynon from Imperial Chemical Industries, UK, first suggested that the spectra could be correlated to structure and outlined the basic rules of MS. MS has now developed into a powerful analytical tool with applications in chemical analysis, drug development, natural products analysis and biomedical applications to name a few. Computer interfacing has added an additional software-driven component, which has brought the instrument within the reach of biologists. In a mass spectrometer, the samples are ionized by different methods. This is usually done in the source part of the mass spectrometer. There are different ionization methods, like electron impact, chemical ionization, ESI, fast atom bombardment, field ionization, field desorption and laser desorption. In electron impact ionization, the samples are ionized by the bombardment of electrons. The ionization is caused by the interaction of the fields of the bombarded electron and the molecule, resulting in the emission of an electron. In an ESI mass spectrometer the sample is sprayed as a fine liquid aerosol. A strong electric field is applied under atmospheric pressure to the liquid passing through a capillary tube, which induces charge accumulation at the liquid surface, which then breaks up to form highly charged droplets. As the solvent evaporates, the droplets explode to give ions. The spectra obtained are usually those of mutiply charged molecular ions owing to protonation. In laser desorption ionization, a laser pulse is focused onto the surface of the sample, some part of the compound gets desorbed and reactions among the molecules in the vapor-phase region result in ions. An extension of this method is the MALDI method. In this technique, samples are mixed with a suitable matrix and allowed to crystallize on grid surfaces. Samples are then irradiated with laser pulses to induce ionization. Most of the energy of the laser pulse is absorbed by the matrix, so unwanted fragmentation of the biomolecule is avoided. Chemical ionization is considered to be "soft ionization" technique as the number of fragment ions produced is less. In this method a reactant gas like methane is passed through the sample and the interaction of the ions with neutral molecules produces new ions. The ions formed by any of the methods are then accelerated through a column and deflected in a magnetic field. In the mass analyzer the ions are separated according to their mass-to-charge ratio (m/z) and finally they are detected by an ion detector. In a triple-quadrupole mass analyzer, the ions from the source are passed between four parallel rods. The motion of the ions depends on the electric field, which allows only ions of the same m/z to be in resonance and to the detector at the same time. Triple-quadrupole MS is most often used for quantification purposes. In the case of an ion-trap mass analyzer, the ions are focused using an electrostatic lensing system into the ion trap. An ion will be stably trapped depending upon the values for the mass and the charge of the ion. A mass spectrum is obtained by changing the electrode voltages to eject the ions from the trap. In a TOF detector the molecules are detected on the basis of the time that each molecule takes to reach the detector.

A schematic representation of the acquisition of mass-spectral data is provided in Fig. 3.1.

Chromatographic separations followed by mass-spectral analysis provide additional separation. This is because the metabolites are first separated on a chromatographic column which partitions the metabolites into different fractions and each of the fractions is further analyzed by a mass spectrometer. The separation of the metabolites into fractions helps in reducing the ion suppression effect and enhances detection and therefore more metabolites can be analyzed from samples. The metabolites can be fragmented for identification purposes using a tandem mass spectrometer. A Tandem mass spectrometer can be considered as two mass spectrometers in series with a collision cell in-between. This type of instrument

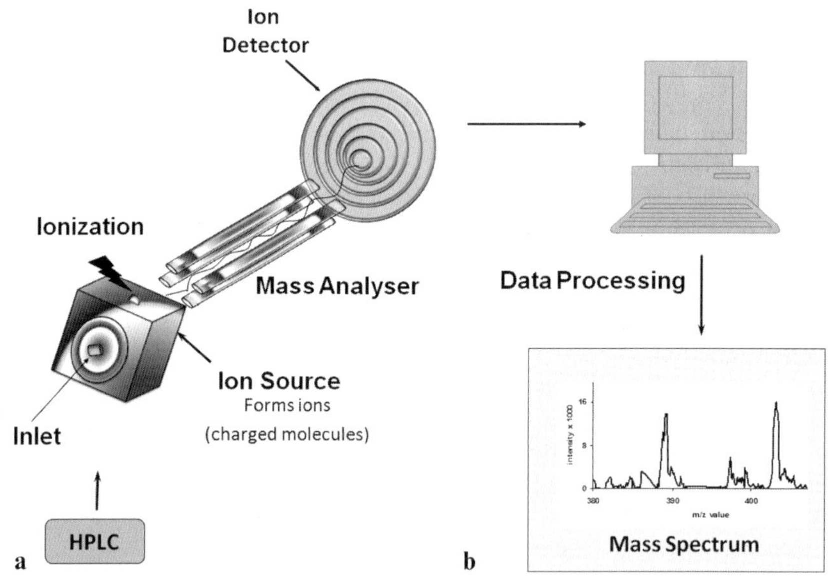

Fig. 3.1 The various steps in acquiring a mass spectrum

helps in fragmenting targeted ions that give rise to daughter ions. These daughter ions form a fingerprint that can then be compared with fingerprints of standards or databases. Mass spectrometers have been used in conjunction with various chromatographic methods. The two commonly used chromatographic methods are LC in conjunction with MS and GC in conjunction with MS. Some of these analytical techniques have been described recently (Dunn et al. 2005; Sumner et al. 2003). MS can also be used to optimize the chromatographic separations. The separations are best if the metabolites are limited to the least number of fractions. Conversely, if the same metabolite is present in more fractions then the separations by HPLC are not very efficient (Fig. 3.2). In this example, each of the metabolites is in just one fraction (no more than one) and 80% belong to this category, indicating an effective HPLC-based separation.

3.3.4.3 Gas Chromatography/Mass Spectrometry

This technique is mostly used to study volatile compounds. As GC/MS relies on the hard ionization methods, ion spectra are highly uniform and reproducible between experiments. Owing to this advantage, standard databases can be created and shared between laboratories. Several examples of GC/MS use are available in the plant metabolomics literature. For example, the GC/MS technique has been used to study the differences in plants of different developmental stages with respect to their day length (Jonsson et al. 2004). GC/MS has been useful in identifying molecules such as those involved in signaling during ectomycorrhizae formation (Menotta et al. 2004). These molecules are exuded during the presymbiotic interaction between *Tuber borchii* (ectomycorrhiza) and the host plant *Tilia americana*. Seventy-three volatile organic compounds (VOCs) could be identified and 29 of these were produced during interaction between the fungi and the host and; therefore, they could possibly be signaling molecules. The technique thus assists in increasing our understanding of rhizosphere signaling.

GC–combustion–isotope-ratio MS (GC/C/IRMS) is another useful technique that has been adopted in rhizosphere metabolomics. An Isotope-ratio mass spectrometer accurately determines the elemental isotope ratios very precisely and accurately. Single focusing magnetic sector mass spectrometers with multiple detectors are used in this technology. The principle of IRMS is that the ratio of isotopes in a compound varies according to its source and forms an isotopic fingerprint, which can be detected using a mass spectrometer. The advent of IRMS has helped in evaluating the interactions between organisms and the environment by studying the variability of the natural abundance of stable isotopes. Stable isotope mass-spectrometric approaches are also useful in understanding biotic interactions in complex ecosystems. Different phenomena involving soil microorganisms and soil invertebrates were recently determined by the $\delta^{13}C$ values of individual compounds (Evans and Evershed 2003). GC/C/IRMS was used in conjunction with ^{13}C labeling to study the nature and dynamics of plant sugars in the

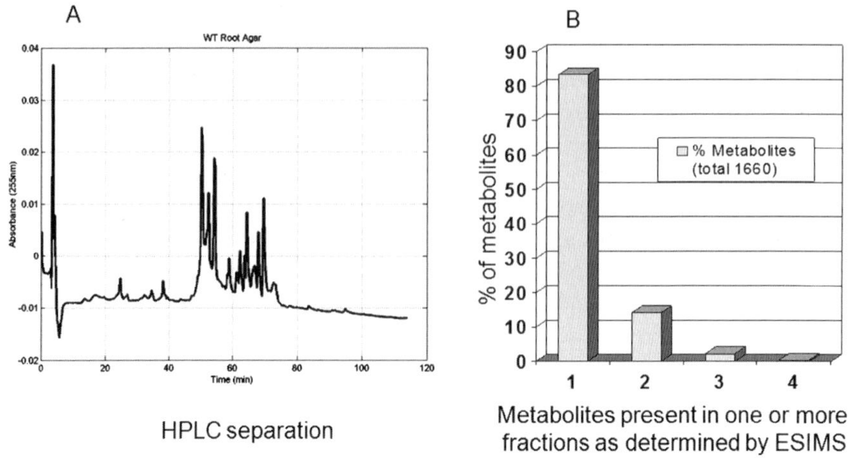

Fig. 3.2 Use of a mass spectrometer to optimize high-performance liquid chromatography (HPLC) methods. **a** Offline analytical HPLC chromatogram of *Arabidopsis* Columbia plant roots grown on agar plates. Metabolites present in plant roots were collected into 16 major reverse-phase HPLC fractions and each fraction is separated by electrospray ionization mass spectrometry. **b** The frequency of each mass/charge (*m*/*z*) value in the different fractions. The frequency helps in determining the efficiency of the HPLC run

rhizosphere (Derrien et al. 2003). In this study, the sugars were hydrolyzed and trimethylsilylated (addition of several carbon atoms per sugar) to derivatize the carbohydrates prior to GC analysis. The polar hydroxyl groups were replaced by nonpolar groups that contain carbon. The isotope excess of each sample was determined using calibration of the number of analyzed added carbon atoms in terms of the ratio of ^{13}C to normal C of individual sugars. The study highlighted the use of this technique and discussed the derivatization aspects and proposed further use of the technique in understanding the sugar dynamics in soil. IRMS has been coupled with the continuous flow mode to understand the C cycling in forest soil (Formanek and Ambus 2004). The efflux of CO_2 is a combination of respiratory activity of roots and associated rhizosphere organisms, soil fauna and soil microorganisms. The contribution of the CO_2 from each group can be analyzed to understand C cycling and sequestration.

3.3.4.4 Liquid Chromatography/Mass Spectrometry

Although numerous metabolites can be identified in a single run using GC/MS, the technique may not always prove useful especially in the case of metabolites that are sequestered in compartments and are labile or degraded in high-temperature regimes; hence, such metabolites are difficult to derivatize. In such cases, LC/MS may be the technique of choice. This technique is very commonly used as it is a very convenient platform especially when used in conjunction with ESI MS. Nearly 13–20 isomeric isoflavone conjugates have been identified from roots of lupine species using ESI MS (Kachlicki et al. 2005). In that study, a comparative analysis of triple-quadruple and ion-trap analyzers was conducted. The study highlighted the utility of these techniques in analyzing metabolites in biological samples. Such techniques can be used to study the role of metabolite conjugations in root–microbe interactions since flavonoids play a major role in plant–microbe interactions as discussed in Sect. 3.4.2.

3.3.4.5 Matrix-Assisted Laser Desorption Ionization Time of Flight

This is a very sensitive method and quantities as low as 10^{-15}–10^{-18} mol can be detected. This method has been useful in the study of aconitum alkaloids from aconite roots (Sun et al. 1998). This kind of analysis often leads to the identification of new metabolites as in this case where three new alkaloids were identified. MALDI TOF is most useful for determining the mass accurately.

3.3.4.6 Proton Transfer Reaction Mass Spectrometry

This new technology allows rapid and real-time analysis of most biogenic VOCs without preconcentration or chromatography. Compounds are ionized by a chemical ionization method using H_3O^+ ions. The H_3O^+ ions transfer their protons to the VOCs, which have higher proton affinities than water. The process is referred to as "soft ionization" as it avoids excessive fragmentation of the biomolecules and allows real-time analysis. The detection limit in proton transfer reaction MS is as low as a few parts per trillion. This technique has been used to study rhizosphere VOCs and their induction by biotic stresses (Steeghs et al. 2004). The VOCs can be analyzed without previous separation by chromatography. VOCs induced specifically as a result of interactions between microbes and insects and *Arabidopsis* roots could be detected. For example, in the abovementioned study, compatible interactions of *Pseudomonas syringae* DC3000 and *Diuraphis noxia* with *Arabidopsis* roots showed rapid release of 1,8-cineole, a monoterpene that was not previously reported in *Arabidopsis*.

3.3.4.7 Spectroscopy Methods

Spectroscopic techniques have been increasingly used in studying and identifying metabolites. FTIR spectroscopy is a technique that is useful in identifying organic and inorganic chemicals. The chemical bonds in a molecule can be determined by interpreting the IR absorption spectrum. Molecular bonds vibrate at various frequencies depending on the elements and the type of bonds and therefore give a specific absorption spectrum. FTIR spectra of pure compounds are so unique that they are like a molecular "fingerprint." Raman spectroscopy is another technique where the observed spectrum is based on the vibration of a scattering molecule. When a photon is incident on a molecule, it interacts with the electric dipole of the molecule. The interaction can be viewed as a perturbation of the molecule's electric field. Both FTIR and Raman spectroscopy are effective in the rapid identification of bacteria and fungi (Goodacre et al. 2000).

3.3.4.8 Nuclear Magnetic Resonance Spectroscopy

NMR spectroscopy is a less sensitive technique than MS; however, it is highly powerful in identification of small molecules and is one of the most used forms of spectroscopy. It helps in accurately determining the structure of a metabolite. Any molecule containing one or more atoms with nonzero moment is detectable by NMR. Biologically important atoms such as ^{1}H, ^{13}C, ^{14}N, ^{15}N and ^{31}P are all detectable by NMR. All biologically important metabolites provide NMR signals. NMR spectra are characterized by the chemical shifts, intensity and fine structure of the signals. These signals help in identification and quantification of the metabolites. The use of NMR in metabolic fingerprinting and profiling of plants has recently been reviewed by Krishnan et al. (2005). These authors reviewed NMR profiling and multivariate data analysis with respect to the effect of stress on wild-type, mutant and transgenic plants. This highlights the potential of NMR in plant metabolomics. NMR has also been used in investigating the operation of networks in plants. Labeling with isotopes in conjunction with metabolite analysis by NMR can help in building flux maps that can be useful in metabolic network modeling (Ratcliffe and Shachar-Hill 2005). ^{133}Cs NMR has been used to examine the intracellular and extracellular pools of Cs^{+}-containing, and CsCl-perfused, excised maize seedling roots. Hence, further insight was gained into the ion transport and subcompartmentalization in the root tissues (Pfeffer et al. 1992). Quantitative NMR is used for quantitative measurements. Further, solid-state NMR has been used in studying plant nitrogen metabolism (Mesnard and Ratcliffe 2005). Solid-state ^{1}H NMR is useful in elucidating the structure as well as the dynamic nature in the solid phase. In this technique, the structure is derived on the basis of the number of H atoms, the neighboring H atoms and the environment of the H atoms. These techniques give a wealth of information on the metabolites in plants and can also be useful in identifying and quantitating root exudation metabolites. The utility of NMR in metabolomic studies has been reviewed by Griffin (2004) and Reo (2002).

3.3.5 *Methods and Tools for Metabolomics Data Handling and Analysis*

Metabolomic studies often lead to huge data sets that need to be analyzed and stored. In typical HPLC-based separation followed by offline MS such as ESI MS of each fraction, ions can be detected per fraction in positive mode and a slightly lesser number in the negative mode. Hence, one sample from a single injection can yield 10,000–12,000 data points. With replications and various samples, the number increases considerably, generating several hundred megabytes of data per experiment.

Databases to store such data as well as for the identification of the metabolites, therefore, become essential. Raw data obtained from the instruments have to be preprocessed by several methods to minimize effects of machine variations or experimental errors during weighing of samples or injections into the instruments. Examples of preprocessing include data normalization, baseline correction and alignment of spectra. Following preprocessing, the data need to be converted to useful biological information by a variety of data analysis techniques. Lastly, the analysis of these data can lead to verification of an original hypothesis or the discovery of new associations, which can be experimentally validated. The following criteria have been proposed for creating robust and interpretable multivariate models for comparison of many samples (Jonsson et al. 2005):

1. Each sample is characterized by the same number of variables.
2. Each of these variables is represented across all observations.
3. A variable in one sample has the same biological meaning or represents the same metabolite in all other samples.

Increasing attention is recently being given to metabolic modeling, which leads to development of metabolic networks. Such networks help in understanding the biochemical behavior at the whole-cell level. Evolutionary computation-based methods such as genetic algorithms and genetic programming are ideal strategies for mining such high-dimensional data to generate useful relationships, rules and predictions (Goodacre 2005).

3.3.5.1 Tools for Data Analysis

Several tools for data analysis have been developed by computational biologists for the preprocessing and analysis of data. One such recent tool is MZmine (Katajamaa and Oresic 2005), which is useful for data generated via LC/MS studies. It contains algorithms useful in data preprocessing such as spectral filtration, peak detection, alignment and normalization. The visualization tools enable comparative viewing of the data across multiple samples and peak lists can be exported to other tools for data analysis. Data obtained from metabolomic studies often are very complex as multiple dimensions may be involved. For example, various dimensions due to different treatments such as doses or time-point-based data may need to be handled. Specific types of biostatistical tools are required to make meaningful conclusions from such data. One such statistical tool is ASCA analysis of variance–simultaneous component analysis) (Smilde et al. 2005). It is a direct generalization of analysis of variance from univariate to multivariate data. This tool helps in analyzing complex data generated by LC/MS involving different parameters such as time and dose factors. Other multivariate data analysis techniques such as principal component analysis (PCA) and partial least squares regression (Martens and Naes 1993) can also be used to analyze metabolomics data. The technique can reduce the number of dimensions to two or three, which can be represented graphically. These representations allow the user to visualize the patterns or clusters in the data sets as hierarchical plots or scatter plots. PCA helps in visualizing the data in a simplified way and helps in extracting meaningful biological interpretations. A variation of PCA is the weighted PCA where spectra of repeated measurements are converted to weights describing the experimental error and it adds interpretation to the metabolomics data (Jansen et al. 2004). Multivariate data analysis has been reviewed by van der Werf et al. (2005).

MSFACT is another metabolite data analysis tool, and consists of spectral formatting, alignment and conversion tools (Duran et al. 2003). This tool helps in reformatting, alignment and export of large chromatographic data sets to allow more rapid visualization and interrogation of metabolomics data. Applications of the tool were illustrated using GC/MS profiles from *Medicago truncatula*. Metabolites from various tissues such as roots, stem and leaves from the same plant were easily differentiated on the basis of metabolite profiles. The tool uses hierarchical clustering, two-dimensional PCA and three-dimensional PCA as visualization tools.

Another recent tool is XCMS, which is suitable for analyses of LC/MS data. It is able to filter and identify relevant peaks and match the peaks in different samples. The tool can also calculate retention time deviations. It is capable of simultaneously preprocessing, analyzing and visualizing the raw data from hundreds of samples. Statistical data analysis can also be performed and it includes functionality for peak picking, nonlinear retention time alignment and relative quantitation. It is freely available at http://metlin.scripps.edu/download/.

A more recent and comprehensive online tool for preprocessing, chemometrics and analysis of LC and MS data is Metabolomics Data Analysis Tool (MetDAT). This tool (http://smbl.nus.edu.sg/METDAT/) is available free online for researchers from academic and nonprofit organizations. MetDAT performs alignment, baseline correction and normalization of data using a number of algorithms. It can calculate log ratios of different treatments with respect to a reference data set. It also allows generation of Venn diagrams that identify common as well as unique molecules in two to four data sets. MetDAT includes chemometric methods like PCA and hierarchal and K-means clustering as well as biostatistical methods such as analysis of variance. The online tool deals with small data sets for rapid analysis. In its offline complete software package, this tool allows user-provided databases to store the data as well as analysis of the data to enable extraction of meaningful biological interpretations. Users are able to upload their data, analyze and store the data, which can later be recalled. The data are also organized according to the project, subproject and experiments. The sample extraction methods, LC methods, buffer system used, the type of instrument and the instrument settings can be input into the database. Hence, this software package can be used for managing large metabolomics projects involving several resources and experiments. Though there are other tools for MS data analysis and HPLC data analysis, MetDAT provides a complete set of programs for data preprocessing and analysis, unique algorithms and programs for data analysis, and a user database in a single package.

Upon metabolite data analysis, novel compounds and metabolic pathway features are frequently discovered. It then becomes essential to mine the literature for reports on such compounds or their enzymes or pathways. Searching databases can often lead to a large number of publications. For example, a simple search for plant sugars yields 43,926 abstracts in the PubMed database. Analysis and integration of knowledge from such abstracts becomes highly cumbersome to an extent such that it takes enormous effort, manpower and time to assimilate and interpret the information. This problem can be minimized by using knowledge mining tools such as Dragon Plant Biology Explorer (DPBE; http://research.i2r.a-star.edu.sg/DRAGON/ME2/) (Bajic et al. 2005). This tool allows plant biologists to mine existing literature and visualize the interconnectedness. DPBE is a system which integrates information on genes from PubMed abstracts with gene functions based on standard gene ontologies and biochemical entity vocabularies, and presents the associations as interactive networks. DPBE complements the existing biological resources for systems biology by identifying potentially novel associations using text analysis between cellular entities based on genome annotation terms. One of the most useful aspects of DPBE to biologists is that it condenses information from a large volume of documents for easy inspection and analysis, thus making it feasible for individual users. Two modules of the explorer, the Metabolome Explorer and the Pharmacology Explorer, are especially relevant to metabolomics researchers (Bhalla et al. 2005). Interconnections among cellular entities such as metabolites, enzymes, genes, mutants, plant anatomical features, cellular components and function can be visualized as networks. Nodes in the networks are hyperlinked to the original abstracts in color-coded forms. Hence, DPBE can be used to interpret novel information generated from metabolomics projects as well as to research new topics by beginners or experienced scientists.

3.3.5.2 Databases

Different kinds of databases are required to efficiently analyze metabolomics data. Some such databases are outlined here briefly:

1. Databases for storing experimental data. These helps to recall data for comparison or different types of analyses and usually such databases are created in-house. Data are stored as flat files for smaller data sets, while for larger complex data, relational databases are used. Relational databases also have functions "built in"

that help them to retrieve, sort and edit the data in many different ways. A universal database for the input of all metabolomics data would be helpful for comparison of results from different experiments by different people. Such databases are at present available for microarray data. These data can then be used as a starting point for experiments by other scientists. The MetDAT software package described previously provides one such option as it incorporates a project and experiment management system.

2. Databases for comparing with other standard or data sets. One of the biggest challenges in MS is to generate reproducible fragmentation patterns especially using soft ionization methods such as ESI MS. Databases then have to include parent, precursor and daughter ion information. Currently, there is very little understanding of the fragmentation patterns of various metabolites; hence, individual laboratories have to generate their own databases based on their methods and instrument settings. Spectral data are available in the public domain for only some metabolites. For example, the National Institute of Standards and Technology (NIST), USA, has a chemistry Web book that provides information on the molecular weight, formula, structure as well as mass spectra. Chemical information on several parameters is available for over 40,000 compounds and mass spectra of 15,000 compounds are available (http://webbook.nist.gov/). Another freely available database for drugs and metabolites is provided by the Mass Spectrometry Database Committee (http://www.ualberta.ca/~gjones/mslib.htm). In addition, a number of commercial databases are available for users. Some of these are the NIST/EPA/NIH Mass Spectral Library for electron impact (ESI) spectra (http://www.nist.gov/srd/nist1a.htm) and the Wiley Registry of Mass Spectral Data (http://www.wileyregistry.com/). Some of the tools and databases available are compiled in Table 3.2.
3. Databases of biochemical reactions and pathways. To understand the role of various metabolites in biological processes it is imperative to understand the biochemical reactions in which such metabolites are involved and the pathways to which they belong. This helps to predict the molecular mechanisms that govern the various processes taking place in the cell as a whole. Consequently several attempts are under way to create large-scale databases on gene-regulatory and biochemical networks. Such databases provide a comprehensive coverage of the chemical reactions. Such database can therefore, be helpful in deducing the reactions that are affected upon treatment or in transgenic, mutant, knockout or knockdown RNA interference plants.

A summary of some of the available databases is provided in Table 3.2 All these databases contain features that make them unique, but none of them singly fulfills all the requirements for a good reference for metabolic pathway studies (Mendes 2002; Wittig and De Beuckelaer 2001). Despite this, such databases provide a wealth of knowledge and play an important role in understanding the complexities and the interrelationships among the genes, proteins and metabolites. Selected tools and databases from this category are described here briefly:

- MetaCyc. This is a database of experimentally elucidated pathways. It has around 700 pathways from 600 organisms. It stores pathways involved in primary

Table 3.2 Summary of tools and databases available for metabolite studies

	Application	Reference and Web site
Tools		
MZmine	Data preprocessing such as spectral filtering, peak detection, alignment and normalization	Katajamaa and Oresic (2005), http://mzmine.sourceforge.net/
ANOVA–simultaneous component analysis (ASCA)	Analysis of complex data generated by liquid chromatography/mass spectrometry involving different parameters like time and dose factors	Smilde et al. (2005), http://www.bdagroup.nl/
Metabolomics Data Analysis Tool (MetDAT)	Data processing such as alignment, baseline correction and normalization of data. It also includes chemometric methods such as principal component analysis and hierarchal clustering and biostatistical methods such as ANOVA	http://smbl.nus.edu.sg/METDAT/
MSFACTs	Includse spectral formatting, aligning and conversion tools	Duran et al. (2003), http://www.noble.org/PlantBio/MS/MSFACTs/MSFACTs.html
XCMS	Incorporates data preprocessing such as nonlinear retention time alignment, matched filtration, peak detection and peak matching	http://metlin.scripps.edu/
Dragon Plant Biology Explorer (DPBE)	A knowledge-mining tool that extracts information and organizes it for easy interpretation	Bajic et al. (2005), http://research.i2r.a-star.edu.sg/DRAGON/ME2/
Databases		
National Institute of Standards and Technology (NIST)	Provides information on the molecular weight, formula, structure as well as mass spectrum	http://webbook.nist.gov
Kyoto Encyclopedia of Genes and Genomes (KEGG)	Molecular interaction networks in biological processes	Kanehisa et al. (2004), http://www.genome.ad.jp/kegg/
MetaCyc	Elucidated pathways from 600 organisms	Zhang et al. (2005), http://metacyc.org/
AraCyc	Biochemical pathways of *Arabidopsis* developed at The *Arabidopsis* Information Resource	Mueller et al. (2003), http://www.arabidopsis.org

(continued)

Table 3.2 (continued)

	Application	Reference and Web site
BioPathAt	Knowledge-based analysis of genome-scale data by integrating biochemical pathway maps	Lange and Ghassemian (2005)
LIGAND	Consists of the following databases: COMPOUND, GLYCAN, REACTION and ENZYME	Goto et al. (2002), http://www.genome.ad.jp/dbget/ligand.html
BRITE	Biomolecular relations in information transmission and expression database	Kanehisa et al. (2006), http://www.genome.ad.jp/kegg/brite.html
Alliance for Cellular Signalling (AFCS)	Provides information on signal transduction	http://www.signaling-gateway.org/
PathDB	A metabolic pathway database	http://www.ncgr.org/pathdb/

ANOVA analysis of variance

metabolism (including photosynthesis), secondary metabolism, as well as associated compounds, enzymes and genes. It is available at http://metacyc.org/.
- AraCyc. This is a database containing biochemical pathways of *Arabidopsis* developed at The *Arabidopsis* Information Resource (http://www.arabidopsis.org) with the aim of representing *Arabidopsis* metabolism using a Web-based interface (Mueller et al. 2003). This database now contains 197 pathways that include information on compounds, metabolic intermediates, cofactors, reactions, genes, proteins and protein subcellular locations. The Web site also has an "omics viewer" that allows users to upload the data onto a pathway chart and then visualize the variations in the data and map the pathway changes in a visual form.
- The Kyoto Encyclopedia of Genes and Genomes (KEGG) (http://www.genome.ad.jp/kegg/) is a complex set of databases, and includes knowledge on molecular interaction networks in biological processes (PATHWAY database), knowledge of genes and proteins (GENES/SSDB/KO databases) and knowledge of the chemical compounds and reactions (COMPOUND/GLYCAN/REACTION databases). KEGG currently covers 15,037 pathways, of which 229 are reference pathways. It has genome information of 181 organisms and catalogs 646,192 genes with ortholog clusters known for 33,305 of the genes. It describes 6,000 chemical reactions and links to 10,000 chemical compounds.
- BioPathAt. This newly developed visual interface allows the knowledge-based analysis of genome-scale data by integrating biochemical pathway maps (BioPathAtMAPS module) with a manually scrutinized gene-function database (BioPathAtDB) for the model plant *Arabidopsis thaliana* (Lange and Ghassemian 2005).

Often there is great discord in the data generation and analysis methods of different laboratories or even during different experiments, which may lead to misinterpretation of the data or irreproducible data. In order to have uniformity in the data published and also for efficient analysis and recording, The Standard Metabolic Reporting Structures (SMRS) group (http://www.smrsgroup.org/) has been set up to provide guidelines for reporting metabolomic studies. The guidelines pertain to three main areas: (1) origin of the biological sample, (2) analytical methods used in the analysis of the material and (3) the multivariate statistical methods (chemometrics) used to retrieve information from the sample data (The Standard Metabolic Reporting Structures Working Group 2005). Further, to enhance the accuracy and descriptions of the methods and experiments, a framework for plant metabolomics called ArMet (Architecture for Metabolomics) has been created (Jenkins et al. 2004). It provides the entire experimental timeline from the sample preparation to data analysis. Such data models will help in comparison of data sets, allow proper interpretation of the results and repetition of results. It gives a basis for storage and transmission of data.

We have provided a brief introduction to the available analytical methods, the bioinformatics tools and the databases available for metabolomics with examples related to rhizosphere metabolomics. We shall now describe the different events in the rhizosphere such as bioconversions that have been deduced with the help of these techniques and tools.

3.4 Bioconversions of Rhizosphere Metabolites

Bioconversion helps in the generation of energy and also signaling molecules for intracellular and intercellular functions. How soil bacteria transform these molecules (bioconversion) from the environment to generate energy and use it for growth and other purposes is very interesting. Bacteria take most of the starting materials from the living environment by involving processes such as biotransformation, biocatalysis or biodegradation. While these terms seem different, they refer to the same group of processes, namely, bioconversion or microbial metabolism. The use of the term depends on what is being studied and more often is based on the intended focus of the study. "Microbial metabolism" from an industrial application viewpoint refers to the process as biotransformation or employing biocatalysis. If the study concerns degrading environmental pollutants or organic compounds, it is commonly referred to as "biodegradation." In this section, we shall refer to the process collectively as "bioconversion" for it may involve one of the abovementioned process or a combination of them, depending on the environment and its inhabitants. Bioconversions have been known since the days of Louis Pasteur (1857); however, a renewed interest in biotransformation was witnessed only in the late twentieth century. This is partly due to interests in developing a sustainable environment coupled with a healthier life.

Plants release chemicals into the rhizosphere; they can positively or negatively regulate growth and development of the microenvironment, including the rhizosphere (Rice 1984). Aromatic hydrocarbons, including many plant phenolics, are ubiquitous in nature. Plant phenolics, including quinines, are the most common class of subterranean allelochemicals (Inderjit 1996). Indeed, next to glucosyl residues, the benzene ring is the most widely distributed chemical structure in nature (Dagley 1981) and therefore they are the second largest group of natural products, including many plant metabolites released into the rhizosphere. Some of the common aromatic hydrocarbons released into the rhizosphere are flavonoids (Narasimhan et al. 2003), members of the phenylpropanoid group of compounds; additionally contamination of vegetated soils can also expose rhizobacterial populations to pollutants such as naphthalene, phenanthrene, fluoranthene and benzo[*a*]pyrene. Although the signature benzene ring of aromatic hydrocarbons is commonly found in many of these compounds, their breakdown involves several complex metabolic networks. These compounds, therefore, signal developmental events in the roots of some plants that enhance their growth, indicating some active events that occur in the rhizosphere. These events are centered on the process of bioconversion/biotransformation. Interestingly, to cope with the complexity of metabolizing these aromatic compounds, microorganisms have adopted two simple but fundamentally different strategies: aerobic and anaerobic mechanisms. Under aerobic conditions, aromatic compounds are generally transformed by monooxygenases and dioxygenases into a few central intermediates such as catechol, protocatechuate and gentisate. This phenomenon is commonly referred to as "funneling pathways." Under anoxic (anaerobic) conditions, aromatic compounds need to be transformed by means other than by oxygenases, more so owing to the absence of oxygen, thus reductively attacking the ring structure via an ATP-dependent mechanism.

3.4.1 Metabolite Conversion Pathways in Bacteria

Bacteria possess the unique ability to break down ring-containing compounds that are resistant to degradation. The first representation of the metabolic pathway and the enzymatic reactions resulting in the mineralization of aromatic compounds (naphthalene in this study) was by Davies and Evans (1964). Since then, metabolite conversion by microbes, including bacteria, fungi and algae, has been extensively studied (Kuiper et al. 2004). Conversion of metabolites, for instance, compounds with more than three fused rings, is less ubiquitous, but the number of organisms found to degrade these aromatics is increasing (Kanaly and Harayama 2000). For some of the aromatic and complex compounds with more than three fused rings, "cometabolism" often serves as the main route for degradation. Another notable aspect of most aromatics that could influence bioconversion includes their hydrophobicity, which in turn depends on the number of fused rings, and their relative water solubility is low. Generally, the uptake of aromatic compounds by bacteria proceeds via the water phase and hence their water solubility may also be an important aspect, at least in the context of bioavailability of metabolites. Some microbes are able to produce biosurfactants that may be involved in enhanced bioavailability of metabolites for further conversion. Bioconversion involves the breakdown of metabolites through one of the abovementioned processes. Metabolites and/or often-recalcitrant compounds are transformed into less complex metabolites or through mineralization into inorganic minerals, water and carbon dioxide (in the case of aerobic pathways) or methane (in anaerobic pathways). The general ability of bacteria to use aromatic or ring-containing compounds such as the plant phenolics, halogenated hydrocarbons and others is related to the fact that most of these compounds are commonly present in the environment as a result of plant-derived material (Harwood and Parales 1996). By contrast, man-made compounds have been in contact for over 100 years only, and as a result their breakdown and properties are less well characterized. Two major routes of bioconversion, i.e., aerobic and anaerobic, and the peripheral and central pathways are discussed in the following sections.

Aerobic and anaerobic pathways have similarities and yet significant differences (Table 3.3). In the aerobic catabolic funnel, most peripheral pathways involve oxygenation reactions carried out by monooxygenases and/or hydroxylating dioxygenases that generate dihydroxy aromatic compounds (such as catechol, protocatechuate, gentisate, homoprotocatechuate, homogentisate, hydroquinone and hydroxyquinol) (Gibson and Subramanian 1984). These intermediate compounds are the substrates of ring-cleavage enzymes that use molecular oxygen to open the aromatic ring between the two hydroxyl groups (*ortho* cleavage, catalyzed by intradiol dioxygenases) or proximal to one of the two hydroxyl groups (*meta* cleavage, catalyzed by extradiol dioxygenases) (Harayama and Timmis 1992). Central pathways, therefore, involve a series of reactions leading to the formation of Krebs cycle intermediates (central metabolism) that are further easily converted to tricarboxylic acid (TCA) cycle intermediates (ven der Meer et al. 1992). In the

Table 3.3 Comparison of aerobic and anaerobic aromatic metabolism pathways

Features	Aerobic	Anaerobic
Channeling reactions	+ O_2	+H_2O, +2[H], −2[H] + H_2O + CO_2, + CoA + ATP
Central intermediates	Catechol, gentisate, protocatechuate	Benzoyl-CoA, resorcinol, phloroglucinol
Properties of central intermediates	Easy to oxidize (cleave)	Easy to reduce (hydrate)
Attack at the ring	O_2	2 or 4[H] (H_2O)
Ring cleavage	Oxygenolysis of aromatic	Hydrolysis of 3-oxo compound
Pathway to central metabolites	3-Oxoadipate pathway, e.g., → succinate + acetyl-CoA	Oxidation, e.g., → glutaryl-CoA → acetyl-CoA

CoA Coenzyme A

anaerobic catabolism of aromatic compounds, the peripheral pathways converge mostly to benzoylcoenzyme A and occasionally to resorcinol and phloroglucinol, which become dearomatized by a specific multicomponent reductase that requires energy in the form of ATP (Gibson and Harwood 2002). At times, the rhizosphere can have partially to completely anaerobic (anoxic) conditions, depending on the soil characteristics (such as compactness and waterlogging conditions). Under anoxic (anaerobic) conditions, aromatic compounds need to be transformed by means other than by oxygenases, more so owing to the absence of oxygen (early studies by Tarvin and Buswell 1934; Dutton and Evans 1967), implying that the aromatic-ring structures are reductively attacked (Dutton and Evans 1969; Evans and Fuchs 1988). It should be noted, however, that there is limited knowledge on anaerobic degradation of polymeric high molecular weight aromatics such as lignins, which could represent probably more than half of the aromatic compounds (Young and Frazer 1987).

The lignin pathway branches out from the initial steps of the phenylpropanoid biosynthesis pathways, which is well known for the production of flavonoids. These groups of compounds are released as rhizosecretion and they influence rhizobacterial populations and competition. The bioconversions of these rhizosecretions are briefly discussed here. Rhizobia and *Agrobacterium* that are capable of degrading *nod* gene-inducing flavonoids have been reported (Rao and Cooper 1994) A *Rhodococcus rhodochrous* strain has been described as being capable of styrene degradation (Warhurst et al. 1994). Microbial enzymes with wide substrate specificity are certain to provide better survival benefits to those harboring the enzymes than those that do not.

3.4.2 *Case Study: Bioconversion of Flavonoids*

Flavonoids are ubiquitous in the plant kingdom and in the rhizosphere and are also an integral part of the human diet (Hollman et al. 1997). Understandably, several microbial genera are known to participate in the breakdown of aromatic compounds, including phenylpropanoids. These have been reported from two ecological niches, viz., soil and intestines, and include rhizobia, *Agrobacterium tumefaciens*, a thermophilic *Bacillius* sp., *Pseudomonas* sp., a *Rhodococcus* strain and a strain of the fungus *Aspergillus niger*. Other degraders are from the anoxic environment of the intestine and include *Clostridium* strains, *Eubacterium* species and *Butyrivibrio* species. Flavonoid degradation pathways have been well studied in the intestinal flora. As in most other cases involving degradation of aromatics, studies on uptake and detection of intermediates and accumulation of end products have generally lead to the elucidation of the pathway (Chang and Zylstra 1998; Bode et al. 2000).

Several rhizobia, including the lotus rhizobia, can degrade quercetin via a novel form of ring cleavage, yielding phloroglucinol and protocatechuic acid (Rao et al. 1991; Rao and Cooper 1994). Hopper and Mahadevan (1991) reported the degradation of catechin by *Bradyrhizobium japonicum*, which was cleaved through an inducible catechin oxygenase to yield phloroglucinolcarboxylic acid and protocatechuic acid as the initial products that were further decarboxylated to phloroglucinol and dehydrated to resorcinol. Phenylpropanoid degradation by a soil pseudomonad and the presence of new oxygenases in the degradation of flavones and flavonones by *Pseudomonas putida* suggests that degradation occurred by a fission in the A-ring, via hydroxylation at C-8 (Shultz et al. 1974). A common pathway for the degradation of flavones and flavonones by *Pseudomonas putida* is generally accepted, where it is converted to protocatechuate and/or catechol, which is further cleaved via the β-ketoadipate pathway, resulting in the formation of oxaloacetic acid. Oxaloacetate could be routed though the TCA cycle for further metabolism and energy generation. A more detailed flavonoid mineralization pathway in a plant-growth-promoting rhizobacterial strain was recently described by Pillai and Swarup (2002) (Fig. 3.3). Using a comparative metabolomics approach with wild-type and flavonoid auxotrophic strains, they elucidated the metabolic pathway. The archetypal flavonoid quercetin was used in the study to monitor the degradation events in the soil pseudomonad. Quercetin was converted to naringenin and then to dihydroxy aromatic compounds that could be cleaved by ring-cleavage oxygenases and led to formation of single-ring compounds such as protocatechuate. As seen from Fig. 3.3, at least two intermediate compounds are produced during the flavonoid bioconversion before production of compounds that enter the TCA cycle. In addition to flavonoids, various intermediates have been identified with different phenolics as carbon sources in the environment. Phenolic degradation pathways produce acids that are partly subjected to further degradation and the phenolics detected over time may not be consistent. Jeffrey et al. (1972a) have reported degradation of taxifolin involving hydroxylation of its A-ring in a pseudomonad, while in another study, also involving a soil pseudomonad, they reported the oxidative fission of the A-ring of dihydrogossypetin (Jeffrey et al. 1972b). This shows the existence of a variety of catabolic metabolisms for different compounds, all in a single bacterial species. Such versatility of soil microbes allows speculation of the existence of novel metabolic regulatory pathways in these strains.

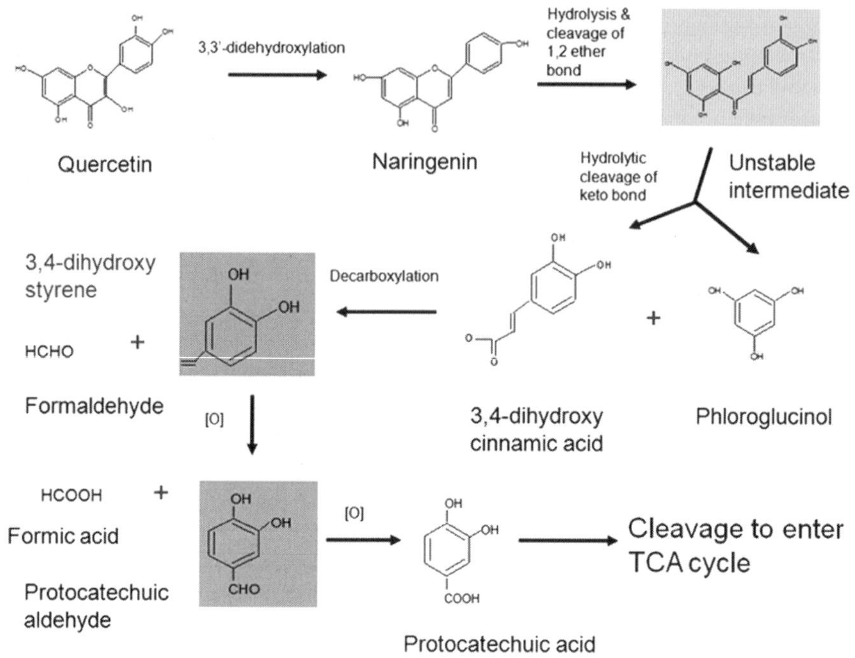

Fig. 3.3 Phenylpropanoid degradation pathway in phenylpropanoid utilizing strains of *Pseudomomas putida*. Quercetin degradation pathway in *P. putida* strain PML2. The identities of all compounds except compounds III, V and VI were confirmed by NMR spectroscopy. All compounds are stably formed except for compound III. Hydrolysis and cleavage of ether and keto bonds and the presence of an unstable intermediate (compound III) were inferred on the basis of the structures of compounds II and IV

3.5 Applications of Rhizosphere Metabolomics

Rhizoremediation involves the use of plants as well as rhizobacteria to clean up contaminated soil and water. Two processes have been described to constitute rhizoremediation, namely, *phytoremediation* and *bioaugmentation* (Kuiper et al. 2004). "Bioaugmentation" refers to enhanced availability of a substrate using specific microbes. Microbial degradation of the pollutants is enhanced owing to stimulation of root exudates. The root system of plants aid in spreading the rhizobacteria through soil and help to penetrate otherwise impermeable soil layers. Pollutant-degrading bacteria can be inoculated on plant seeds to improve the efficiency of phytoremediation or bioaugmentation.

Phytoremediation of polyaromatic hydrocarbons (PAHs) is driven by root–microbe interactions (Rugh et al. 2005). Bacterial degradation has been shown to be the dominant pathway for environmental PAH dissipation. The authors tested various plant species and the efficacy in degrading PAHs. It was found that in soils that were planted, there was an increase in heterotrophic and biodegradative cell numbers compared with the situation in unplanted soils. The study showed that the expanded metabolic range of the rhizosphere bacterial community would contribute more to effective degradation of PAHs.

Plants can be genetically engineered so as to create a biased rhizosphere. This is possible by enhancing the growth of selected microbial species which can help in increasing the biodegradation capacity of the soil. Plants can be engineered such that they are resistant to soil-borne pathogens, are better hosts to beneficial microorganisms, can remediate toxic waste or can attract communities of soil microorganisms that enhance plant health (O'Connel et al. 1996). Engineering plants which exudate specific nutrients that enhance the growth of specific microorganisms helps in creating a biased rhizosphere that is more efficient in biodegradation (Fig. 3.4). For example, the genetically engineered plants that produce opines could change the bacterial populations in soil (Oger et al. 1997). Degradation of environmental pollutants can be enhanced in the rhizosphere by microorganisms that can utilise root exudates as carbon source. For example, biodegradation of polychlorinated biphenyls (PCBs) can be enhanced by growth of PCB-degrading rhizobacteria along with plants that can exude phenylpropanoids. The rhizobacteria are able to utilize phenylpropanoids and hence are able to grow better in the rhizosphere as there is less competition for compounds like phenylpropanoids from other

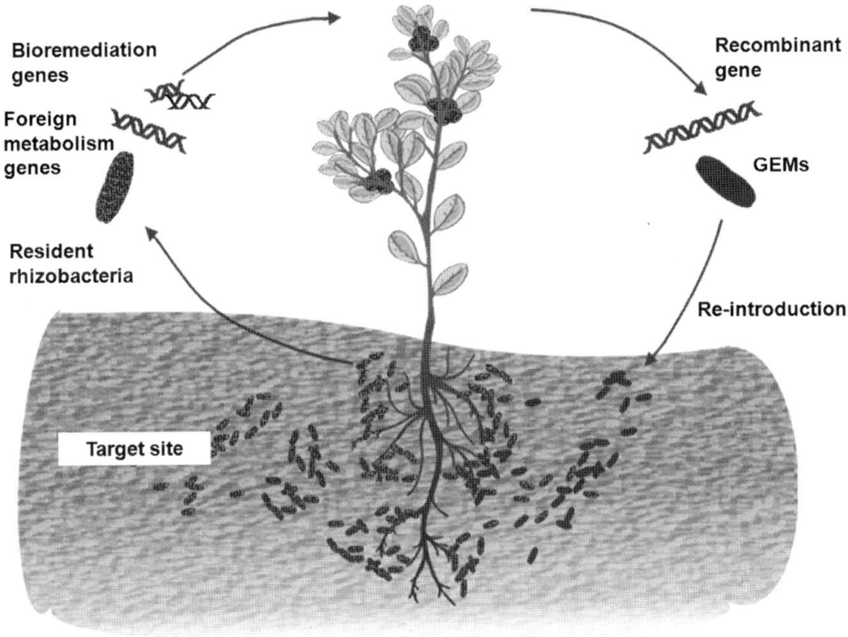

Fig. 3.4 Rhizoengineering approach to improve bioremediation efficiency. Resident rhizobacteria that can mineralize phenylpropanoids for their growth can be isolated from polluted site. They can be modified by transfer of bioremediation genes/pathway. The genetically modified microorganisms (*GEMs*) thus produced can be reintroduced to the rhizosphere, where they will be competitive for growth and perform bioremediation efficiently

microorganisms. Since they are able to grow better, the efficiency of PCB degradation is increased (Narasimhan et al. 2003). This method has the advantage that it does not rely on genetic engineering of plants as in the case of the previous methods. Plant-assisted rhizoremediation in the long run will turn into an effective mode for rhizoremediation of toxic organic pollutants. At petroleum hydrocarbon contaminated sites, two genes encoding hydrocarbon degradation, alkane monooxygenase (*alkB*) and naphthalene dioxygenase (*ndoB*), were 2 and 4 times more prevalent in bacteria extracted from the root interior (endophytic) than from the bulk soil and sediment, respectively (Siliciano et al. 2001). These results indicate that the enrichment of catabolic genotypes in the root interior is both plant-dependent and contaminant-dependent.

3.6 Conclusions

Metabolomics has emerged as the final frontier in functional genomics. The field has broad applications in understanding the composition and interactions of the rhizosphere. Although there are certain limitations in rhizosphere metabolomics in its present state, these are likely to be addressed as the field becomes more widely appreciated. Some of the techniques used for studying plant or animal metabolism can be extended to the rhizosphere as well. A number of analytical tools for the separation of metabolites are available for metabolomics researchers. However, metabolite databases and tools for processing and analyzing the data need further improvement. Future directions in this field are likely to be (1) in methods development, (2) identifying signaling molecules that originate from both plants as well as rhizosphere microbiota and (3) understanding the role of rhizosphere metabolites in affecting plant growth and physiological functioning.

References

Bajic VB, Veronika M, Veladandi PS, Meka A, Heng M-W, Rajaraman K, Pan H, Swarup S (2005) Dragon Plant Biology Explorer. A text-mining tool for integrating associations between genetic and biochemical entities with genome annotation and biochemical terms lists. Plant Physiol 138:1914–1925

Bais HP, Weir TL, Perry LG, Gilroy S, Vivanco JM (2006) The role of root exudates in rhizosphere interactions with plants and other organisms. Annu Rev Plant Biol 57:233–266

Bending GD, Poole EJ, Whipps JM, Read DJ (2002) Characterisation of bacteria from Pinus sylvestris-Suillus luteus mycorrhizas and their effects on root-fungus interactions and plant growth. FEMS Microbiol Ecol 39:219–227

Bhalla R, Narasimhan K, Swarup S (2005) Metabolomics and its role in understanding cellular responses in plants. Plant Cell Rep 24:562–571

Bode HB, Zeeck A, Plückhahn K, Jendrossek D (2000). Physiological and chemical investigtions into microbial degradation of synthetic poly (cis-1,4-isoprene). Appl Environ Microbiol 66:3680–3685

Boersma MG, Solyanikova IP, Van Berkel WJ, Vervoort J, Golovleva LA, Rietjens IM (2001) 19F NMR metabolomics for the elucidation of microbial degradation pathways of fluorophenols. J Ind Microbiol Biotechnol 26:22–34

Bonnington L, Eljarrat E, Guillamón M, Eichhorn P, Taberner A, Barceló D (2003) Development of a liquid chromatography-electrospray-tandem mass spectrometry method for the quantitative determination of benzoxazinone derivatives in plants. Anal Chem 75:3128–3136

Cataldi TRI, Margiotta G, Iasi L, Di Chio B, Xiloyannis C, Bufo SA (2000) Determination of Sugar Compounds in olive plant extracts by anion-exchange chromatography with pulsed amperometric detection. Anal Chem 72:3902–3907

Chang H-K, Zylstra GJ (1998) Novel organization of the genes for phthalate degradation from *Burkholderia cepacia* DBO1. J Bacteriol 180:6529–6537

Chin-A-Woeng TFC, van den Broek D, Lugtenberg BJJ, Bloemberg GV (2005) The *Pseudomonas* chlororaphis PCL1391 sigma regulator psrA represses the production of the antifungal metabolite phenazine-1-carboxamide. Mol Plant Microbe Interact 18:244–253

Czarnota MA, Rimando AM, Weston LA (2003). Evaluation of root exudates of seven sorghum accessions. J Chem Ecol 29(9):2073–2083

Dagley S (1981) New perspectives in aromatic catabolism. In: Leisinger T, Cook AM, Hütter R, Nüesch J (eds) Microbial degradation of xenobiotics and recalcitrant compounds. Academic, New York, pp 181–186

Davies JI, Evans WC (1964) Oxidative metabolism of naphthalene by soil pseudomonads. Biochem J 91:251–261

Derrien D, Balesdent J, Marol, Santaella C (2003) Measurement of the 13C/12C ratio of soil-plant individual sugars by gas chromatography/combustion/isotope-ratio mass spectrometry of silylated derivatives. Rapid Commun Mass Spectrom 17:2626–2631

Desbrosses GG, Kopka J, Udvardi MK (2005) Lotus japonicus metabolic profiling. Development of gas chromatography-mass spectrometry resources for the study of plant-microbe interactions. Plant Physiol 137:1302–1318

Dunn WB, Ellis DI (2005) Metabolomics: current analytical platforms and methodologies. Trends Anal Chem 24:285–294

Dunn WB, Bailey NJ, Johnson HE (2005) Measuring the metabolome: current analytical technologies. Analyst 130:606–625

Duran AL, Yang J, Wang L, Sumner LW (2003) Metabolomics spectral formatting, alignment and conversion tools (MSFACTs). Bioinformatics 19:2283–2293

Dutton PL, Evans WC (1967). Dissimilation of aromatic substrates by *Rhodopseudomonas palustris*. Biochem J 104:30–31

Dutton PL, Evans WC (1969) The metabolism of aromatic compounds by *Rhodopseudomonas palustris*: a new reductive method of aromatic ring metabolism. Biochem J 113:525–536

Eljarrat E, Barcelo D (2001) Sample handling and analysis of allelochemical compounds in plants. Trends Anal Chem 20:584–590

Evans CJ, Evershed RP (2003) Compound-specific stable isotope analysis of soil mesofauna using thermally assisted hydrolysis and methylation for ecological investigations Anal Chem 75:6056–6062

Evans WC, Fuchs G (1988) Anaerobic degradation of aromatic compounds. Annu Rev Microbiol 42:289–317

Fan TWM, Lane AN, Shenker M, Bartley JP, Crowley D, Higashi RM (2001) Comprehensive chemical profiling of gramineous plant root exudates using high resolution NMR and MS. Phytochemistry 57:209–221

Formanek P, Ambus P (2004) Assessing the use of delta C-13 natural abundance in separation of root and microbial respiration in a Danish beech (*Fagus sylvatica* L.) forest. Rapid Commun Mass Spectrom 18:897–902

Gibson DT, Subramanian V (1984) Microbial degradation of aromatic hydrocarbons. In Gibson DT (ed) Microbial degradation of organic compounds. Dekker, New York, pp 181–252

Gibson J, Harwood CS (2002) Metabolic diversity in aromatic compound utilization by anerobic microbes. Annu Rev Microbiol 56:345–369

Gleye C, Laurens A, Hocquemiller R, Cave A, Laprevote O, Serani L (1997) Isolation of montecristin, a key metabolite in biogenesis of acetogenins from *Annona muricata* and its structure elucidation by using tandem mass spectrometry. J Org Chem 62:510–513

Goodacre R (2005) Making sense of the metabolome using evolutionary computation: seeing the wood with the trees. J Exp Bot 56:245–254

Goodacre R, Shann B, Gilbert RJ, Timmins EM, McGovern AC, Alsberg BK, Kell DB, Logan NA (2000) Detection of the dipicolinic acid biomarker in Bacillus spores using Curie-point pyrolysis mass spectrometry and Fourier transform infrared spectroscopy. Anal Chem 72:119–127

Goto S, Okuno Y, Hattori M, Nishioka T, Kanehisa M (2002) LIGAND: database of chemivcal compounds and reactions in biological pathways. Nucleic Acids Res 30:402–404

Griffin JL (2004) Metabolic profiles to define the genome: can we hear the phenotypes? Philos Trans R Soc Lond B Biol Sci 359:857–871

Harayama S, Timmis KN (1992) Aerobic biodegradation of aromatic hydrocarbons by bacteria. In: Sigel H, Sigel A (eds) Metal ions in biological systems, vol 28. Dekker, New York, pp 99–156

Harwood CS, Parales RE (1996) The B-ketoadipate pathway and the biology of self-identity. Annu Rev Microbiol 50:553–590

Hollman PCH, van Trijp JMP, Buysman MNCP, Gaga MSVD, Mengelers MJB, Vries JHM, Katan MB (1997) Relative bioavailability of the antioxidant flavonoid quercetin from various foods in man. FEBS Lett 418:152–156

Hopper W, Mahadevan A (1991) Utilization of catechin and its metabolites by *Bradyrhizobium japonicum*. Appl Microbiol Biotechnol 35:411–415

Huang WE, Griffiths RI, Thompson IP, Bailey MJ, Whiteley AS (2004) Raman microscopic analysis of single microbial cells. Anal Chem 76:4452–4458

Inderjit (1996) Plant phenolics in allelopathy. Bot Rev 62:186–202

Inderjit, Duke SO (2003) Ecophysiological aspects of allelopathy. Planta 217:529–539

Jansen JJ, Hoefsloot HCJ, Boelens HFM, van der Greef J, Smilde AK (2004) Analysis of longitudinal metabolomics data. Bioinformatics 20:2438–2446

Jeffrey AM, Knight M, Evans WC (1972a) The bacterial degradation of flavonoids: hydroxylation of the A-ring of taxifolin by a soil pseudomonad. Biochem J 130:373–380

Jeffrey AM, Jerina DM, Self R, Evans WC (1972b) The bacterial degradation of flavonoids: oxidative fission of the A-ring of dihydrogossypetin by a *Pseudomonas* sp. Biochem J 130:383–390

Jenkins H, Hardy N, Beckmann M, Draper J, Smith A, Taylor J et al. (2004) A proposed framework for the description of plant metyabolomics experiments and their results. Nat Biotechnol 22:1601–1607

Jonsson P, Gullberg J, Nordstrom A, Kusano M, Kowalczyk M, Sjostrom M, Moritz T (2004) A strategy for identifying differences in large series of metabolomic samples analyzed by GC/MS. Anal Chem 76:1738–1745

Jonsson P, Bruce SJ, Moritz T, Trygg J, Sjöström M, Plumb R, Granger J, Maibaum J, Nicholson JK, Holmes E, Antti H (2005) Extraction, interpretation and validation of information for comparing samples in metabolic LC/MS data sets. Analyst 130:701–707

Kachlicki P, Marczak L, Kerhoas L, Einhorn J, Stobiecki M (2005) Profiling isoflavone conjugates in root extracts of lupine species with LC/ESI/MSn systems. J Mass Spectrom 40:1088–1103

Kanaly RA, Harayama S (2000) Biodegradation of high-molecular-weight polycyclic aromatic hydrocarbons by bacteria. J Bacteriol 182:2059–2067

Kanehisa M, Goto S, Kawashima S, Okuno Y, Hattori M (2004) The KEGG resource for deciphering the genome Nucleic Acids Rese 32:D277–D280

Kanehisa M, Goto S, Hattori M, Aoki-Kinoshita KF, Itoh M, Kawashima S, Katayama T, Araki M, Hirakawa M (2006) From genomics to chemical genomics: new developments in KEGG. Nucleic Acids Res 34:D354–D357

Katajamaa M, Oresic M (2005) Processing methods for differential analysis of LC/MS profile data. BMC Bioinformatics 6:179

Kerry BR (2000) Rhizosphere interactions and the exploitation of microbial agents for the biological control of plant parasitic nematodes. Annu Rev Phytopathol 38:423–441

Krishnan P, Kruger NJ, Ratcliffe RG (2005) Metabolite fingerprinting and profiling in plants using NMR. J Exp Bot 56:255–265

Kuiper I, Lagendijk EL, Bloemberg GV, Lugtenberg BJJ (2004) Rhizoremediation: a beneficial plant-microbe interaction. Mol Plant Microbe Interact 17 6–15

Lange BM, Ghassemian M (2005) Comprehensive post-genomic data analysis approaches integrating biochemical pathway maps. Phytochemistry 66:413–451

López-Díez EC, Goodacre R (2004) Characterization of microorganisms using UV resonance raman spectroscopy and chemometrics. Anal Chem 76:585–591

Martens H, Naes T (1989) Multivariate calibration. Wiley, New York

Mendes P (2002) Emerging bioinformatics for the metabolome. Brief Bioinformatics 3:134–145

Menotta M, Gioacchini AM, Amicucci A, Buffalini M, Sisti D, Stocchi V (2004) Headspace solid-phase microextraction with gas chromatography and mass spectrometry in the investigation of volatile organic compounds in an ectomycorrhizae synthesis system. Rapid Commun Mass Spectrom 18:206–210

Mesnard F, Ratcliffe RG (2005) NMR analysis of plant nitrogen metabolism. Photosynth Res 83:163–180

Mueller LA, Zhang P, Rhee SY (2003) AraCyc: a biochemical pathway database for *Arabidopsis*. Plant Physiol 132:453–460

Mukerji KG, Manoharachary C, Singh J (eds) (2006) Microbial activity in the rhizosphere. Soil biology, vol 7, Springer, Heidelberg

Narasimhan K, Basheer C, Bajic VB, Swarup S (2003) Enhancement of plant-microbe interactions using a rhizosphere metabolomics-driven approach and its application in the removal of polychlorinated biphenyls. Plant Physiol 132:146–153

O'Connell KP, Goodman RM, Handelsman J (1996) Engineering the rhizosphere: expressing a bias. Trends Biotechnol 14:83–88

Oger P, Petit A, Dessaux Y (1997) Genetically engineered plants producing opines alter their biological environment. Nature biotechnology 15:369–372

Pasteur L (1857) Mémoire sur la fermentation appelée lactique. Mém Soc Sci Agric Arts 5:13–37

Perera MR, Vanstone VA, Jones MGK (2005) A novel approach to identify plant parasitic nematodes using matrix-assisted laser desorption/ionization time-of-flight mass spectrometry. Rapid Commun Mass Spectrom 19:1454–1460

Pfeffer PE, Rolin DB, Schimdt JH, Tu SI, Kumosinski TF, Douds DD (1992) Ion transport and subcellular compartmentation in maize root tissue as examined by in vivo CS-133 NMR spectroscopy. J Plant Nutr 15:913–927

Pillai BVS, Swarup S (2002) Elucidation of the flavonoid catabolism pathway in *Pseudomonas putida* PML2 strain by comparative metabolic profiling. Appl Environ Microbiol 68:143–151

Pinton R, Varanni Z, Nannipier Pi, Willig W (eds) (2000) The rhizosphere: biochemistry and organic dubstance at the soil-plant interface. Dekker, New York

Rao RJ, Cooper JE (1994) *Rhizobia* catabolize nod gene-inducing flavonoids via C-ring fission mechanisms. J Bacteriol 176:5409–5413

Rao RJ, Sharma ND, Hamilton JTG, Boyd DR, Cooper JE (1991) Biotransformation of the pentahydroxy flavone quercetin by *Rhizobium loti* and *Bradyrhizobium* strains (Lotus). Appl Environ Microbiol 57:1563–1565

Ratcliffe RG, Shachar-Hill Y (2005) Revealing metabolic phenotypes in plants: inputs from NMR analysis. Biol Rev Camb Philos Soc 80:27–43

Reo NV (2002) NMR-based metabolomics. Drug Chem Toxicol 25:375–382

Rice EL (1984) Allelopathy 2nd edn. Academic, Orlando

Rugh CL, Susilawati E, Kravchenko AN, Thomas JC (2005) Biodegrader metabolic expansion during polyaromatic hydrocarbons rhizoremediation. Z Naturforsch C 60:331–339

Ryals J (2004) Drug discovery metabolomics. Metabolomics–an important emerging science. Business briefing: Pharmatech 51–54

Shultz E, Engle FE, Wood JM (1974) New oxygenases in the degradation of flavones and flavonones by *Pseudomonas putida*. Biochemistry 13:1768–1776

Siciliano SD, Fortin N, Mihoc A, Wisse G, Labelle S, Beaumier D, Ouellette D, Roy R, Whyte LG, Banks MK, Schwab P, Lee K, Greer CW (2001) Selection of specific endophytic bacterial genotypes by plants in response to soil contamination. Appl Environ Microbiol 67:2469–2475

Smilde AK, Jansen JJ, Hoefsloot HCJ, Lamers R-J AN, van der Greef J, Timmerman ME (2005) ANOVA-Simultaneous component analysis (ASCA): a new tool for analyzing designed metabolomics data. Bioinformatics 21:3043–3048

Spaink HP, Wijfjes AHM, Vanvliet TB, Kijne JW, Lugtenberg BJJ (1993) Rhizobial lipo-oligosaccharide signals and their role in plant morphogenesis are analogous lipophilic chitin derivatives produced by the plant. Aust J Plant Physiol 20:381–392

Steeghs M, Bais HP, de Gouw J et al. (2004) Proton-transfer-reaction mass spectrometry as a new tool for real time analysis of root-secreted volatile organic compounds in *Arabidopsis*. Plant Physiol 135:47–58

Sumner LW, Mendes P, Dixon RA (2003) Plant metabolomics: large-scale phytochemistry in the functional genomics era Phytochemistry 62:817–836

Tarvin D, Buswell AM (1934) The methane fermentation of organic acids and carbohydrates. J Am Chem Soc 56:1751–1755

The Standard Metabolic Reporting Structures Working Group (2005) Summary Recommendations for standardization and reporting of metabolic analyses. Nat Biotechnol 23:833–839

van der Meer JR, de Vos WM, Harayama S, Zehnder AJB (1992) Molecular mechanisms of genetic adaptation to xenobiotic compounds. Microbiol Rev 56:677–694

van der Werf MJ, Jellema RH, Hankemeier T (2005) Microbial metabolomics: replacing trial-and-error by the unbiased selection and ranking of targets. J Ind Microbiol Biotechnol 32:234–252

Villas-Boas SG, Rasmussen S, Lane GA (2005) Metabolomics or metabolite profiles? Trends Biotechnol 23:385–386

Walker TS, Bais HP, Halligan KM, Stermitz FR, Vivanco JM (2003) Metabolic profiling of root exudates of *Arabidopsis thaliana*. J Agric Food Chem 51:2548–2554

Warhurst AM, Clarke KF, Hill RA, Holt RA, Fewson CA (1994) Metabolism of styrene by *Rhodococcus rhodochrous* NCIMB 13259. Appl Environ Microbiol 60:1137–1145

Weir TL, Park SW, Vivanco JM (2004) Biochemical and physiological mechanisms mediated by allelochemicals. Curr Opin Plant Biol 7:472–479

Weidenhamer JD (2005). Biomimetic measurement of allelochemical dynamics in the rhizosphere. J Chem Ecol 31(2):221–236

Sun W, Liu S, Liu Z, Song F, Fang S (1998) A study of *Aconitum* alkaloids from aconite roots in *Aconitum carmichaeli* Debx using matrix-assisted laser desorption/ionization mass spectrometry. Rapid Commun Mass Spectrom 12:821–824

Wittig U, De Beuckelaer A (2001) Analysis and comparison of metabolic pathway databases. Brief Bioinformatics 2:126–142

Young LY, Frazer AC (1987) The fate of lignin and lignin-derived compounds in anaerobic ecosystems. Geomicrobiol J 5:261–293

Zhang P, Foerster H, Tissier CP, Mueller L, Paley S, Karp PD, Rhee SY (2005) MetaCyc and AraCyc. Metabolic pathway databases for plant research. Plant Physiol 138:27–37

Chapter 4
N-Acyl Homoserine Lactone Quorum Sensing in Gram-Negative Rhizobacteria

Sara Ferluga, Laura Steindler, and Vittorio Venturi(✉)

4.1 Introduction

In the last 15 years microbiologists have become aware that in most bacteria a major level of regulation exists which involves intercellular communication via the production and response to signal molecules. The concentration of the signal molecules increases alongside the bacterial population density and when it reaches a critical level, when a sufficient number of cells are present, bacteria respond and modulate gene expression. This cell-density-dependent modulation of gene expression has been termed quorum sensing (QS) (Fuqua et al. 1994). This allows bacteria to modify their behavior and act as multicellular entities; it is believed that in natural ecosystems bacteria are always aiming at establishing communities rather than choosing to exist as solitary cells. The reason being that intercellular communication provides significant advantages to a group of bacteria such as improving access to environmental niches, enhancing its defense capabilities against other microorganisms or eukaryotic host-defense mechanisms, and facilitating the adaptation to changing environmental conditions.

Bacterial QS signaling compounds at present can be broadly divided in two main classes, one being produced by Gram-positive bacteria and the other by Gram-negative bacteria. Gram-positive bacteria produce short, usually modified peptides processed from precursors which are then exported out of the cell and are then sensed by the bacterium through a signal transduction cascade (Bassler 2002; Sturme et al. 2002). A typical Gram-negative QS system, on the other hand (Fig. 4.1), involves the production of an acylated homoserine lactone (AHL) which was first described in the marine bioluminescent bacterium *Vibrio fischeri* in which QS regulates light production (Ruby 1996). Other types of less common signaling molecules have also been identified (Barber et al. 1997; Flavier et al. 1997a), including a furanosyl borate diester which appears to be employed by bacteria for interspecies communication (Chen et al. 2002).

Vittorio Venturi
Plant Bacteriology Group, International Centre for Genetic Engineering & Biotechnology, Biosafety Outstation, Via Piovega 23, 31050 Ca' Tron di Roncade, Treviso, Italy
e-mail: venturi@icgeb.org

P. Karlovsky (ed.), *Secondary Metabolites in Soil Ecology. Soil Biology 14.*

Several AHL QS systems have been described for Gram-negative plant-associated bacteria, including *Pseudomonas putida, P. chlororaphis/P. aureofaciens, P. syringae, Burkholderia cepacia, B. glumae, Erwinia carotovora, E. chrysanthemi, E. stewartii, Ralstonia solanacearum, Agrobacterium tumefaciens, Rhizobium etli, R. leguminosarum*, and *Sinorhizobium meliloti*. Among them, QS is involved in the regulation of antibiotic biosynthesis, extracellular enzymes, antifungal production, plasmid conjugation, biofilm formation, virulence factors, and rhizosphere gene

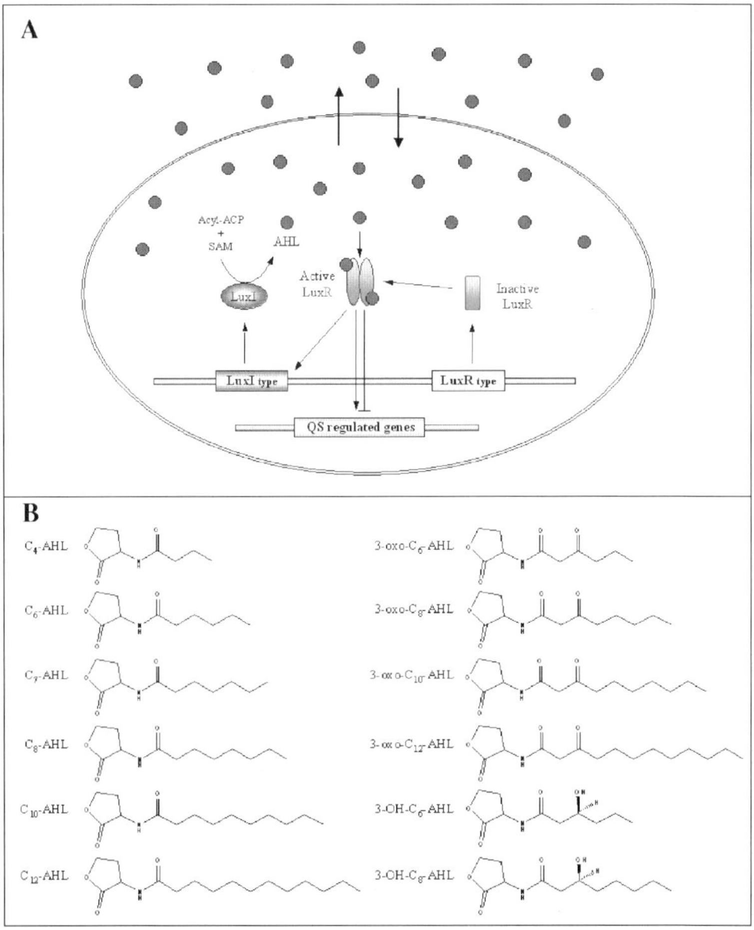

Fig. 4.1 **a** A typical *N*-acyl homoserine lactone (AHL) dependent quorum sensing (QS) system in Gram-negative bacteria. The LuxI-type proteins are the main class of enzymes capable of synthesizing AHLs and they use the cellular metabolites *S*-adenosyl-methionine (SAM) and acetylated acyl carrier proteins (ACP) to form AHLs. At high cell density, the AHL signal accumulates and interacts directly with the LuxR-type protein, inducing a conformational change (usually allowing multimerization) altering the affinity for specific DNA sequences (known as *lux* boxes) at target gene promoters changing gene expression (see text for all details). **b** Some common AHL signal molecules

expression (Loh et al. 2002; Pierson et al. 1998b; von Bodman et al. 2003). The scope of this review is to outline only the current knowledge on the AHL QS systems of rhizosphere bacteria, discussing recent advances in the role of gene regulation by QS and potential functions in bacteria–bacteria and plant–bacteria interactions. More general excellent reviews on QS in bacteria have recently been published (Fuqua et al. 2001; Miller and Bassler 2001; Whitehead et al. 2001).

4.2 AHL-Mediated QS Regulation

The typical model system for AHL QS regulation is rather simple, being most commonly mediated by two proteins belonging to the LuxI/LuxR protein families (Fig. 4.1). These families originate from the LuxI–AHL synthase and LuxR–AHL-response regulator, which is the first AHL QS system discovered involved in regulating light production in *V. fischeri* (Ruby 1996). LuxI-type proteins are the main class of enzymes capable of synthesizing AHLs and they use the cellular metabolites *S*-adenosyl-methionine and acetylated acyl carrier proteins to form AHLs. Many different Gram-negative bacteria have been reported to produce AHLs via LuxI-type proteins, differing only in the length of the acyl-chain moiety and substitution at position C3, which can be either unmodified or carries an oxo or hydroxyl group (Fig. 4.1b). Most AHLs are believed to diffuse across the cell wall; however, long-chain AHLs also utilize efflux pumps for translocation (Kohler et al. 2001; Pearson et al. 1999). The AHLs then interact directly at quorum concentration with the cognate LuxR-type protein and this protein–AHL complex can then bind at specific gene promoter sequences called *lux* boxes affecting expression of QS target genes. In most cases, the LuxR–AHL complex positively regulates the *luxI* family gene, creating a positive induction loop resulting in significant signal amplification. LuxR-type proteins display preferential binding for the AHL produced by the cognate LuxI-family protein, guaranteeing a good degree of selectivity; however, LuxR-family proteins can also to some extent respond to AHLs of different length and substitution of the acyl-chain moiety, raising important implications for the role of AHLs in interspecies communication. In some cases, for example, in *P. aeruginosa, P. aureofaciens*, and *Rhizobium* spp., bacteria possess multiple LuxI/LuxR systems producing multiple AHLs which can be hierarchically organized (see later). Recent reviews have appeared which cover in depth information on AHL structure and synthesis, LuxI- and LuxR-type proteins, LuxR–AHL interactions, and AHL QS regulons (Bassler 2002; Fuqua et al. 2001; Miller and Bassler 2001; Whitehead et al. 2001).

4.3 AHL QS and the Rhizosphere

The rhizosphere is the environment which surrounds and is influenced by the root system and has an important impact on the health and yield of crops. Plants release many compounds in the rhizosphere and microbial communities establish themselves creating a microenvironment of plant–microbe associations (Bais et al. 2004). Some of these interactions are beneficial to both plants and microbes, involving nutrient exchange, and are encouraged, for example, in the case of nitrogen-fixing bacteria or plant-growth-promoting rhizobacteria (PGPR). On the other hand, the rhizosphere can be an environment of growth, establishment, and attack of disease-causing microorganisms, resulting in crop damage and loss. Plants therefore have evolved strategies to defend themselves from pathogens, one of which is to favor the colonization of the rhizosphere by PGPR which will then exclude deleterious pathogenic microorganisms from this environment. The generally accepted mechanisms of biocontrol of phytopathogens by PGPR are competition for a substrate, production of inhibitory substances, and induction of systematic resistance (Compant et al. 2005; Haas and Defago 2005).

In the last 10 years it has become apparent that a diversity of *Proteobacteria* isolated from the rhizosphere use AHL signal molecules for QS-dependent gene expression. Among these are strains belonging to the species or genera of *P. aureofaciens/P. chlororaphis, P. putida, P. syringae, Burkholderia, Serratia, Erwinia, Ralstonia*, and *Rhizobium* and related genera involved in legume symbiosis. These bacteria employ AHL QS, for example, to regulate the production of biologically active secondary metabolites, enzymes, or exopolysaccharide which can improve the biological control activities of PGPR or are virulence determinants for plant pathogens. In then following we describe and discuss the current knowledge of AHL QS control in these beneficial or pathogenic rhizosphere-associated bacteria.

4.4 AHL QS in *Pseudomonas*

Pseudomonads can colonize several environmental niches and *P. aeruginosa* is also an important and dangerous human opportunistic pathogen as it can infect and chronically colonize the lungs of humans suffering from cystic fibrosis. In fact in the scientific community, pseudomonads are studied (1) for their role as human and plant pathogens, (2) for their remarkable catabolic potential, metabolism, and physiological versatility, and (3) for rhizosphere colonization and potential biological control agents. AHL QS has been most extensively studied in *P. aeruginosa*, making it one of the most studied systems in bacteria (Juhas et al. 2005; Smith and Iglewski 2003). Two systems, LasI/LasR and RhlI/RhlR, are present in *P. aeruginosa*; in the LasI/LasR system, LasI synthesizes *N*-(3-oxo-dodecanoyl)-L-homoserine lactone (3-oxo-C_{12}-AHL), which interacts with LasR and regulates target promoters. In the RhlI/RhlR system, RhlI directs the synthesis of *N*-butanoyl-L-homoserine lactone (C_4-AHL), which interacts with the cognate regulator RhlR and regulates target gene promoters. The two systems are under positive feedback control and are intimately connected forming a regulatory cascade; LasR/3-oxo-C_{12}-AHL positively regulates the *lasI* AHL synthase, creating a positive induction loop, and also activates *rhlR* expression initiating the RhlI/RhlR systems (Latifi et al. 1996; Seed et al. 1995). The LasI/LasR and RhlI/RhlR regulons have been extensively studied and have been found to regulate the production of multiple virulence factors, including elastase, alkaline protease, exotoxin A, rhamnolipids, pyocyanin, lectins, superoxide dismutases, and biofilm formation (Smith and Iglewski 2003). The effects of the two AHL QS systems have been tested in various models of *P. aeruginosa* infection, including several mouse models and alternative infection models of *Caenorhabditis elegans, Arabidopsis thaliana*, and *Dictyostelium discoideum*, all of which have shown that AHL QS MUTANTS showed reduction in virulence (reviewed recently in Juhas et al. 2005; Smith and Iglewski 2003). Genetic and microrray studies on regulons of *P. aeruginosa* have shown that the expression of over 300 genes is affected by AHL QS; thus, it is a major global regulatory response/regulation system (Hentzer et al. 2003; Schuster et al. 2003; Wagner et al. 2003; Whiteley et al. 1999). It is not intended to review and discuss in detail AHL QS in *P. aeruginosa* since studies have been focused on its pathogenicity to humans and because it is not regarded as a predominant rhizosphere bacterial species as is the case for *P. putida, P. fluorescens*, and *P. chlororaphis/P. aureofaciens*.

One of the first reports of AHL QS in *Pseudomonas* was the PhzI/PhzR of the wheat plant growth promoting rhizosphere colonizing *P. aureofaciens* (synonym of *P. chlororaphis*) strain 30-84 producing and responding to *N*-hexanoyl-L-homoserine lactone (C_6-AHL) (Pierson et al. 1994, 1995; Wood et al. 1997). PhzR/C_6-AHL regulates transcription, through binding to *lux*-box-like sequences in the promoter region, of the phenazine operon *phzXYFABCD* (Wood et al. 1997). The production of phenazine antibiotics in the wheat rhizosphere by *P. aureofaciens* strain 30-84 is important for its biocontrol properties by antagonizing the fungus *Gaeumannomyces graminins* var. *tritici*, which is the causal agent of take-all disease of wheat. A second AHL QS system is present in *P. aureofaciens* strain 30-84 which has been designated CsaI/CsaR (Zhang and Pierson 2001). The two systems are not organized in a hierarchical way (as is the case for the Las and Rhl systems of *P. aeruginosa*) and appear to function independently. CsaI/CsaR is not involved in the regulation of phenazine production, whereas it regulates exoprotease production in a synergistic way together with the PhzI/PhzR system; the precise molecular mechanism by which this occurs is still unknown (Zhang and Pierson 2001). In addition, CsaI/CsaR is involved in the regulation of cell-surface properties and is important for rhizosphere colonization. Interestingly, the most notable reduction in rhizosphere colonization in this strain was observed when both the PhzI/PhzR and CsaI/CsaR systems were inactivated (Zhang and Pierson 2001). The role of AHL QS in phenazine regulation has also been investigated in the plant-beneficial rhizobacterium *P. chlororaphis* strain PCL1391 (Chin et al. 2001, 2005). Similarly to *P. aureofaciens* 30-84, also in strain PCL1391, phenazine-1-carboxamide is regulated by a PhzI/PhzR QS system which produces and responds to C_6-AHL (Chin et al. 2001, 2005). The two PhzI/PhzR systems are highly identical and initial studies have shown that both *phzI/phzR* systems are themselves under considerable regulation (Chancey et al. 1999; Chin et al. 2005). *P. chlororaphis* PCL1391 has been shown to produce other types of AHL molecules in addition to C_6-AHL; however, the genetic determinants as well as the possible roles in gene regulation are currently unknown (Chin et al. 2001, 2005). Recently a PhzI/PhzR system has also been reported in rhizosphere *P. fluorescens* 2-79 in which it is also involved in the regulation of the antifungal secondary metabolite phenzine-1-carboxamide (Khan et al. 2005). Unlike the PhzI/PhzR systems of *P. chlororaphis/P. aurefaciens* 30-84 and PCL1391, the Phz/PhzR of *P. fluorescens* 2-79 produces and responds to *N*-(3-oxo-hexanoyl)-L-homoserine lactone (3-oxo-C_6-AHL) despite it being almost 90% identical to the other two PhzI/PhzR systems. Genetic and molecular studies have shown that in *P. fluorescens* PhzI/PhzR regulates the transcription of the phenazine biosynthesis operon in response to 3-oxo-C_6-AHL and hence to cell density (Khan et al. 2005).

AHL QS has been studied in two strains of plant-beneficial *P. putida* rhizobacteria (Bertani and Venturi 2004; Steidle et al. 2002). *P. putida* strain IsoF produces and responds to a 3-oxo-C_{12}-AHL via the PpuI/PpuR AHL QS system. This system has been shown to be important for biofilm formation; an important trait for colonization when growing on surfaces. Similarly, in *P. putida* strain WCS358 a PpuI/PpuR system identical to that of strain IsoF has been identified and characterized; however, no phenotypes have yet been identified which are regulated by this system. Interestingly, the PpuI/PpuR system is highly identical to the LasI/LasR system of *P. aeruginosa*, both systems responding to the same AHL molecule and being regulated in a similar way (Bertani and Venturi 2004).

The plant pathogen *P. syringae* has been reported to be able to synthesize AHLs (Dumenyo et al. 1998; Elasri et al. 2001). *P. syringae* pv. *syringae*, the causal agent of brown spot of bean, has an AHL QS system designated AhlI/AhlR which produces and responds to 3-oxo-C_6-AHL and was shown to be important for cell aggregation and epiphytic fitness for *in planta* growth and disease (Quinones et al. 2004). Interestingly, an extensive study analyzing the AHL production ability of 137 soil-borne and plant-associated *Pseudomonas* sp. bacterial strains revealed that AHL production was more common among plant-associated bacteria than among free-living soil-borne ones (Elasri et al. 2001). This study involved strains belonging to *P. syringae, P. chlororaphis, P. fluorescens*, and *P. putida*. It was observed that none of the last three *Pseudomonas* sp. isolated from the soil produced AHLs, whereas it was a very common feature if they were isolated from the rhizosphere. This raises the question of the possible important role of AHL QS in plant–bacteria interaction.

4.5 AHL QS in *Burkholderia*

Just like in the *Pseudomonas* genus, *Burkholderia* species can populate very different niches, including plants, soil, water, and rhizosphere; they may have both pathogenic and symbiotic interactions with plants and are also pathogenic to humans (Coenye and Vandamme 2003). The species *B. cepacia* was originally described by Burkholder (1950) in 1950 as the causative agent of bacterial rot of onions causing a disease called sour skin. In the last 10 years several taxonomic studies resulted in the classification of the *Burkholderia* genus, also creating a group of nine closely species, designated as the *B. cepacia* complex (BCC), isolated from both clinical and environmental sources (Coenye and Vandamme 2003). BCC strains have emerged as opportunistic pathogens in patients with cystic fibrosis causing serious chronic infections. Some species of the BCC are potential biocontrol agents as they can efficiently colonize the root rhizosphere of several important crops and antagonize growth of microbial plant pathogens (Coenye and Vandamme 2003; O'Sullivan and Mahenthiralingam 2005).

Several studies involving cell–cell communication in *Burkholderia* have been performed mainly involving species of the BCC where the AHL QS systems are very well conserved. All reported QS systems of *Burkholderia* consist of the AHL synthase CepI, which mainly directs the synthesis of *N*-octanoyl-L-homoserine lactone (C_8-AHL), which then interacts with LuxR-family member CepR, leading to induction or repression of gene expression (Venturi et al. 2004a). CepI/CepR systems are highly conserved and thus very homologous within the *Burkholderia* genus (Aguilar et al. 2003a, b; Gotschlich et al. 2001; Lutter et al. 2001; Yao et al. 2002). A second AHL QS system (called BviI/BviR), in addition to the CepI/CepR system, has been reported in some strains of *B. vietnamiensis* involving *N*-decanoyl-L-homoserine lactone (C_{10}-AHL) (Conway and Greenberg 2002; Venturi et al. 2004a); strains belonging to this species have been the subject of bioremediation studies. The role of CepI/CepR and BviI/BviR and whether the two systems interact in *B. vietnamiensis* are currently unknown. Another system, called CciI/CciR, has been reported in a highly transmissible opportunistic human pathogen of *B. cenocepacia* (Mahenthiralingam et al. 2005). CciI/CciR is part of a 31-kb pathogenicity island and recently it has been demonstrated that the CepI/CepR and the CciI/CciR systems are interacting with each other (Malott et al. 2005).

The AHL QS system of *Burkholderia*, just like in *P. aeruginosa*, contributes to the virulence as determined using various infection models including plants, nematodes, and murines (Bernier et al. 2003; Huber et al. 2004; Sokol et al. 2003). In the environmental isolate *B. cepacia* ATCC 25419^T, the CepI/CepR system is associated with onion pathogenicity; *cepI* and *cepR* mutants are less virulent in onion rot since attenuated tissue maceration was observed (Aguilar et al. 2003a). This reduction in maceration is mainly due to the lower levels of extracellular polygalacturonase activity since in *cepI/cepR* knockout mutants display 40% of enzyme activity when compared with the parent strain. A systematic study on the CepI/CepR regulon of a *B. cenocepacia* strain has revealed that just like in *P. aeruginosa*, also in *Burkholderia*, QS is a global regulatory system modulating gene expression of approximately 6% of the loci present in the genome (Aguilar et al. 2003b; Riedel et al. 2003).

AHL QS plays an important role in virulence in the soil- and seed-borne rice grain rot (also known as panicle blight) pathogen *B. glumae* (Kim et al. 2004). *B. glumae* contains a system called TofI/TofR which has high identity (75%) to the CepI/CepR system, also producing and responding to C_8-AHL (Kim et al. 2004). The TofI/TofR system of *B. glumae* has been implicated in the regulation of toxoflavin, an essential toxin for the rice pathogenicity of this organism (Kim et al. 2004). Toxoflavin production occurs best in the late-exponential phase and employs QS through TofR/C_8-AHL, ensuring high expression of the toxoflavine biosynthesis genes at high cell densities. It is currently unknown if this occurs directly through activation by TofR/C_8-AHL of the toxoflavin genetic loci or whether the TOF system regulates another regulator called ToxJ which then activates the toxoflavin operons (Kim et al. 2004).

4.6 AHL QS in *Erwinia*

Erwinia spp. are Gram-negative bacterial necrotrophic plant pathogens. They are the causative agents of plant diseases such as soft rots, the potato disease blackleg (stem rot), and Stewart's wilt. AHL QS regulation of pathogenicity factors has been studied in *E. carotovora, E. stewartii* subsp. *stewartii* (Ess; synonym *P. stewartii*), and in *Erwinia chrysnathemi*. In *E. carotovora*, AHL QS systems have been reported in three subspecies: *E. carotovora* subsp. *carotovora* (Ecc), which is pathogenic to many different crops, *E. carotovora* subsp. *atroseptica* (Eca), whose genome has been sequenced and which attacks mainly potato (reviewed in Perombelon 2002), and *E. carotovora* spp. *betavasculorum*, which is pathogenic to sugar beet (Costa and Loper 1997). The major AHL signal molecule produced by *Erwinia* spp. is 3-oxo-C_6-AHL, which in Ecc and Eca is generated by CarI (Swift et al. 1993) and in *E. stewartii* by EsaI. *E. amylovora*, a bacterial pathogen that causes fire blight in plants, produces both 3-oxo-C_6-AHL and *N*-(3-hydroxyhexanoyl)-L-homoserine lactone (3-OH-C_6-AHL) (Venturi et al. 2004b) and the AHL synthase gene, *eamI*, and its putative activator gene, *eamR*, were recently identified and found to be involved in disease symptom development (Molina et al. 2005). *Erwinia* spp. produce an array of exoenzymes (Barras et al. 1994) and some subspecies produce the antibiotic 1-carbapen-2-em-3-carboxylic acid (carbapenem). The production of carbapenem is directly regulated by QS since CarR/AHL binds upstream of the CarA-H biosynthetic operon, resulting in transcriptional activation (Bainton et al. 1992; McGowan et al. 1995; Williams et al. 1992). Mutations in either *carI* or *carR* block carbapenem synthesis (McGowan et al. 1995). Self-resistance to carbapenem antibiotic is obtained by expressing CarF and CarG, which are expressed on a basal level in an AHL-independent manner, while their upregulated expression is CarR/AHL-dependent (McGowan et al. 2005). AHLs are also required to induce the production of plant cell wall degrading exoenzymes (PCWDEs; Jones et al. 1993). CarR is not the AHL receptor activator driving PCWDE production, as exoenzyme production is unaltered in Ecc harboring a disrupted *carR* gene, whereas *carI* mutations affect exoenzyme production. The LuxR-family regulator involved in this regulation is currently unidentified. As the regulators characterized in different *Erwinia* spp. share a higher degree of amino acid identity with each other than with other members of the LuxR-like proteins, Andersson et al. (2000) proposed that they may form a distinct subfamily.

The involvement of QS in plant infection by *Erwinia* spp. is manifested also in experiments with transgenic tobacco plants producing AHLs (Mae et al. 2001) or transgenic potato plants expressing the lactonase enzyme AiiA (Dong et al. 2001), which show increased resistance to *Erwinia* infections. The "quenching" of the QS of *Erwinia* spp. can also be achieved by coinfecting *Erwinia* spp. with bacterial strains that are capable of degrading the AHL signals; recently *Bacillus thuringiensis*, the most widely used biocontrol agent for insect control, was found to effectively stop the otherwise rapid spread of *E. carotovora* cells in plant tissues by the production of AHL lactonases (Dong et al. 2004).

The search for additional QS-regulated genes that are related to pathogenicity of *Erwinia* spp. is ongoing, and has revealed such genes both in Eca and in Ecc. Seven novel genes that are either activated or repressed by the presence of AHLs were found in Ecc (Pemberton et al. 2005). One of these genes, Nip_{Ecc}, was found to be a member of the Nep-1-like (NPL) proteins family (Nep1 is an elicitor of plant necrosis from *Fusarium oxysporum*; Bailey 1995). Nip_{Ecc} was found to cause necrosis when infiltrated into *Nicotiana tabacum* leaves. The Eca homologue of Nip_{Ecc}, Nip_{Eca}, was found to be also involved in pathogenicity. As for exoenzyme production, also nip expression was not affected by *carR* mutations, and *eccR* mutation (*eccR* codes for EccR which is a second AHL LuxR-family response protein) led to a slight increase in nip transcription (Pemberton et al. 2005). A novel gene, *svx*, encoding for a virulence factor which is also regulated by QS was recently found in Eca, the mutant of the *svx* gene was found to have reduced virulence, and *carI* mutants did not produce Svx (Corbett et al. 2005).

E. chrysanthemi strain 3937 produces three different AHLs: 3-oxo-C_6-AHL, C_6-AHL and *N*-decanoyl-L-homoserine lactone (C_{10}-AHL). The genes for the QS signal generator (*expI*) and a response regulator (*expR*) were identified and shown to have high similarity to the *expI/expR* genes of *E. carotovora* (Nasser et al. 1998). ExpI is responsible for only two of the AHLs produced. Disruption of *expI* had no apparent effect on the growth-phase-dependent expression of *hrpN* and *pelE*, or on the virulence of *E. chrysanthemi* in witloof chicory leaves (Ham et al. 2004).

4.7 AHL QS in Rhizobia

Bacteria belonging to the genera *Rhizobium, Mesorhizobium, Sinorhizobium, Bradyrhizobium*, and *Azorhizobium* (collectively referred to as rhizobia) grow in the soil as free-living organisms and can also live as nitrogen-fixing symbionts inside root nodule cells of legume plants playing important roles in agriculture by inducing nitrogen-fixing nodules on the roots of legumes such as peas, beans, clover, and alfalfa (Gage 2004). Several species of rhizobia have been shown to produce AHLs playing important roles in plant–bacteria interactions (Gonzalez and Marketon 2003).

R. leguminosarum bv. *viciae* has a genome consisting of a circular chromosome and six plasmids and possesses several AHL QS systems designated Rhi, Cin, Tra, and Rai (reviewed in Gonzalez and Marketon 2003). The Rhi system is located on the symbiotic plasmid pRL1JI and is composed of *rhiI* and *rhiR* genes responsible for producing and responding to C_6-AHL and C_8-AHL (Rodelas et al. 1999). RhiI/RhiR has been shown to regulate the expression of the *rhiABC* operon, the function of which is currently unknown but it is believed to be involved in the early stages of the symbiotic process. Moreover it was demonstrated that the plant signal compounds known as flavonoids inhibit the expression of both *rhiR* and the *rhiABC* operon (Cubo et al. 1992). The TraI/TraR system is present on the plasmid pRL1JI, it produces and responds to *N*-(3-oxo-octanoyl)-L-homoserine lactone (3-oxo-C_8-AHL), and is involved in the regulation of plasmid transfer (Danino et al. 2003). The RaiI/RaiR system mainly produces and responds to *N*-(3-hydroxyoctanoyl)-L-homoserine lactone (3-OH-C_8-AHL) and is intimately connected with the CinI/CinR system (Wisniewski-Dye et al. 2002; see later). The CinI/CinR system is present on the chromosome and is responsible for the production and response to *N*-(3-hydroxy-7-*cis*-tetradecenoyl)-L-homoserine lactone (3-OH-$C_{14:1}$-AHL) (Lithgow et al. 2000). This signaling molecule is rather unusual in that it inhibits the growth of several strains of *R. leguminosarum* and was previously known as a small bacteriocinin. The CinI/CinR system appears to be at the top of the AHL regulatory cascade since it influences several AHL QS systems, including the RhiI/RhiR and TraI/TraR systems, as well as being involved in the transfer of pRL1JI (Gonzalez and Marketon 2003; Lithgow et al. 2000).

R. etli strain CNPAF512 differs from *R. leguminosarum*, since it possesses only the RaiI/RaiR and the CinI/CinR AHL QS systems. Both systems are present on the chromosome in this strain and are important for growth inhibition and nitrogen fixation (Daniels et al. 2002). *R. etli* strain CFN42 contains one chromosome and six plasmids (p42a to p42f) and possesses the Tra and part of the Cin AHL QS systems, both being involved in the mobilization of the p42a symbiotic plasmid (Tun-Garrido et al. 2003).

S. meliloti is a free-living soil bacterium capable of establishing a symbiotic relationship with the alfalfa plant (*Medicago sativa*). Several strains of *S. meliloti* have been reported to produce one or more AHLs, suggesting the presence of QS systems in this species (Cha et al. 1998; Gonzalez and Marketon 2003; Shaw et al. 1997). The well-characterized *S. meliloti* strain Rm1021 contains two different AHL QS systems on its chromosome: the Sin and the Mel systems (Marketon and Gonzalez 2002).The SinI/SinR system is responsible for the production of long-chain AHLs, ranging from *N*-dodecanoyl-L-homoserine lactone (C_{12}-AHL) to *N*-octadecanoyl-L-homoserine lactone (C_{18}-AHL); *sinI* and *sinR* mutants lead to a decrease in the number of pink nodules during nodulation assays, suggesting a role for QS in symbiosis (Marketon and Gonzalez 2002). In addition, SinI/SinR is necessary for the synthesis of EPSII, an exopolysaccharide important for the nodule invasion process (Marketon et al. 2003). The Mel system appears to be responsible for the production of short-chain AHLs, but the genetic loci as well as its function have not yet been identified (Marketon and Gonzalez 2002). A third AHL QS system has been identified in *S. meliloti* strain Rm41. This system, named TraI/TraR for its homology to the QS system in *A. tumefaciens* and *Rhizobium*, is present on a plasmid called pRme41a and has been shown to be controlling conjugal plasmid transfer (Marketon and Gonzalez 2002).

4.8 AHL QS in Other Gram-Negative Rhizobacteria

The production of exopolysaccharides and plant cell wall degrading enzymes by the phytopathogen *Ralstonia solanacearum* contributes significantly to its virulence and they are produced maximally at high cell densities. *R. solanacearum* contains an AHL QS system designated SolI/SolR which produces and responds to C_6-AHL and/or C_8-AHL (Flavier et al. 1997b). At present there is no evidence that SolI/SolR is directly involved in virulence gene expression; however SolI/SolR is part of the regulatory cascade as it is regulated by a "higher-level" autoinduction system responsive to 3-hydroxypalmitic acid methyl ester via the LysR family regulator PchA (Flavier et al. 1997b, 1998). In addition, *solI/solR* is also additionally regulated by the stationary phase RpoS sigma factor. The QS system is therefore regulated by two other global regulatory systems which are both required for the expression of virulence factors.

Members of the *Serratia* genus are able to colonize a wide variety of surfaces in water and soils and are opportunistic pathogens for plants, insects, fish, and humans (Grimont and Grimont 1978). In *S. liquefaciens* MG1 the AHL QS system SwrI/SwrR produces and responds to C_4-AHL and is involved in (1) the regulation of swarming motility through the direct control of the *swrA* gene which encodes a peptide synthetase responsible for the synthesis of the biosurfactant serraweetin 2 which reduces surface tension and allows swarming motility to occur (Lindum et al. 1998) and (2) mature biofilm formation through the regulation of two loci (called *bsmA* and *bsmB*) responsible for the formation of cell aggregates at a specific time point in biofilm development (Labbate et al. 2004). In *Serratia* sp. ATCC 39006 the AHL QS system SmaI/SmaR produces and responds to C_4-AHL and is involved in the regulation of the antibiotic prodigiodin, of the secondary metabolite carbapenem (a broad spectrum β-lactam antibiotic also produced by *Erwinia* sp.), and of the exoenzymes pectate lyase and cellulose (Fineran et al. 2005; Slater et al. 2003; Thomson et al. 2000). The biocontrol strain *S. plymutica* IC1270 has been reported to produce AHLs; however, the genetic loci have not yet been isolated (Ovadis et al. 2004).

4.9 Interspecies Signaling via AHLs Among Bacteria in the Rhizosphere

Most studies involving production and response to AHL of plant-associated bacteria have been performed in the laboratory and might not reflect what occurs in vivo in close proximity to the plant or *in planta*. Scientists need to focus more attention on the *in situ* production and ability to respond to AHL signal molecules in order to understand when bacteria are coordinating their gene expression in response to cell density or if they are subjected to interference by other bacteria or by the plant. Using green fluorescent protein based AHL sensor plasmids (which are able to respond to the presence of AHLs by producing the easily detectable green fluorescent protein) Steidle et al. (2001) have demonstrated that *P. putida* and *S. liquefaciens* can perceive AHL signals in the rhizosphere of tomato plants when coinoculated with an AHL-producing strain in axenically grown tomato plants. *P. putida* can also perceive AHL signals produced by the indigenous rhizosphere community as it responds to AHLs when inoculated in nonsterile soil (Steidle et al. 2001). This latter result clearly demonstrates that AHL molecules are produced at quorum concentrations in the rhizosphere and that they can be utilized/perceived by the bacterial consortium, implicating interspecies communication. Interspecies communication via AHLs has also been demonstrated between *Burkholderia* and *Pseudomonas* including in the biofilm mode of growth (Lewenza et al. 2002; McKenney et al. 1995; Riedel et al. 2001). Similarly, cross-talk via AHLs has also been demonstrated in the rhizosphere of wheat as a naturally coexisting nonisogenic bacterial population exchange AHL signal with phenazine-producing *P. aureofaciens* strain 30-84 (Pierson et al. 1998a).

Interspecies communication can also be significantly affected by microorganisms which have the capability of degrading the AHLs. Over the past 5 years scientists have reported that a diversity of soil microbes are capable of biodegrading AHLs by cleaving either the amide or the lactone bonds. These enzyme activities could have potent negative effects on AHL signal accumulation as has been demonstrated in pure culture laboratory studies, in soil microcosms, and in transgenic plants expressing bacterial proteins (reviewed in Dong and Zhang 2005).

4.10 AHL Interference, Coordination, and Response by the Plant

A question which is now beginning to be addressed by the scientific community is how eukaryotic hosts are responding to and/or defending themselves against bacterial AHL signal molecules. Plants have been shown to produce chemical compounds that can interfere with QS systems in bacteria by acting as agonists or antagonists of AHL signaling pathways. The chemical structure of these AHL mimics is currently not known and they are referred to as mimics because of their functional interference with bacterial AHLs. Pea seedling exudates inhibited AHL QS in *Chromobacterium violaceum*, whereas they induced the system in *S. liquefaciens* and in LuxR-, LasR-, and AhyR-based engineered *Escherichia coli* AHL sensor systems (Teplitski et al. 2000). In addition, extracts from rice, soybean, tomato, *Medicago truncatula*, and the green alga *Chlamydomonas reinhardtii* all contain AHL mimic molecules (Bauer and Mathesius 2004; Gao et al. 2003; Teplitski et al. 2000, 2004). The implications of a plant interfering with the AHL QS system can be the following: (1) if a plant pathogenic bacterium employs QS in order to prevent activation of virulence gene expression at low cell densities, the interference by the plant to prematurely express these genes would result in the bacterium revealing its presence at a time during the infection at which the plant can effectively prevent its establishment, and (2) on the other hand, in symbiotic plant–bacteria interactions direct signaling via AHLs may permit the coordination of gene expression, resulting in a beneficial interaction for both partners. Interestingly, tobacco plants genetically modified to produce AHLs could induce AHL QS target gene expression in bacteria and consequently restore biocontrol activity of a *P. aureofaciens* AHL-deficient mutant. These transgenic plants could also reestablish pathogenicity to an AHL-defective *E. carotovora* mutant (Fray et al. 1999). Similarly, plants genetically modified to produce a bacterial lactonase enzyme able to degrade AHL molecules displayed significant resistance to disease caused by *E. carotovora* (Dong et al. 2001). AHLs have also been shown to be able to modulate gene expression in plant cells as shown in a proteomic study. Three-day-old roots of *M. truncatula* were exposed to 3-oxo-C_{12}-AHL or *N*-(3-oxo-9-*cis*-hexadecanoyl)-L-homoserine lactone (3-oxo-$C_{16:1}$-AHL) and protein expression was examined by two-dimensional gel electrophoresis and the abundance of 150 proteins showed altered levels depending on the identity, time of exposure, and concentration of the AHL (Mathesius et al. 2003). Our understanding of how plants interfere, coordinate, or respond to bacterial AHL signal molecules is at an early stage and these are important questions to address in the future.

4.11 Conclusions

Several AHL QS systems belonging to rhizobacteria have been reported and studied (summarized in Table 4.1); most systems have been isolated in the course of deciphering the regulation of particular target genes. From studies in other bacteria, it is very likely that also in rhizobacteria AHL QS is a global regulatory network controlling the expression of several hundred genes, thus changing the gene expression profile of bacteria. It will be important to determine if bacterial species which have evolved to particularly adapt to colonize the rhizosphere display unique features with respect to AHL QS. In order to establish this, more *in situ* studies are required especially to determine if AHL QS is influenced by the plant or the microbial consortium present in the rhizosphere. Initial investigations have shown that the plant responds to AHLs, synthesizes AHL analogues and that in the rhizosphere interspecies communication via AHL takes place. These are very important observations indicating that AHL QS could be a way to communicate in the bacterial community and across kingdoms. These results will encourage the scientific community to dedicate more attention to *in situ* studies; understanding the role of these systems will most probably have an impact on more appropriate and effective bioinoculants as well as designing efficient strategies for combating bacterial plant pathogens.

Acknowledgements We thank coresearchers I. Bertani, G. Degrassi, and G. Devescovi for their useful discussions and for reading the manuscript. We thank ICGEB and Fondazione Cassamarca (Treviso, Italy) for their support.

Table 4.1 *N*-Acyl homoserine lactone (*AHL*) quorum sensing (*QS*) systems of rhizobacteria

Species	AHL-QS	AHL molecule	Regulated phenotype	References
Pseudomonas aeruginosa	*lasI, lasR* *rhlI, rhlR*	3-oxo-C_{12}-AHL C_4-AHL	Elastase, alkaline protease, exotoxin A, rhamnolipids, pyocyanin, lectins, superoxide dismutases, biofilm formation	Juhas et al. (2005) Smith and Iglewski (2003)
Pseudomonas aureofaciens 30-84	*phzI, phzR* *csaI, csaR*	C_6-AHL Unknown	Phenazine antibiotics exoprotease, cell surface components, rhizosphere colonization	Pierson et al. (1995) Wood et al. (1997)
Pseudomonas chlororaphiss PCL1391	*phzI, phzR*	C_6-AHL Others	Phenazine antibiotics Unknown	Chin et al. (2001) Chin et al. (2005)
Pseudomonas fluorescens 2-79	*phzI, phzR*	3-oxo-C_6-AHL	Phenazine-1-carboxamide regulation	Khan et al. (2005)
Pseudomonas putida IsoF	*PpuI, PpuR*	3-oxo-C_{12}-AHL	Biofilm development	Steidle et al. (2002)
Pseudomonas putida WCS358	*PpuI, PpuR*	3-oxo-C_{12}-AHL	Unknown	Bertani and Venturi (2004)
Pseudomonas syringiae	*ahlI, ahlR*	3-oxo-C_6-AHL	Cell aggregation, epiphytic fitness	Dumenyo et al. (1998), Elasri et al. (2001)
Burkholderia cenocepacia	*cciI, cciR*	Unknown	Unknown	Malott et al. (2005)
Burkholderia cepacia	*cepI, cepR*	C_8-AHL	Unknown	Aguilar et al. (2003a)
Burkholderia glumae	*tofI, tofR*	C_8-AHL	Toxoflavin regulation	Kim et al. (2004)
Burkholderia vietmamiensis	*bviI, bviR*	C_{10}-AHL	Unknown	Conway and Greenberg (2002), Venturi et al. (2004a)
Erwinia amylovora Ea02	*eamI, eamR*	3-oxo-C_6-AHL 3-oH-C_6-AHL	Virulence Extracellular polysaccharide production, hydrogen peroxide tolerance	Venturi et al. (2004b) Molina et al. (2005)
Erwinia carotovora subsp. *betavasculorum* Ecb168	*ecbI, ecbR*	Unknown	Antibiotic production, extracellular protease	Costa and Loper (1997)

(continued)

Table 4.1 (continued)

Species	AHL-QS	AHL molecule	Regulated phenotype	References
Erwinia carotovora subsp. *atrseptica* SCRI1043	*carI, carR*	3-oxo-C_6-AHL 3-oxo-C_8-AHL, C_8-AHL, C_{10}-AHL	Carbapenem production Exoenzyme synthesis	Smadja et al. (2004)
Erwinia carotovora subsp. *carotovora*	*carI, carR* *expI, expR* *hslI, hslR*	3-oxo-C_6-AHL	Carbapenem production Exoenzyme synthesis Hrp secretion system	Bainton et al. (1992) Jones et al. (1993) McGowan et al. (1995), Swift et al. (1993)
Erwinia chrysanthemi 3937	*expI, expR*	3-oxo-C_6-AHL, C_6-AHL, C_{10}-AHL	Unknown	Nasser et al. (1998)
Erwinia stewartii subsp. *stewartii* SS104	*esaI, esaR*	3-oxo-C_6-AHL, 3-oxo-C_8-AHL	Exopolysaccharide production	Beck von Bodman and Farrand (1995), Beck von Bodman et al. (1998)
Rhizobium etli CNPAF512	*cinI, cinR* *raiI, raiR*	OH-long-chain AHLs Short-chain AHLs	Nitrogen fixation, growth inhibition, symbiosome development Nitrogen fixation, growth inhibition	Daniels et al. (2002)
Rhizobium etli CNPAF512	*traI, traR* Unknown	3-oxo-C_8-AHL 3-OH-C_8-AHL	Conjugal plasmid transfer Unknown	Tun-Garrido et al. (2003)
Rhizobium legumino-sarum bv. *viciae*	*rhiI, rhiR* *cinI, cinR* *traI, traR* *raiI, raiR*	C_6-AHL, C_7-AHL, C_8-AHL 3-OH-$C_{14:1}$-AHL 3-oxo-C_8-AHL 3-OH-C_8-AHL	Influences nodulation Mediates growth inhibition Regulation of plasmid transfer Unknown	Rodelas et al. (1999) Lithgow et al. (2000) Danino et al. (2003) Wisniewski-Dye et al. (2002)
Sinorhizobium meliloti Rm1021	*sinI, sinR* *melI, melR*	Long-chain AHL Short-chain AHLs	Exopolysaccharide EPSII synthesis Unknown	Marketon et al. (2003) Marketon and Gonzalez (2002)
Sinorhizobium meliloti Rm41	*traI, traR* *sinI, sinR*	3-oxo-C_8-AHL, 3-oxo-$C_{16:1}$-AHL, $C_{16:1}$-AHL, C_{16}-AHL 3-oxo-C_{14}-AHL	Conjugal plasmid transfer Exopolysaccharide EPSII synthesis	Marketon and Gonzalez (2002) Marketon and Gonzalez (2002)

(continued)

Table 4.1 (continued)

Species	AHL-QS	AHL molecule	Regulated phenotype	References
Ralstonia solanacearum	*solI, solR*	3-Hydroxypalmitic acid methyl ester	Virulence gene expression	(Flavier et al. (1998)
Serratia liquefaciens MG1	*swrI, swrR*	C_4-AHL	Regulation of swarming motility Mature biofilm formation	Lindum et al. (1998) Labbate et al. (2004)
Serratia ATCC 39006	*smaI, smaR*	C_4-AHL	Regulation of antibiotic prodigiodin, exoenzymes pectate lyase, cellulose	Fineran et al. (2005)

References

Aguilar C, Bertani I, Venturi V (2003a) Quorum-sensing system and stationary-phase sigma factor (*rpoS*) of the onion pathogen *Burkholderia cepacia* genomovar I type strain, ATCC 25416. Appl Environ Microbiol 69:1739–1747

Aguilar C, Friscina A, Devescovi G, Kojic M, Venturi V (2003b) Identification of quorum-sensing-regulated genes of *Burkholderia cepacia*. J Bacteriol 185:6456–6462

Andersson RA, Eriksson AR, Heikinheimo R, Mae A, Pirhonen M, Koiv V, Hyytiainen H, Tuikkala A, Palva ET (2000) Quorum sensing in the plant pathogen *Erwinia carotovora* subsp. *carotovora*: the role of *expR*(Ecc). Mol Plant Microbe Interact 13:384–393

Bailey BA (1995) Purification of a protein from culture filtrates of *Fusarium oxysporum* that induces ethylene and necrosis in leaves of *Erythroxylum coca*. Phytopathology 85:1250–1255

Bainton NJ, Stead P, Chhabra SR, Bycroft BW, Salmond GP, Stewart GS, Williams P (1992) N-(3-Oxohexanoyl)-L-homoserine lactone regulates carbapenem antibiotic production in *Erwinia carotovora*. Biochem J 288:997–1004

Bais HP, Park SW, Weir TL, Callaway RM, Vivanco JM (2004) How plants communicate using the underground information superhighway. Trends Plant Sci 9:26–32

Barber CE, Tang JL, Feng JX, Pan MQ, Wilson TJ, Slater H, Dow JM, Williams P, Daniels MJ (1997) A novel regulatory system required for pathogenicity of *Xanthomonas campestris* is mediated by a small diffusible signal molecule. Mol Microbiol 24:555–566

Barras F, van Gijsegem F, Chatterjee AK (1994) Extracellular enzymes and pathogenesis of soft-rot *Erwinia*. Annu Rev Phytopathol 32:201–234

Bassler BL (2002) Small talk. Cell-to-cell communication in bacteria. Cell 109:421–424

Bauer WD, Mathesius U (2004) Plant responses to bacterial quorum sensing signals. Curr Opin Plant Biol 7:429–433

Beck von Bodman S, Farrand SK (1995) Capsular polysaccharide biosynthesis and pathogenicity in Erwinia stewartii require induction by an N-acylhomoserine lactone autoinducer. J Bacteriol 177:5000–5008

Beck von Bodman SB, Majerczak DR, Coplin DL (1998) A negative regulator mediates quorum-sensing control of exopolysaccharide production in Pantoea stewartii subsp. stewartii. Proc Natl Acad Sci USA 95(13):7687–7692

Bernier SP, Silo-Suh L, Woods DE, Ohman DE, Sokol PA (2003) Comparative analysis of plant and animal models for characterization of *Burkholderia cepacia* virulence. Infect Immun 71:5306–13

Bertani I, Venturi V (2004) Regulation of the N-acyl homoserine lactone-dependent quorum-sensing system in rhizosphere *Pseudomonas putida* WCS358 and cross-talk with the stationary-phase RpoS sigma factor and the global regulator GacA. Appl Environ Microbiol 70:5493–502

Burkholder WH (1950) Sour skin, a bacterial rot of onion bulbs. Phytopathology 40:115–117

Cha C, Gao P, Chen YC, Shaw PD, Farrand SK (1998) Production of acyl-homoserine lactone quorum-sensing signals by gram-negative plant-associated bacteria. Mol Plant Microbe Interact 11:1119–1129

Chancey ST, Wood DW, Pierson LS 3rd (1999) Two-component transcriptional regulation of N-acyl-homoserine lactone production in *Pseudomonas aureofaciens*. Appl Environ Microbiol 65:2294–2299

Chen X, Schauder S, Potier N, Van Dorsselaer A, Pelczer I, Bassler BL, Hughson FM (2002) Structural identification of a bacterial quorum-sensing signal containing boron. Nature 415:545–549

Chin AWTF, van den Broek D, de Voer G, van der Drift KM, Tuinman S, Thomas-Oates JE, Lugtenberg BJ, Bloemberg GV (2001) Phenazine-1-carboxamide production in the biocontrol strain *Pseudomonas chlororaphis* PCL1391 is regulated by multiple factors secreted into the growth medium. Mol Plant Microbe Interact 14:969–979

Chin AWTF, van den Broek D, Lugtenberg BJ, Bloemberg GV (2005) The *Pseudomonas chlororaphis* PCL1391 sigma regulator *psrA* represses the production of the antifungal metabolite phenazine-1-carboxamide. Mol Plant Microbe Interact 18:244–253

Coenye T, Vandamme P (2003) Diversity and significance of *Burkholderia* species occupying diverse ecological niches. Environ Microbiol 5:719–729

Compant S, Duffy B, Nowak J, Clement C, Barka EA (2005) Use of plant growth-promoting bacteria for biocontrol of plant diseases: principles, mechanisms of action, and future prospects. Appl Environ Microbiol 71:4951–4659

Conway BA, Greenberg EP (2002) Quorum-sensing signals and quorum-sensing genes in *Burkholderia vietnamiensis*. J Bacteriol 184:1187–1191

Corbett M, Virtue S, Bell K, Birch P, Burr T, Hyman L, Lilley K, Poock S, Toth I, Salmond G (2005) Identification of a new quorum-sensing-controlled virulence factor in *Erwinia carotovora* subsp. *atroseptica* secreted via the type II targeting pathway. Mol Plant Microbe Interact 18:334–342

Costa JM, Loper JE (1997) EcbI and EcbR: homologs of LuxI and LuxR affecting antibiotic and exoenzyme production by *Erwinia carotovora* subsp. *betavasculorum*. Can J Microbiol 43:1164–1171

Cubo MT, Economou A, Murphy G, Johnston AW, Downie JA (1992) Molecular characterization and regulation of the rhizosphere-expressed genes *rhiABCR* that can influence nodulation by *Rhizobium leguminosarum* biovar viciae. J Bacteriol 174:4026–4035

Daniels R, De Vos DE, Desair J, Raedschelders G, Luyten E, Rosemeyer V, Verreth C, Schoeters E, Vanderleyden J, Michiels J (2002) The *cin* quorum sensing locus of *Rhizobium etli* CNPAF512 affects growth and symbiotic nitrogen fixation. J Biol Chem 277:462–468

Danino VE, Wilkinson A, Edwards A, Downie JA (2003) Recipient-induced transfer of the symbiotic plasmid pRL1JI in *Rhizobium leguminosarum* bv. viciae is regulated by a quorum-sensing relay. Mol Microbiol 50:511–525

Dong YH, Zhang LH (2005) Quorum sensing and quorum-quenching enzymes. J Microbiol 43:101–109

Dong YH, Wang LH, Xu JL, Zhang HB, Zhang XF, Zhang LH (2001) Quenching quorum-sensing-dependent bacterial infection by an N-acyl homoserine lactonase. Nature 411:813–817

Dong YH, Zhang XF, Xu JL, Zhang LH (2004) Insecticidal *Bacillus thuringiensis* silences *Erwinia carotovora* virulence by a new form of microbial antagonism, signal interference. Appl Environ Microbiol 70:954–960

Dumenyo CK, Mukherjee A, Chun W, Chatterjee AK (1998) Genetic and physiological evidence for the production of *N*-acyl homoserine lactones by *Pseudomonas syringae* pv. *syringae* and other fluorescent plant pathogenic *Pseudomonas* species. Eur J Plant Pathol 104:569–582

Elasri M, Delorme S, Lemanceau P, Stewart G, Laue B, Glickmann E, Oger PM, Dessaux Y (2001) Acyl-homoserine lactone production is more common among plant-associated *Pseudomonas* spp. than among soilborne *Pseudomonas* spp. Appl Environ Microbiol 67:1198–1209

Fineran PC, Slater H, Everson L, Hughes K, Salmond GP (2005) Biosynthesis of tripyrrole and beta-lactam secondary metabolites in *Serratia*: integration of quorum sensing with multiple new regulatory components in the control of prodigiosin and carbapenem antibiotic production. Mol Microbiol 56:1495–1517

Flavier AB, Clough SJ, Schell MA, Denny TP (1997a) Identification of 3-hydroxypalmitic acid methyl ester as a novel autoregulator controlling virulence in *Ralstonia solanacearum*. Mol Microbiol 26:251–259

Flavier AB, Ganova-Raeva LM, Schell MA, Denny TP (1997b) Hierarchical autoinduction in *Ralstonia solanacearum*: control of acyl-homoserine lactone production by a novel autoregulatory system responsive to 3-hydroxypalmitic acid methyl ester. J Bacteriol 179: 7089–7097

Flavier AB, Schell MA, Denny TP (1998) An RpoS (sigmaS) homologue regulates acylhomoserine lactone-dependent autoinduction in *Ralstonia solanacearum*. Mol Microbiol 28:475–486

Fray RG, Throup JP, Daykin M, Wallace A, Williams P, Stewart GS, Grierson D (1999) Plants genetically modified to produce N-acylhomoserine lactones communicate with bacteria. Nat Biotechnol 17:1017–1020

Fuqua WC, Winans SC, Greenberg EP (1994) Quorum sensing in bacteria: the LuxR-LuxI family of cell density-responsive transcriptional regulators. J Bacteriol 176:269–275

Fuqua C, Parsek MR, Greenberg EP (2001) Regulation of gene expression by cell-to-cell communication: acyl-homoserine lactone quorum sensing. Annu Rev Genet 35:439–468

Gage DJ (2004) Infection and invasion of roots by symbiotic, nitrogen-fixing rhizobia during nodulation of temperate legumes. Microbiol Mol Biol Rev 68:280–300

Gao M, Teplitski M, Robinson JB, Bauer WD (2003) Production of substances by *Medicago truncatula* that affect bacterial quorum sensing. Mol Plant Microbe Interact 16:827–834

Gonzalez JE, Marketon MM (2003) Quorum sensing in nitrogen-fixing rhizobia. Microbiol Mol Biol Rev 67:574–592

Gotschlich A, Huber B, Geisenberger O, Togl A, Steidle A, Riedel K, Hill P, Tummler B, Vandamme P, Middleton B, Camara M, Williams P, Hardman A, Eberl L (2001) Synthesis of multiple N-acylhomoserine lactones is wide-spread among the members of the *Burkholderia cepacia* complex. Syst Appl Microbiol 24:1–14

Grimont PA, Grimont F (1978) The genus *Serratia*. Annu Rev Microbiol 32:221–248

Haas D, Defago G (2005) Biological control of soil-borne pathogens by fluorescent pseudomonads. Nat Rev Microbiol 3:307–319

Ham JH, Cui Y, Alfano JR, Rodriguez-Palenzuela P, Rojas CM, Chatterjee AK, Collmer A (2004) Analysis of *Erwinia chrysanthemi* EC16 4 *pelE:uidA, pelL:uidA*, and *hrpN:uidA* mutants reveals strain-specific atypical regulation of the Hrp type III secretion system. Mol Plant Microbe Interact 17:184–194

Hentzer M, Wu H, Andersen JB, Riedel K, Rasmussen TB, Bagge N, Kumar N, Schembri MA, Song Z, Kristoffersen P, Manefield M, Costerton JW, Molin S, Eberl L, Steinberg P, Kjelleberg S, Hoiby N, Givskov M (2003) Attenuation of Pseudomonas aeruginosa virulence by quorum sensing inhibitors. EMBO J 22:3803–3815

Huber B, Feldmann F, Kothe M, Vandamme P, Wopperer J, Riedel K, Eberl L (2004) Identification of a novel virulence factor in *Burkholderia cenocepacia* H111 required for efficient slow killing of *Caenorhabditis elegans*. Infect Immun 72:7220–7230

Jones S, Yu B, Bainton NJ, Birdsall M, Bycroft BW, Chhabra SR, Cox AJ, Golby P, Reeves PJ, Stephens S, Winson MK, Salmond GP, Stewart G, Williams P (1993) The *lux* autoinducer regulates the production of exoenzyme virulence determinants in *Erwinia carotovora* and *Pseudomonas aeruginosa*. EMBO J 12:2477–2482

Juhas M, Eberl L, Tummler B (2005) Quorum sensing: the power of cooperation in the world of *Pseudomonas*. Environ Microbiol 7:459–471

Khan SR, Mavrodi DV, Jog GJ, Suga H, Thomashow LS, Farrand SK (2005) Activation of the phz operon of *Pseudomonas fluorescens* 2-79 requires the LuxR homolog PhzR, N-(3-OH-hexanoyl)-L-homoserine lactone produced by the LuxI homolog PhzI, and a *cis*-acting *phz* box. J Bacteriol 187:6517–6527

Kim J, Kim JG, Kang Y, Jang JY, Jog GJ, Lim JY, Kim S, Suga H, Nagamatsu T, Hwang I (2004) Quorum sensing and the LysR-type transcriptional activator ToxR regulate toxoflavin biosynthesis and transport in *Burkholderia glumae*. Mol Microbiol 54:921–934

Kohler T, van Delden C, Curty LK, Hamzehpour MM, Pechere JC (2001) Overexpression of the MexEF-OprN multidrug efflux system affects cell-to-cell signaling in *Pseudomonas aeruginosa*. J Bacteriol 183:5213–5222

Labbate M, Queck SY, Koh KS, Rice SA, Givskov M, Kjelleberg S (2004) Quorum sensing-controlled biofilm development in *Serratia liquefaciens* MG1. J Bacteriol 186:692–698

Latifi A, Foglino M, Tanaka K, Williams P, Lazdunski A (1996) A hierarchical quorum-sensing cascade in *Pseudomonas aeruginosa* links the transcriptional activators LasR and RhlR (VsmR) to expression of the stationary-phase sigma factor RpoS. Mol Microbiol 21:1137–1146

Lewenza S, Visser MB, Sokol PA (2002) Interspecies communication between *Burkholderia cepacia* and *Pseudomonas aeruginosa*. Can J Microbiol 48:707–716

Lindum PW, Anthoni U, Christophersen C, Eberl L, Molin S, Givskov M (1998) N-Acyl-L-homoserine lactone autoinducers control production of an extracellular lipopeptide biosurfactant required for swarming motility of *Serratia liquefaciens* MG1. J Bacteriol 180:6384–6388

Lithgow JK, Wilkinson A, Hardman A, Rodelas B, Wisniewski-Dye F, Williams P, Downie JA (2000) The regulatory locus *cinRI* in *Rhizobium leguminosarum* controls a network of quorum-sensing loci. Mol Microbiol 37:81–97

Loh J, Pierson EA, Pierson LS 3rd, Stacey G, Chatterjee A (2002) Quorum sensing in plant-associated bacteria. Curr Opin Plant Biol 5:285–290

Lutter E, Lewenza S, Dennis JJ, Visser MB, Sokol PA (2001) Distribution of quorum-sensing genes in the *Burkholderia cepacia* complex. Infect Immun 69:4661–4666

Mae A, Montesano M, Koiv V, Palva ET (2001) Transgenic plants producing the bacterial pheromone N-acyl-homoserine lactone exhibit enhanced resistance to the bacterial phytopathogen *Erwinia carotovora*. Mol Plant Microbe Interact 14:1035–1042

Mahenthiralingam E, Urban TA, Goldberg JB (2005) The multifarious, multireplicon *Burkholderia cepacia* complex. Nat Rev Microbiol 3:144–156

Malott RJ, Baldwin A, Mahenthiralingam E, Sokol PA (2005) Characterization of the *cciIR* quorum-sensing system in *Burkholderia cenocepacia*. Infect Immun 73:4982–4992

Marketon MM, Gonzalez JE (2002) Identification of two quorum-sensing systems in *Sinorhizobium meliloti*. J Bacteriol 184:3466–3475

Marketon MM, Glenn SA, Eberhard A, Gonzalez JE (2003) Quorum sensing controls exopolysaccharide production in *Sinorhizobium meliloti*. J Bacteriol 185:325–331

Mathesius U, Mulders S, Gao M, Teplitski M, Caetano-Anolles G, Rolfe BG, Bauer WD (2003) Extensive and specific responses of a eukaryote to bacterial quorum-sensing signals. Proc Natl Acad Sci USA 100:1444–1449

McGowan S, Sebaihia M, Jones S, Yu B, Bainton N, Chan PF, Bycroft B, Stewart GS, Williams P, Salmond GP (1995) Carbapenem antibiotic production in *Erwinia carotovora* is regulated by CarR, a homologue of the LuxR transcriptional activator. Microbiology 141:541–550

McGowan SJ, Barnard AM, Bosgelmez G, Sebaihia M, Simpson NJ, Thomson NR, Todd DE, Welch M, Whitehead NA, Salmond GP (2005) Carbapenem antibiotic biosynthesis in *Erwinia carotovora* is regulated by physiological and genetic factors modulating the quorum sensing-dependent control pathway. Mol Microbiol 55:526–545

McKenney D, Brown KE, Allison DG (1995) Influence of *Pseudomonas aeruginosa* exoproducts on virulence factor production in *Burkholderia cepacia*: evidence of interspecies communication. J Bacteriol 177:6989–6992

Miller MB, Bassler BL (2001) Quorum sensing in bacteria. Annu Rev Microbiol 55:165–199

Molina L, Rezzonico F, Defago G, Duffy B (2005) Autoinduction in *Erwinia amylovora*: evidence of an acyl-homoserine lactone signal in the fire blight pathogen. J Bacteriol 187:3206–3213
Nasser W, Bouillant ML, Salmond G, Reverchon S (1998) Characterization of the *Erwinia chrysanthemi expI-expR* locus directing the synthesis of two N-acyl-homoserine lactone signal molecules. Mol Microbiol 29:1391–1405
O'Sullivan LA, Mahenthiralingam E (2005) Biotechnological potential within the genus *Burkholderia*. Lett Appl Microbiol 41:8–11
Ovadis M, Liu X, Gavriel S, Ismailov Z, Chet I, Chernin L (2004) The global regulator genes from biocontrol strain *Serratia plymuthica* IC1270: cloning, sequencing, and functional studies. J Bacteriol 186:4986–4993
Pearson JP, Van Delden C, Iglewski BH (1999) Active efflux and diffusion are involved in transport of *Pseudomonas aeruginosa* cell-to-cell signals. J Bacteriol 181:1203–1210
Pemberton CL, Whitehead NA, Sebaihia M, Bell KS, Hyman LJ, Harris SJ, Matlin AJ, Robson ND, Birch PR, Carr JP, Toth IK, Salmond GP (2005) Novel quorum-sensing-controlled genes in *Erwinia carotovora* subsp. carotovora: identification of a fungal elicitor homologue in a soft-rotting bacterium. Mol Plant Microbe Interact 18:343–353
Perombelon MCM (2002) Potato diseases caused by soft rot erwinias: an overview of pathogenesis. Plant Pathol 51:1–12
Pierson LS 3rd, Keppenne VD, Wood DW (1994) Phenazine antibiotic biosynthesis in *Pseudomonas aureofaciens* 30-84 is regulated by PhzR in response to cell density. J Bacteriol 176:3966–3674
Pierson LS 3rd, Gaffney T, Lam S, Gong F (1995) Molecular analysis of genes encoding phenazine biosynthesis in the biological control bacterium *Pseudomonas aureofaciens* 30-84. FEMS Microbiol Lett 134:299–307
Pierson EA, Wood DW, Cannon JA, Blachere FM, Pierson LS (1998a) Interpopulation signaling via *N*-acyl-homoserine lactones among bacteria in the wheat rhizosphere. Mol Plant Microbe Interact 11:1078–1084
Pierson LS 3rd, Wood DW, Pierson EA (1998b) Homoserine lactone-mediated gene regulation in plant-associated bacteria. Annu Rev Phytopathol 36:207–225
Quinones B, Pujol CJ, Lindow SE (2004) Regulation of AHL production and its contribution to epiphytic fitness in *Pseudomonas syringae*. Mol Plant Microbe Interact 17:521–531
Riedel K, Hentzer M, Geisenberger O, Huber B, Steidle A, Wu H, Hoiby N, Givskov M, Molin S, Eberl L (2001) N-Acylhomoserine-lactone-mediated communication between *Pseudomonas aeruginosa* and *Burkholderia cepacia* in mixed biofilms. Microbiology 147:3249–3262
Riedel K, Arevalo-Ferro C, Reil G, Gorg A, Lottspeich F, Eberl L (2003) Analysis of the quorum-sensing regulon of the opportunistic pathogen *Burkholderia cepacia* H111 by proteomics. Electrophoresis 24:740–750
Rodelas B, Lithgow JK, Wisniewski-Dye F, Hardman A, Wilkinson A, Economou A, Williams P, Downie JA (1999) Analysis of quorum-sensing-dependent control of rhizosphere-expressed (rhi) genes in *Rhizobium leguminosarum* bv. viciae. J Bacteriol 181:3816–3823
Ruby EG (1996) Lessons from a cooperative, bacterial-animal association: the *Vibrio fischeri-Euprymna scolopes* light organ symbiosis. Annu Rev Microbiol 50:591–624
Schuster M, Lostroh CP, Ogi T, Greenberg EP (2003) Identification, timing, and signal specificity of *Pseudomonas aeruginosa* quorum-controlled genes: a transcriptome analysis. J Bacteriol 185:2066–2079
Seed PC, Passador L, Iglewski BH (1995) Activation of the *Pseudomonas aeruginosa* lasI gene by LasR and the *Pseudomonas* autoinducer PAI: an autoinduction regulatory hierarchy. J Bacteriol 177:654–659
Shaw PD, Ping G, Daly SL, Cha C, Cronan JE Jr, Rinehart KL, Farrand SK (1997) Detecting and characterizing N-acyl-homoserine lactone signal molecules by thin-layer chromatography. Proc Natl Acad Sci USA 94:6036–6041
Slater H, Crow M, Everson L, Salmond GP (2003) Phosphate availability regulates biosynthesis of two antibiotics, prodigiosin and carbapenem, in *Serratia* via both quorum-sensing-dependent and -independent pathways. Mol Microbiol 47:303–320

Smadja B, Latour X, Faure D, Chevalier S, Dessaux Y, Orange N (2004) Involvement of N-acylhomoserine lactones throughout plant infection by *Erwinia carotovora* subsp. *atroseptica* (*Pectobacterium atrosepticum*). Mol Plant Microbe Interact 17:1269–1278

Smith RS, Iglewski BH (2003) *P. aeruginosa* quorum-sensing systems and virulence. Curr Opin Microbiol 6:56–60

Sokol PA, Sajjan U, Visser MB, Gingues S, Forstner J, Kooi C (2003) The CepIR quorum-sensing system contributes to the virulence of *Burkholderia cenocepacia* respiratory infections. Microbiology 149:3649–3658

Steidle A, Sigl K, Schuhegger R, Ihring A, Schmid M, Gantner S, Stoffels M, Riedel K, Givskov M, Hartmann A, Langebartels C, Eberl L (2001) Visualization of N-acylhomoserine lactone-mediated cell-cell communication between bacteria colonizing the tomato rhizosphere. Appl Environ Microbiol 67:5761–5770

Steidle A, Allesen-Holm M, Riedel K, Berg G, Givskov M, Molin S, Eberl L (2002) Identification and characterization of an N-acylhomoserine lactone-dependent quorum-sensing system in *Pseudomonas putida* strain IsoF. Appl Environ Microbiol 68:6371–6382

Sturme MH, Kleerebezem M, Nakayama J, Akkermans AD, Vaugha EE, de Vos WM (2002) Cell to cell communication by autoinducing peptides in gram-positive bacteria. Antonie Van Leeuwenhoek 81:233–243

Swift S, Winson MK, Chan PF, Bainton NJ, Birdsall M, Reeves PJ, Rees CE, Chhabra SR, Hill PJ, Throup JP et al (1993) A novel strategy for the isolation of *luxI* homologues: evidence for the widespread distribution of a LuxR:LuxI superfamily in enteric bacteria. Mol Microbiol 10:511–520

Teplitski M, Robinson JB, Bauer WD (2000) Plants secrete substances that mimic bacterial N-acyl homoserine lactone signal activities and affect population density-dependent behaviors in associated bacteria. Mol Plant Microbe Interact 13:637–648

Teplitski M, Chen H, Rajamani S, Gao M, Merighi M, Sayre RT, Robinson JB, Rolfe BG, Bauer WD (2004) *Chlamydomonas reinhardtii* secretes compounds that mimic bacterial signals and interfere with quorum sensing regulation in bacteria. Plant Physiol 134:137–146

Thomson NR, Crow MA, McGowan SJ, Cox A, Salmond GP (2000) Biosynthesis of carbapenem antibiotic and prodigiosin pigment in *Serratia* is under quorum sensing control. Mol Microbiol 36:539–556

Tun-Garrido C, Bustos P, Gonzalez V, Brom S (2003) Conjugative transfer of p42a from *Rhizobium etli* CFN42, which is required for mobilization of the symbiotic plasmid, is regulated by quorum sensing. J Bacteriol 185:1681–1692

Venturi V, Friscina A, Bertani I, Devescovi G, Aguilar C (2004a) Quorum sensing in the *Burkholderia cepacia* complex. Res Microbiol 155:238–244

Venturi V, Venuti C, Devescovi G, Lucchese C, Friscina A, Degrassi G, Aguilar C, Mazzucchi U (2004b) The plant pathogen *Erwinia amylovora* produces acyl-homoserine lactone signal molecules in vitro and in planta. FEMS Microbiol Lett 241:179–83

Von Bodman SB, Bauer WD, Coplin DL (2003) Quorum sensing in plant-pathogenic bacteria. Annu Rev Phytopathol 41:455–482

Wagner VE, Bushnell D, Passador L, Brooks AI, Iglewski BH (2003) Microarray analysis of *Pseudomonas aeruginosa* quorum-sensing regulons: effects of growth phase and environment. J Bacteriol 185:2080–2095

Whitehead NA, Barnard AM, Slater H, Simpson NJ, Salmond GP (2001) Quorum-sensing in Gram-negative bacteria. FEMS Microbiol Rev 25:365–404

Whiteley M, Lee KM, Greenberg EP (1999) Identification of genes controlled by quorum sensing in *Pseudomonas aeruginosa*. Proc Natl Acad Sci USA 96:13904–13909

Williams P, Bainton NJ, Swift S, Chhabra SR, Winson MK, Stewart GS, Salmond GP, Bycroft BW (1992) Small molecule-mediated density-dependent control of gene expression in prokaryotes: bioluminescence and the biosynthesis of carbapenem antibiotics. FEMS Microbiol Lett 79:161–167

Wisniewski-Dye F, Jones J, Chhabra SR, Downie JA (2002) *railR* genes are part of a quorum-sensing network controlled by *cinI* and *cinR* in *Rhizobium leguminosarum*. J Bacteriol 184:1597–1606

Wood DW, Gong F, Daykin MM, Williams P, Pierson LS 3rd (1997) N-Acyl-homoserine lactone-mediated regulation of phenazine gene expression by *Pseudomonas aureofaciens* 30-84 in the wheat rhizosphere. J Bacteriol 179:7663–7670

Yao F, Zhou H, Lessie TG (2002) Characterization of N-acyl homoserine lactone overproducing mutants of *Burkholderia multivorans* ATCC 17616. FEMS Microbiol Lett 206:201–207

Zhang Z, Pierson LS 3rd (2001) A second quorum-sensing system regulates cell surface properties but not phenazine antibiotic production in *Pseudomonas aureofaciens*. Appl Environ Microbiol 67:4305–4315

Part II
Bacterial Metabolites

Chapter 5
The Effect of Bacterial Secondary Metabolites on Bacterial and Fungal Pathogens of Rice

P. Velusamy(✉) and S.S. Gnanamanickam

5.1 Introduction

Certain antagonistic bacteria are considered ideal biological control agents owing to their rapid growth, easy handling and aggressive colonization of the rhizosphere. These bacteria may mediate biocontrol by one or more of the several mechanisms of disease suppression (Weller 1988). A primary mechanism of pathogen inhibition is by the production of secondary metabolites and other factors such as siderophore production and microbial cyanide, and lytic enzymes may also play a role (Fravel 1988; Keel et al. 1992; O'Sullivan and O'Gara 1992). These bacteria are involved in the biological control of bacterial, fungal, and viral diseases of plants.

The antagonistic fluorescent pseudomonads produce one or more metabolites, such as phenazine-1-carboxylicacid (PCA), 2,4-diacetylphloroglucinol (DAPG), pyoluteorin, pyrrolnitrin, and oomycin A. Among these, DAPG is an antibiotic produced by fluorescent *Pseudomonas* spp. of diverse geographic origin (Dowling and O'Gara 1994; Keel et al. 1996; Thomashow and Weller 1995; Raaijmakers et al. 1997). It has been implicated as the mechanism involved in the biological control of some of the most important crop diseases, such as the root rot of wheat caused by *Fusarium oxysporum* f. sp. *graminis* (Garagulya et al. 1974), black root rot of tobacco caused by *Thielaviopsis basicola* (Defago et al. 1990; Keel et al. 1992), damping-off of sugarbeet caused by *Pythium ultimum* and *Rhizoctonia solani* (Nowak-Thompson et al. 1994), and the "take-all" of wheat caused by *Gaeumannomyces graminis tritici* (Defago et al. 1990; Keel et al. 1992). Strains of *Pseudomonas fluorescens* that produce DAPG also have had a key role in the natural biological control of "take-all" known as "take-all decline" (Raaijmakers et al. 1997).

DAPG is a bacterial and plant metabolite (Bangera and Thomashow 1996, 1999; Keel et al. 1992), phenolic in nature, probably of polyketide origin with a broad spectrum of antifungal, antibacterial, antiviral, and antihelminthic properties (Garagulya

P. Velusamy
Centre for Advanced Studies in Botany, University of Madras, Guindy Campus, Chennai-600025, India
e-mail: velusamy22973@yahoo.com

P. Karlovsky (ed.), *Secondary Metabolites in Soil Ecology. Soil Biology 14.*

et al. 1974; Keel et al. 1992; Nowak-Thompson et al. 1994; Levy et al. 1992; Reddy and Borovko 1970). The production of DAPG has recently been recognized as an important feature in the biological control of plant diseases by antagonistic bacteria, and a number of researchers are now investigating the genetics of this metabolite production.

The first report on the cloning of the antibiotic genes was by Gutterson et al. (1986). Their work outlined the isolation of the mutants of *Pseudomonas fluoresencs* strain HV37a which had lost the ability to inhibit *Pythium ultimum* on iron-rich media. Complementation analysis of such mutants indicated that at least five genes in this strain encoded the antibiotic production. Subsequently it was determined that at least three different antibiotics were produced by the strain HV37a (James and Gutterson 1986). One of these compounds was identified as oomycin A and this was found to be responsible for about 70% of the ability of this antagonistic strain to reduce *Pythium* root infection of cotton and about 50% of the ability to increase cotton seed emergence (Howie and Suslow 1991). Production of these compounds was found to be regulated differentially by glucose (James and Gutterson 1986). This regulation occurs at the transcriptional level and it was dependent on the products of the *afuA* and *afuB* genes. Four transcriptional units, *afuE, afuR, afuAB*, and *afuP*, were found to be involved in antibiotic production (Gutterson et al. 1988). Although the *afuAB* operon is involved in regulation, the precise functions of the other units are not yet clearly known. The regulation of *afuE* transcriptional units appears to be more complex and autoregulated. Expression of *afuE* has been increased by cloning it downstream from the *tac* promoter and this regulated increased production of oomycin A by the genetically manipulated strains (Gutterson 1990). Preliminary experiments on the oomycin A overproducing strain suggest an improved inhibitory capacity against *Pythium ultimum.*

The antibiotic DAPG is produced by many strains of fluorescent *Pseudomonas* spp. with biocontrol activity against many soil-borne bacterial and fungal plant pathogens. Genes required for DAPG synthesis by *Pseudomonas fluorescens* Q2-87 are encoded by a 6.5-kb fragment of genomic DNA that can transfer production of DAPG to nonproducing recipient *Pseudomonas* strains. The nucleotide sequence was determined for the 6.5-kb fragment and its flanking region of genomic DNA in strain Q2-87 (Bangera and Thomashow 1999). Six open reading frames were identified, four of which (*phlABCD*) comprised an operon that includes a set of three genes (*phlACB*) conserved between eubacteria and archaebacteria and a gene (*phlD*) encoding a polyketide synthase with homology to chalcone and stilbene synthases from plants. The biosynthetic operon is flanked on either side by *phlE* and *phlF*, which code, respectively, for putative efflux and regulatory proteins. Expression in *Escherichia coli* of *phlA, phlC, phlB*, and *phlD*, individually or in combination, identified a novel polyketide biosynthetic pathway in which *phlD* is responsible for the production of monoacetylphloroglucinol (MAPG). *phlA, phlC*, and *phlB* are necessary to convert MAPG to DAPG, and they also may function in the synthesis of MAPG (Bangera and Thomashow 1996, 1999). Cloning of a 6.5-kb genomic DNA fragment from

Pseudomonas fluorescens Q2-87 conferred production of DAPG upon DAPG-nonproducing recipient biocontrol strains and increased the biocontrol efficiency (Bangera and Thomashow 1999).

In this review we describe how we have used a rapid PCR-based assay method to identify a plant-associated *Pseudomonas* strain from India that produces DAPG and its role for biological suppression of rice bacterial blight. We also discuss how we have used a number of very efficient *Pseudomonas fluorescens* and *Bacillus* strains which produce different antibacterial and antifungal metabolites (other than DAPG) to suppress rice pathogens such as *Xanthomonas oryzae* pv. *oryzae* (rice bacterial blight), *Magnaporthe grisea* (rice blast), and *Rhizoctonia solani* (rice sheath blight).

5.2 A Critical Review of the Work Done in India

Rice is the most widely cultivated crop in the world and is very important to the economy of India. Annually, more than 40% of the world's rice crop is lost as a result of biotic stresses like insects, pests, pathogens, and weeds (Hossain 1996). Among several diseases caused by bacterial, fungal, and viral pathogens that devastate rice yields all over the world, bacterial blight caused by *Xanthomonas oryzae* pv. *oryzae*, blast caused by *Magnaporthe grisea*, sheath blight caused by *Rhizoctonia solani*, sheath rot caused by *Sarocladium oryzae*, and tungro virus are most important.

5.3 Bacterial Blight of Rice

5.3.1 Causal Organism: Xanthomonas oryzae pv. oryzae

Bacterial blight caused by *Xanthomonas oryzae* pv. *oryzae* is one of the most important and very serious diseases of rice (Swings et al. 1990). Bacterial blight is also one of the oldest known diseases and was first noticed by the farmers of the Fukuko area, Kyushu, Japan, as early as 1884 (Tagami and Mizukami 1962).

5.3.2 Symptoms

Bacterial blight is a vascular disease resulting in a systemic infection of rice and it produces tannish-gray to white lesions along the veins (Mew 1987). Symptoms are observed at the tillering stage, and disease incidence increases with plant growth, peaking at the flowering stage (Fig. 5.1).

Fig. 5.1 Symptom of rice bacterial blight in the field

5.3.3 Morphology and External Appearance

The causal bacterium of rice bacterial leaf blight *Xanthomonas oryzae* pv. *oryzae* has cells that are short rods with round ends, 1–2 µm × 0.8–1 µm, with a monotrichous flagellum of 6–8 µm. The organism is gram-negative and non-spore-forming (Ishiyama 1922). Bacterial cells are surrounded by mucous capsules. Colonies are circular, convex, and whitish to straw yellow with a smooth surface and an entire

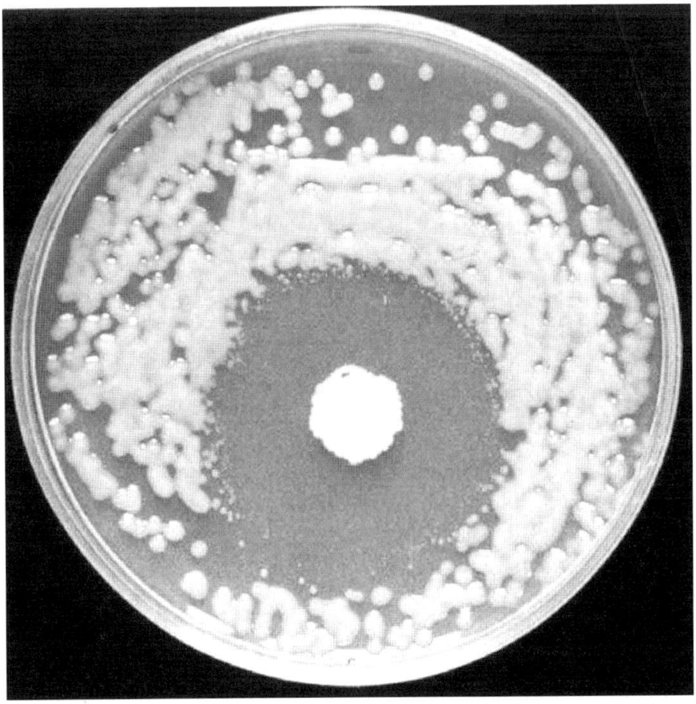

Fig. 5.2 Identification of a bacterial antagonist against *Xanthomonas oryzae* pv. *oryzae* by dual plate assay on peptone sucrose agar

margin and are opaque against transmitted light. The flagellum is 8.75 μm × 30 nm (Yoshimura and Tahara 1960).

Xanthomonas oryzae pv. *oryzae* is a yellow, gram-negative bacterium producing copious amounts of extracellular polysaccharides (EPS) on peptone sucrose agar (PSA) medium (Fig. 5.2).

5.3.4 *Yield Losses*

Bacterial blight is found worldwide and is particularly destructive in Asia during the heavy rains of the monsoon season. In many Asian countries, the disease has become endemic on rice following repeated cultivation. The disease can reduce grain yields to varying levels, depending on the stage of the crop at the time of infection, the degree of cultivar susceptibility, and to a great extent the conduciveness of the environment in which it occurs. Severe crop losses of 10–20% in moderate conditions (Ou 1985), or up to 50% in highly conducive conditions (Mew et al. 1993), have been recorded in several parts of Asia and Southeast Asia (Fig. 5.1).

5.3.5 Disease Cycle

The soil is not considered as an important source of inoculla (Tagami et al. 1963; Srivastava 1967). The bacterium can survive in soil only for 1–2 months (Wakimoto 1956). It can survive in dry form on seeds from infected plants, stored rice straw, and rice stubble. The dry form of the bacterium normally becomes activated by moisture. The growth form of the bacterium is normally found in stubble and in some susceptible grasses, especially *Leersia* sp., *Leptocloa chinensis*, and *Cyperus rotundus*, which serve as alternative hosts.

5.3.6 Bacterial Blight Management

Bacterial blight management tactics, such as us of chemicals, are harmful to the environment, while others, such as host plant resistance based on single genes, may not be durable in the field and might lead to frequent varietal breakdowns. Biological control, therefore, assumes special significance in being an ecology-conscious, cost-effective alternative strategy for bacterial blight management. This can also be used in integration with other strategies to afford greater levels of protection and sustain rice yields. Antagonistic bacteria are considered ideal biological control agents for obvious reasons, like rapid growth, easy handling, and aggressive colonization of the rhizosphere (Weller 1988). Bacterial antagonists have been evaluated with various degrees of success for the suppression of rice diseases of fungal origin (Vasudevan et al. 2002). However, there has been no detailed study on the use of antagonistic bacteria for suppression of bacterial blight except for a recent study with *Bacillus* spp. in our laboratory (Vasudevan 2002). The present study focuses on the use of *Pseudomonas fluorescens* strains from India that produce DAPG to suppress rice bacterial blight.

5.4 Investigation on the Production of DAPG and Its Role in Disease Management

The discovery of plant growth promoting bacteria offered renewed hopes of developing effective biological control agents, which would be ecology-conscious, environment-friendly, and cost-effective. Of the various antagonistic microbes reported so far, *Bacillus* spp. and fluorescent pseudomonads appear most promising. Among these organisms, those *Pseudomonas fluorescens* strains that produce DAPG have been implicated in the dramatic "take-all" decline in wheat in the American Pacific northwest (Weller and Cook 1983), and also in the suppression of damping-off of cotton (Howell and Stipanovic 1979; Howie and Suslow 1991) and black root rot of tobacco (Defago et al. 1990). These studies have clearly demonstrated the possibilities of using microbial metabolites to control major plant diseases.

Since there was no report on the production of DAPG from Indian soils, we started screening plant-associated *Pseudomonas fluorescens* strains for production of this important metabolite which is known to have antifungal, antibacterial, antiviral, and antihelminthic properties. In our recent work, we (Velusamy and Gnanamanickam 2003; Velusamy et al. 2006) have isolated a large number of fluorescent pseudomonads strains associated with roots of rice and other crops in India (Table 5.1).

5.5 Screening for Efficient Biocontrol Strains Against *Xanthomonas oryzae* pv. *oryzae*

5.5.1 Dual Plate Assay

Dual plate assays was performed for the identification of bacterial antagonists against the rice bacterial blight pathogen *Xanthomonas oryzae* pv. *oryzae* on PSA medium. These assays led to the identification of 278 bacterial antagonists out of 637 screened (44%). The antagonist strains showed zones of inhibition whose diameter ranged from 0.5 to 4.7 cm (Table 5.1).

5.5.2 PCR-Based Screening of DAPG Producers from Antagonistic Bacteria

Identification of the *phlD* gene for the production of DAPG from the antagonistic bacteria was carried out by a PCR method by using forward (*phl2a*: 20-mer 5′-GAGGACGTCGAAGACCACCA-3′) and reverse (*phl2b*: 20-mer 5′-ACCGCAGCATCGTGTATGAG-3′) primers. By this method, a total of 27 strains out of 278 antagonistic strains showed a characteristic 745-bp DNA fragment amplification (Table 5.1). This was confirmed from the PCR products of a DAPG-producing strain.

5.6 Evaluation of DAPG Producers for Bacterial Blight Suppression—In Vivo

All 27 DAPG producers were evaluated in net-house and field experiments for performance in bacterial blight suppression at the Regional Agricultural Research Station, Pattambi, Kerala, India. In these experiments *Xanthomonas oryzae* pv. *oryzae* was clip-inoculated on the leaf of rice cultivar IR24 plants.

Table 5.1 List of rhizobacteria isolated from various rice-growing regions of India, their antibiosis towards *Xanthomonas oryzae* pv. *oryzae* (*Xoo*), and detection of 2,4-diacetylphloroglucinol (*DAPG*) by a PCR-based method

Place of collection		Code	Host/crop	Number of strains obtained	No. of strains showing Xoo inhibition (% of bacterial strain in particular host)[a]	No. of strains that produce DAPG[b]
Karnataka	Mandya	MAD	Rice	36	9 (25.0)	–
	Immavur	IMV	Rice	22	5 (22.7)	2
	Nanjangud	NJD	Rice	19	6 (31.6)	–
	Malavali	BGR	Sunnhemp	15	7 (46.7)	1
	Srirangapattinam	RDR	Finger millet	10	7 (70.0)	–
	Mysulipattinum	MLP	Sunnhemp	39	10 (25.6)	–
	Chamrajnagar	CJR	Sunnhemp	12	3 (25.0)	1
	Kanakapura	KKP	Rice	15	2 (13.3)	–
Andhra Pradesh	Nellur	NEL	Rice	35	14 (40.0)	–
	Vijayavada	VYA	Rice	20	9 (45.0)	1
	Sulur	SLU	Rice	21	9 (42.9)	–
	Cuddappah	CUD	Rice	16	6 (37.5)	–
	Nandyal	NAD	Rice	17	10 (58.8)	1
	Tada	TAD	Rice	22	8 (36.4)	–
Kerala	Pattambi	PTB	Rice	11	6 (54.5)	1
	Mannur	MNR	Rice	13	4 (30.8)	–
	Palghat	PAL	Rice	21	11 (52.4)	–
	Mancombu	MON	Rice	13	7 (53.8)	1
	Cheramangalam	CHE	Rice	12	9 (75.0)	–
Tamil Nadu	Thiruvannamalli	TVM	Rice	17	7 (41.2)	1
	Vellore	VEL	Finger millet	10	7 (70.0)	2
	Thiruvercadu	TVR	Finger millet	31	2 (6.5)	–
	Gudiyatham	GDY	Rice	12	8 (66.7)	2
	Tiruchirappalli	TRP	Rice	10	3 (30.0)	–
	Maduri	MDR	Rice	14	9 (64.3)	3
	Sethuvali	STV	Sorghum	12	9 (75.0)	1
	Verunchipuram	VGP	Sorghum	13	4 (30.8)	1
	Puducherry	PDY	Finger millet	12	8 (66.7)	1
	Valajabath	VLB	Rice	10	3 (30.0)	–
	Kavanur	KVR	Black gram	11	7 (63.6)	2
	Theni	TNI	Black gram	15	6 (40.0)	1
	Kovilpatti	KOV	Green gram	11	7 (63.6)	2
	Rajapalayam	RJP	Rice	17	11 (64.7)	1
	Pudukkoittai	PDU	Rice	11	11 (100.0)	1
Maharashtra	Loanavola	LVA	Rice	12	4 (33.3)	–
	Khade	KAD	Black gram	13	8 (61.5)	1
	Dapholi	DPI	Rice	10	9 (90.0)	–
	Miraj	KRA	Rice	15	8 (53.3)	–
	Karhad	KAH	Rice	12	5 (41.7)	–
Total number of strains				637	278	27

[a] Of 637 strains, 278 inhibited Xoo in laboratory dual-plate assays

[b] Production of DAPG was identified through a PCR-based screening procedure that amplified a 745-bp DNA fragment in 27 out of 278 strains tested

5.6.1 *Net-House Experiment*

Six of the 27 strains afforded more than 55% of bacterial blight suppression in this net-house experiment. Three of the six strains were from the rice rhizosphere. Treatments with the rice-associated strains IMV 14, PTB 9, and MDR 7 resulted in suppression of bacterial blight of 58.7, 58.8, and 57.1%, respectively. The other three non-rice-associated DAPG-producer strains, KAD 7, VGP 13, and PDY 7, also significantly reduced bacterial blight disease by 56.9, 55.4, and 58.8%, respectively (Table 5.2).

5.6.2 *Field Experiment*

At least seven of the 27 strains afforded more than 50% bacterial blight suppression in the field experiment. These included six strains which had performed well in the net house. The rice-associated strains IMV 14, PTB 9, and MDR 7 showed significant levels (56.8, 64.5, and 54.4%, respectively) of bacterial blight suppression (Fig. 5.3, Table 5.2).

Both in the net house and in the field, the rice-associated bacterial strain PTB 9 performed consistently well along with two other rice-associated strains (Table 5.2).

5.7 Mechanism Mediating Biocontrol Activity of Purified DAPG

We have analyzed the biological activity of a compound extracted and purified from the superior strains of *Pseudomonas fluorescens* PTB 9 which we supposed was DAPG. The results showed strong inhibition of the growth of *Xanthomonas oryzae* pv. *oryzae* in the PSA plate-well diffusion method, corroborating a causal relationship between the production of this compound and suppression of bacterial blight in rice observed in net-house and field experiments.

The purified compound was cochromatographed with an authentic sample of DAPG (gift sample provided by G. Defago, Switzerland). The thin layer chromatogram showed an identical R_f value of 0.54 for the DAPG extracted from efficient *Pseudomonas fluorescens* PTB 9 and for the authentic sample (result not shown). The presence of DAPG in the condensed extract of *Pseudomonas fluorescens* strain PTB 9 was also confirmed by high-performance liquid chromatography (HPLC). Analysis of the purified compound of the strain PTB 9 showed a major peak which corresponded to the peak obtained for authentic DAPG.

Further confirmation was obtained by proton NMR. The spectrum exhibited a 6H singlet at δ 2.84 due to two acetyl methyl hydrogens, a 1H singlet at δ 5.99 due to aromatic hydrogen, and also a broad singlet at δ 3.56 due to the three hydroxyl hydrogens. Fourier transform infrared spectroscopy (generated a spectrum with a broad band at 3,304 cm^{-1} due to three hydroxyl groups and slightly broadband at

Table 5.2 Evaluation of *Pseudomonas fluorescens* strains producing DAPG in the biological control of rice bacterial blight (*BB*)

Name of strain	Diameter of inhibition of *Xoo* in dual-plate assay (cm)	Net-house experiments		Field experiments	
		Mean BB lesion length (cm)[a]	BB suppression (%)	Mean BB lesion length (cm)[a]	BB suppression (%)
KAD7	3.6	9.04**	56.87	10.15**	53.93
IMV14	4.1	8.66**	56.68	9.53**	56.74
IMV2	2.8	13.25**	36.78	14.72**	33.18
BGR19	3.0	15.20**	27.48	12.62**	42.71
PTB9	4.4	8.64**	58.78	7.83**	64.46
MON1	4.2	16.83**	19.70	13.85**	37.13
TVM8	1.2	17.54*	16.32	16.85**	23.51
VEL17	3.5	16.07**	23.33	11.07**	50.75
VEL10	2.8	17.39*	17.03	18.17*	17.52
GDY4	2.1	14.52**	30.73	12.90**	41.44
GDY7	2.3	10.15**	51.57	12.05**	45.30
TRP5	1.7	13.90**	33.68	14.42**	34.54
TRP18	2.3	16.87**	53.15	11.38**	48.34
MDR9	2.1	14.53**	30.68	16.90**	23.29
MDR7	4.1	8.99**	57.11	10.04**	54.43
STR7	1.8	9.82**	19.51	14.09**	36.04
VGP13	3.5	9.35**	55.39	11.19**	51.21
MDR16	1.4	13.08**	37.60	11.87**	46.12
PDY7	3.8	8.63**	58.83	9.50**	51.88
VLB7	1.7	13.91**	33.64	20.03 (NS)	9.08
KVR5	2.5	9.24**	55.92	13.42**	39.08
TNI13	3.6	15.50**	26.05	15.82**	23.19
KOV8	3.0	12.85**	38.69	19.89 (NS)	9.71
RJP31	2.4	18.80 (NS)	10.31	21.92 (NS)	0.50
KOV3	1.8	16.52**	21.18	20.13 (NS)	8.62
PDU1	2.9	19.25 (NS)	8.16	19.43 (NS)	11.80
PDU9	2.7	17.41*	16.94	21.58 (NS)	2.04
Untreated check	0.0	20.96	0.0	22.03	0.0
LSD 0.05		3.0		3.4	
LSD 0.01		4.0		4.5	

Experiments were performed either in a net house or in the field at the Regional Agriculture Research Station, Pattambi, Kerala

NS not significant, *LSD* least significant difference

[a]Each value is a mean of 40 observations

* Reduction in lesion length significant at the 5% level

** Reduction in lesion length significant at the 1% level

1,620 cm^{-1} due to the two acetyl carbonyl groups. Also there were bands at 1,403 and 1,365 cm^{-1} due to methyl C–H bending vibrations. These results proved beyond doubt that the compound purified from *Pseudomonas fluorescens* PTB 9 was indeed DAPG.

Fig. 5.3 Biological suppression of bacterial blight lesion length in rice cultivar IR24 owing to treatments with *Pseudomonas fluorescens* strains. The leaves detached from field plots raised with rice plants that were treated with six strains show a maximum of 64.5% reduction. A leaf from the untreated control plot showing a longer spreading bacterial blight lesion of more than 22 cm is on the *far left*. Bacterial strains *from left to right*: : control (untreated check), MDR7, PDY7, PTB9, KAD7, IMV14, VGP13

5.8 Conclusions

The inconsistent performance of biocontrol agents in the field developed thus far has plagued efforts to exploit them for commercial applications. There is a compelling need to identify efficient and dependable biocontrol agents to be used singly or as mixtures, so as to ensure consistent performance in the farmer's field.

We screened bacterial antagonists against *Xanthomonas oryzae* pv. *oryzae* by dual plate assay, evaluated *Pseudomonas fluorescens* strains as bacterial agents for the reduction of bacterial blight severity in a field experiment, and corroborated the role for a bacterial metabolite, DAPG, in disease suppression through laboratory dual plate assay.

Biological control offers exciting possibilities for the future. The *Pseudomonas fluorescens* strain PTB 9 described in this study and producing DAPG can be formulated with other strains (*Bacillus* spp., non-DAPG producers) and used as "superstrains of biocontrol agents" for the management of rice diseases and to increase productivity. Opportunities for creating superior strains of biocontrol agents and transgenic crops which express microbial secondary metabolites such as an antibacterial antibiotic (DAPG) and other antifungal proteins are endless.

The choice of the right microbial candidates is one of the most important factors governing the success of biocontrol programs on a commercial basis. Ideal biocontrol agents would reduce the severity of more than one pathogen, as this will make their application cost-effective. It needs to be remembered that most of the world's rice farmers, who live in Asia, are resource-poor. Therefore, only cost-effective formulations of biocontrol agents that perform consistently in the field, either by themselves or as part of an integrated disease management package, will benefit low-income rice growers. In this lies the key to the ultimate success of biocontrol research for rice disease management.

References

Bangera GM, Thomashow LS (1996) Characterization of a genomic locus required for synthesis of the antibiotic 2,4-diacetylphloroglucinol by the biological control agent *Pseudomonas fluorescens* Q2-87. Mol Plant Microbe Interact 9:3155–3163

Bangera GM, Thomashow LS (1999) Identification and characterization of a gene cluster for synthesis of the polyketide antibiotic 2,4-diacetylphloroglucinol from *Pseudomonas fluorescens* Q2-87. J Bacteriol 181:3155–3163

Barry GF (1988) A broad-host-range shuttle system for gene insertion into the chromosomes of Gram-negative bacteria. Gene 71:75–84

Chatterjee A, Valasubramanian R, Ma WL, Vacchani AK Gnanamanickam SS, Chatterjee AK (1996) Isolation of mutants of *Pseudomonas fluorescens* strain Pf7-14 altered in antibiotic production, cloning of ant+ DNA and evaluation of antibiotic production in the control of blast and sheath blight of rice. Biol Control 7:185–195

Defago G, Berling CH, Burger U, Haas D, Kahr G, Keel C, Voisard C, Wirthner P, Wuthrich B (1990) Suppression of black root rot of tobacco and other diseases by strains of *Pseudomonas fluorescens*: potential applications and other mechanisms. In: Hornby D, Cook RJ, Henis Y, Ko WH, Rovira AD, Schippers B, Scott BR (eds) Biological control of soilborne plant pathogen. CAB International, Wallingford, pp 93–108

Dowling DN, O'Gara F (1994) Metabolites of *Pseudomonas* involved in the biocontrol of plant disease. Trends Biotechnol 12:133–141

Fravel DC (1988) Role of antibiotics in the biocontrol of plant diseases. Annu Rev Phytopathol 26:75–91

Garagulya AD, Kiprianova EA, Boiko IO (1974) Antibiotic effect of bacteria from the genus *Pseudomonas* on phytopathogenic fungi. Mikrobiol Zh 36:197–202

Gutterson N (1990) Microbial fungicides: recent approaches to elucidating mechanisms. Crit Rev Biotechnol 10:69–91

Gutterson NL, Layton TJ, Warren GJ (1986) Molecular cloning of genetic determinants for inhibition of fungal growth by a fluorescent pseudomonad. J Bacteriol 165:696–703
Gutterson N, Ziegle JS, Warren GJ, Layton TJ (1988) Genetic determinants for catabolic induction of antibiotic biosynthesis in *Pseudomonas fluorescens* HV37a. J Bacteriol 170:380–385
Hossain M (1996) Recent developments in the Asian rice economy: challenge for the rice research. In: Evenson RE, Herdt, RW, Hossain M (eds) Rice research in Asia: progress and priorities. CAB International, Wallingford, pp 11–17
Howell CR, Stipanovic RD (1979) Control of *Rhizoctonia solani* on cotton seedlings with *Pseudomonas fluorescens* and with an antibiotic produced by the bacterium. Pytopathology 69:480–482
Howie WJ, Suslow TV (1991) Role of antibiotic biosynthesis in the inhibition of *Pythium ultimum* in the cotton spermosphere and rhizosphere by *Pseudomonas fluorescens*. Mol Plant Microb Interact 4:393–399
Ishiyama S (1922) Studies on the white leaf disease of rice plants. Rep Agric Exp Stn Tokyo 47:233–251
James DJ, Gutterson NI (1986) Multiple antibiotics produced by *Pseudomonas fluorescens* HV37a and their differential regulation by glucose. Appl Environ Microbiol 52:1183–1189
Kavitha S (2002) Strategies for management of rice blast and sheath blight with bacterial biocontrol agents in combination with major genes for disease resistance. PhD thesis, University of Madras, Chennai
Keel C, Schinder U, Maurhofer M, Voisard C, Laviller J, Burger U, Wirthner P, Haas D, Defago G (1992) Suppression of root diseases by *Pseudomonas fluorescens* CHAO: Importance of the bacterial secondary metabolite 2,4-diacetylphloroglucinol. Mol Plant Microbe Interact 5:4–13
Keel C, Weller DM, Natsch A, Defago G, Cook RJ, Thomashow LS (1996) Conservation of the 2,4-diacetylphloroglucinol biosynthesis locus among fluorescent *Pseudomonas* strains from diverse geographic locations. Appl Environ Microbiol 62:552–563
Krishnamurthy K (1997) Biological control of rice blast and sheath blight by *Pseudomonas* spp.: survival and migration of the biocontrol agents and the induction of systemic resistance in biological disease suppression. PhD thesis, University of Madras, Chennai
Levy E, Gough FJ, Berlin KD, Guiana PW, Smith TJ (1992) Inhibition of *Septoria tritici* and other phytopathogenic fungi and bacteria by *Pseudomonas fluorescens* and its antibiotics. Plant Pathol 41:335–341
Mew TW (1987) Current status and future prospects of research on bacterial blight of rice. Annu Rev Phytopathology 25:59–382
Mew TW, Alvarez AM, Leach JE, Swings J (1993) Focus on bacterial blight of rice. Plant Dis 77:5–11
Morrison DA (1979) Transformation and preservation of competent bacterial cells by freezing. Methods Enzymol 68:326–331
Narayanan NN, Baisakh N, Vera Cruz CM, Gnanamanickam SS, Datta K, Datta SK (2002) Molecular breeding for the development of blast and bacterial blight resistance in rice cv. IR50. Crop Sci 50:2072–2079
Narayanan NN (2001) Molecular breeding for the development of transgenic rice resistant to blast and bacterial blight diseases. PhD thesis, University of Madras, Chennai
Nowak-Thompson B, Gould SJ, Kraus J, Loper JE (1994) Production of 2,4-diacetylphloroglucinol by the biocontrol agent *Pseudomonas fluorescens* Pf-5. Can J Microbiol 40:1064–1066
O'Sullivan DJ, O'Gara F (1992) Traits of fluorescent *Pseudomonas* spp. involved in the suppression of plant root pathogens. Microbiol Rev 56:662–676
Ou SH (1985) Rice diseases, 2nd edn. Commonwealth Mycological Institute, Kew
Raaijmakers JM, Weller DM, Thomashow LS (1997) Frequency of antibiotics-producing *Pseudomonas spp.* in natural environments. Appl Environ Microbiol 63:881–887
Reddy TKK, Borovkov AV (1970) Antibiotic properties of 2,4-diacetylphloroglucinol produced by strain *Pseudomonas fluorescens* strain 26-0. Antibiotiki 15:19–21
Srivastava DN (1967) Epidemiology and control of bacterial blight of rice in India. Proceedings of a symposium on rice diseases and their control by growing resistant varieties and other measures, Japan. Agric For Fis Res Counc Agric For Tokyo Jpn 11–18

Swings J, Van den Mooter M, Vauterin L, Hoste B, Gillis M, Mew TW, Kersters K (1990) Reclassification of the causal agents of bacterial blight (*Xanthomonas campestris* pv. *oryzae*) and bacterial leaf streak (*Xanthomonas campestris* pv. *oryzicola*) of rice as pathovars of *Xanthomonas oryzae* (ex Ishiyama 1922). sp. nov., nom. rev. Int J Sys Bacteriol 40:309–331

Tagami Y, Mizukami T (1962) Historical review of the researches on bacterial blight of rice caused by *Xanthomonas oryzae* (Uyede and Ishiyama) Dowson. Spec Rep Plant Dis Insect Pests Forecast Serv 10:112

Tagami Y, Kuhara S, Kurita T, Fujii H, Sekiya N, Yoshimura S, Sato T, Watanabe B (1963) Epidemiological studies on the bacterial leaf blight of rice, *Xanthomonas oryzae* (Uyeda et Ishiyama) Dowson. I. The overwintering of the pathogen. Bull Kyushu Agric Exp Stn Japan 9:89–122

Thara KV (1994) Biological control of rice sheath blight by bacterial antagonists: mechanisms of disease suppression. PhD thesis, University of Madras, Chennai

Thomashow LS, Weller DM (1988) Role of phenazine antibiotic from *Pseudomonas fluorescens* in biological control of *Gaeumannomyces graminis* var. *tritici*. J Bacteriol 170:3499–3508

Thomashow LS, Weller DM (1995) Current concepts in the use of introduced bacteria for biological disease control: mechanisms and antifungal metabolitesm. In: Stacey G, Keen N (eds) Plant-microbe interactions, vol 1. Chapman and Hall, New York, pp 187–235

Valasubramanian R (1994) Biological control of rice blast with *Pseudomonas fluorescens* Migula: role of antifungal antibiotics in disease suppression. PhD thesis, University of Madras, Chennai

Vasudevan P (2002) Isolation and characterization *Bacillus* spp. from the rice rhizosphere and their role in biological control of bacterial blight of rice caused by *Xanthomonas oryzae* pv. *oryzae*. Ph.D thesis, University of Madras, Chennai

Vasudevan P, Kavitha S, Priyadarisini VB, Babujee L, Gnanamanickam SS (2002) Biological control of rice diseases. In: Gnanamanickam SS (ed) Biological control of crop diseases. Decker, New York, pp 11–32

Velusamy P (2003) Biological control of bacterial blight of rice by plant-associated bacteria: Molecular and genetic analyses of metabolites for their role in disease control. PhD thesis, University of Madras, Chennai

Velusamy P, Gnanamanickam SS (2003) Plant-associated bacteria, 2,4-diacetylphloroglucinol (DAPG) production and suppression of rice bacterial blight in India. Curr Sci 85:1270–1273

Velusamy P, Immanuel E, Gnanamanickam SS, Thomashow L (2006) Biological control of rice bacterial blight by plant-associated bacteria producing 2,4-diacetylphloroglucinol. Can J Microbiol 52:56–65

Wakimoto S (1956) Hibernation of leaf blight disease bacteria in soil. Agric Hortic Jpn 31:413–1414

Weller DM (1988) Biological control of soilborne plant pathogens in the rhizosphere with bacteria. Annu Rev Phytopathol 26:379–407

Weller DM, Cook RJ (1983) Suppression of take-all of wheat by seed treatments with fluorescent pseudomonads. Phytopathology 73:463–469

Yoshimura S, Tahara SK (1960) Morphology of bacterial leaf blight organism under (electron) microscope. Ann Phytopathol Soc Jpn 26:61

Chapter 6
Secondary Metabolites of Soil Streptomycetes in Biotic Interactions

Mika Tarkka and Rüdiger Hampp(✉)

6.1 Introduction

Streptomyces spp. are ubiquitous in soil microbial communities, and more than 500 species have been described thus far. The streptomycetes are generally saprophytic organisms which spend the majority of their life cycles as semidormant spores (Mayfield et al. 1972). During the life cycle, streptomycete spores germinate to produce substrate mycelium, which during maturation fragments into chains of spores. The substrate mycelium uses extracellular hydrolytic enzymes to gain nutrition from organic compounds that resist degradation by many other microbial groups, e.g. plant and fungal cell wall polysaccharides and insect exoskeletons.

The members of *Streptomyces* are distinguished by their ability to produce an array of secondary metabolites (Goodfellow and Williams 1983; Berdy 2005). The biosynthesis of these substances is influenced by physiological and environmental signals. The production of secondary metabolites commonly precedes the development of aerial hyphae, when the growth rate of bacterial filaments has decreased and sporulation starts (Bibb 2005). Much of the published data indicate that the most important environmental signal triggering secondary metabolism is nutrient starvation, particularly that of phosphate (Sola-Landa et al. 2003). The signalling networks behind the regulation of secondary metabolism in streptomycetes have recently been reviewed by Bibb (2005).

To date, approximately 17% of biologically active secondary metabolites (7,600 out of 43,000; Berdy 2005) have been characterized from streptomycetes. The main source for the bioactive secondary metabolites is soil streptomycetes, but a wide variety of structurally unique and biologically active secondary metabolites have recently been isolated from marine actinomycetes, including those from the genus *Streptomyces* (Cho et al. 2001; Lee et al. 2005; Jensen et al. 2005; Sanchez-Lopez

R. Hampp
Botanical Institute, Physiological Ecology of Plants, University of Tübingen,
Auf der Morgenstelle 1, D-72076 Tübingen, Germany
e-mail: ruediger.hampp@uni-tuebingen.de

et al. 2003). The compounds synthesized by streptomycetes show an extreme chemical diversity. These include substances as diverse as peptide compounds from simple amino acid derivatives to high molecular weight proteides, and macrolactones from simple eight-membered lactones to various condensed macrolactones. Berdy (1974) introduced the first classification scheme for antibiotics (antimicrobial substances) according to the chemical structure. On the basis of Berdy's scheme, Sanglier et al. (1996) recognized that both low and high molecular weight compounds from 63 different chemical classes are produced by streptomycetes, including many molecular skeletons that have not been reported from other organisms or cannot be chemically synthesized. The number of new secondary metabolites increased until the 1990s, but since then no new chemical classes have been reported. This may indicate that all possible types of carbon skeletons have been exploited (Berdy 2005).

Besides antibiotics, which present the largest group of bioactive secondary metabolites, the streptomycete compounds show several other biological activities. The secondary metabolites from streptomycetes can be broadly separated into four classes according to their biological activity: (1) antagonistic agents, including antibacterials, antifungals, antiprotozoans as well as antivirals, (2) pharmacological agents, including antitumorals, immunomodulators, neurological agents and enzyme inhibitors, (3) agrobiologicals, including insecticides, pesticides and herbicides, and (4) compounds with regulatory activities, such as growth factors, siderophores or morphogenic agents. To detect simultaneous bioactivities for a given compound, pharmacological and agricultural screens are increasingly being used in combination with antimicrobial tests. This has revealed several novel therapeutic and agrobiological agents and previously unknown biological activities for antibiotics (Sanglier et al. 1996; Berdy 1995, 2005). Many reports have shown that since streptomycetes are frequently screened for antimicrobial activity, the existence of secondary metabolites with other activities may have been missed (Garcia et al. 2000; Nunes et al. 2005).

6.2 Why Is Multiple Secondary Substance Production in Streptomycetes So Common?

Screening programmes for bioactive substances provide strong evidence that most chemicals do not possess any activity against specific target molecules unless tested at high concentrations (Firn and Jones 2000). Streptomycete species often produce simultaneously several bioactive secondary metabolites that, in combination, act in a synergistic way (Challis and Hopwood 2003). As was suggested by Firn and Jones (2000), secondary substance production must be based on the selection for traits that enhance the generation and retention of chemical diversity, and traits that reduce the fitness costs for this. In streptomycetes this seems to be accomplished by combinatorial biosynthesis, synergism between the secondary compounds and horizontal gene transfer, whereas the retention of this diversity is probably maintained through microbial competition in the heterogeneous soil substrate (Firn and Jones 2000; Challis and Hopwood 2003; Davelos et al. 2004; Weissman and Leadlay 2005; Weisst and Süssmuth 2005).

6.3 Combinatorial Biosynthesis: Low Investment with Good Profit

To lower the costs for secondary substance production, streptomycetes have evolved a genuine way to cope with hostile environments, namely the combinatorial biosynthesis of secondary metabolites. This permits the production of structurally diverse secondary metabolites with small changes in the common synthesis machinery. Polyketides represent an extremely rich source of biologically active compounds produced through combinatorial chemistry (Weissman and Leadlay 2005). Recent research on streptomycete polyketide synthases (PKS) has given insight into how the structural diversity of these compounds has been achieved in streptomycetes. Here, the modular PKS systems contain one to six subunits with differing functions. The total number of modules controls the length, whereas the catalytic domains of each module control the level of oxidation of the polyketide chain (McDaniel et al. 2005). The PKS system has proven to be an efficient and energy-saving way of producing a wide variety of substances, as small differences in, for example, the module hierarchy can cause an alteration in the molecular structure of the final product (Firn and Jones 2000).

6.4 Synergism and Combinatory Action Between Secondary Metabolites

Streptomyces clavuligerus produces β-lactamase inhibitors, β-lactams and cephalosporin-like antibiotics (Fig. 6.1). Only the combination of these three antimicrobial substance groups possesses strong antimicrobial activity against β-lactam-resistant bacteria (Jensen and Paradkar 1999; Liras 1999). Synergistic action has also been shown among the streptogramins A and B that serve as important inhibitors of bacterial protein synthesis (Cocito et al. 1997). Only the production of both substances causes bacteriocidal effects in target organisms (Cocito 1969). Furthermore, the presence of both compounds in all studied culture extracts of streptogramin-producing isolates suggests strong selection for the simultaneous production of the A and B subtypes (Challis and Hopwood 2003).

The combination of different iron-chelating compounds may also be an important fitness factor in an iron-poor soil substrate. Besides the detection of the common siderophores desferroxiamine B and E, Fiedler et al. (2001) detected an untraditional iron-chelating substance from the culture extracts of two streptomycete strains. This additional product was enterobactin, a characteristic siderophore of *Enterobacteriaceae* spp. (Fig. 6.1). Challis and Hopwood (2003) suggested that the ecological function of the multiple siderophore production is based on competition. Desferroxiamines are commonly scavenged by streptomycete strains, even by those that do not produce this kind of siderophore. The streptomycete strains reported by Fiedler et al. (2001) would therefore produce enterobactins to secure their iron source.

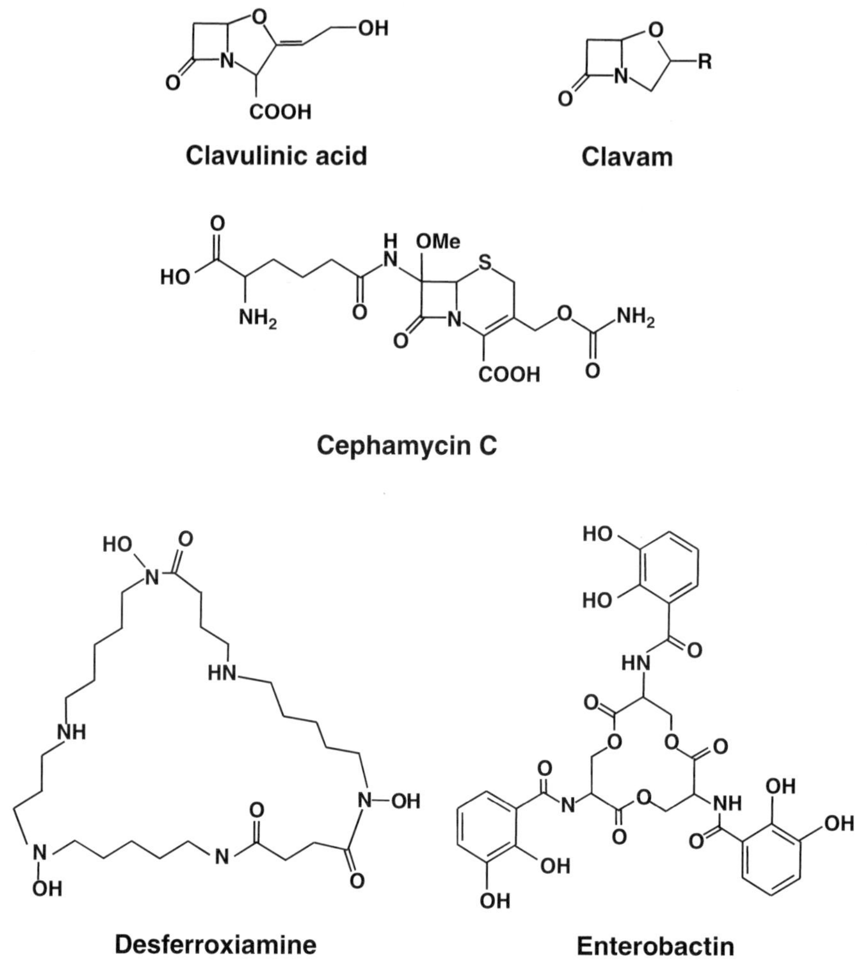

Fig. 6.1 Antibiotics and metal chelating agents from streptomycetes. Structures of the β-lactamase inhibitor clavulinic acid, the β-lactam clavam, cephalosporin C and the siderophores desferroxiamine and enterobactin. (*R* is a variable group)

6.5 Competition Between Streptomycete Populations in Soil: Antibiosis and Resistance

To determine the biological activity of a secondary metabolite under natural circumstances, its activity on species coexisting in the environment of a given streptomycete isolate has to be addressed (Bibb 2005). Natural soil is an inherently heterogeneous environment, consisting of nutrient-poor areas, nutrient patches of dead organic matter and mineral salts, and of areas surrounding plant root surfaces rich in organic matter. The irregular distribution of nutrients creates an uneven distribution of microbes, leading to ‘hot spots’, areas where bacterial and fungal colonies compete for the resources. Streptomycetes are capable of decomposing a variety of macromolecules, making them important soil nutrient recycling agents (Katsifas et al. 2000). An important benefit for streptomycetes competing with other microbes is possibly based on their antagonistic warfare, secondary metabolites and cell wall degrading agents, and on their resistance against toxic compounds produced by interacting microbes (Huddleston et al. 1997; Wiener 2000; Challis and Hopwood 2003). Davelos et al. (2004) surveyed antibiotic production and resistance of 153 prairie soil streptomycetes, isolated from three locations and four soil depths. The location of the individual isolates did not affect antimicrobial resistance, but an interesting correlation existed between antimicrobial metabolite production and spatial occurrence in the soil. The observations by Davelos et al. (2004) corroborated the results of Wiener (2000), who suggested that antibiotic production by streptomycetes is most important in a spatially structured environment. These observations indicate that selection pressure leads to the development of spatial hot spots for antibiotic production by the bacteria (Wiener 2000; Davelos et al. 2004). Both reports indicate that the production of antibiotics indeed plays an important role in the natural environment. Under the selection pressure, antibiotic production might thus be expected to be altered through horizontal transfer of antibiotic gene clusters. Phylogenetic analysis of streptomycin biosynthetic gene clusters has been used to evaluate the hypothesis of horizontal gene transfer in streptomycetes. Both the scattered distribution of streptomycin producers with respect to the overall phylogeny (Wiener et al. 1998) and the detection of streptomycete isolates which contain only part of the streptomycin biosynthesis clusters (Egan et al. 1998; Wiener 2000) indicate that horizontal gene transfer events are common among these filamentous bacteria. Horizontal transfer of the basic secondary metabolite gene clusters, followed by modification of basic modules, and incorporation of novel genes might thus explain the diversity of secondary metabolic pathways in streptomycete species (Stone and Williams 1992; Egan et al. 1998; 2001).

6.6 Chemical Ecology of Rhizosphere Streptomycetes

Plant roots constitute important organs for water and nutrient uptake, but also release a wide range of carbon compounds of low molecular weight. These compounds form the basis for an environment with rich microbial diversity, the rhizosphere (Hiltner 1904). The rhizosphere has been defined as the narrow zone of soil which is influenced by living roots. Bacteria are an important part of these populations. It has been shown that microbial communities within the rhizosphere are distinct from those of non-rhizosphere soil (Curl and Truelove 1986; Whipps and Lynch 1986; Frey-Klett et al. 2005). As a result of their ability to consume a variety of organic carbon sources, and owing to their capacity for antagonism against other microbial species, streptomycetes are often present in the rhizospheres of plants (Huddleston et al. 1997; Smalla et al. 2001; Weller et al. 2002). A substantial fraction of the rhizosphere streptomycetes are beneficial for plants, owing to their ability to control plant pathogens, to promote plant symbioses, to mineralize nutrients and to split biopolymers. There are, however, also streptomycetes which elicit severe plant diseases. The ecological role of the rhizosphere streptomycetes has recently received increased attention, and a combination of biochemical and molecular biological techniques with seminatural culture systems has revealed new aspects about streptomycete–plant interactions. The following part of this review will thus address how the secondary substance production in rhizosphere soil streptomycetes relates to streptomycete–plant interactions.

6.7 Plant Protection Through Secondary Metabolite Production

Streptomycete species are among the most promising biocontrol agents of plant diseases. Not only are the members of *Streptomyces* so effective because of secondary metabolite production, they are also ubiquitous in the rhizosphere, and frequent colonizers of plant tissues. Their ability to exude a variety of fungal cell wall and insect exoskeleton degrading enzymes has also been well documented (Emmert and Handelsman 1999; Siddiqui and Mahmood 1999; Doumbou et al. 2001; Paulitz and Belanger 2001; Whipps 2001; Weller et al. 2002).

6.8 Rhizosphere Streptomycetes as Biocontrol Agents

Antibiosis presents an important factor in the biocontrol effect (Fravel 1988). Some reports have clearly shown how the biological activity of secondary metabolites from streptomycetes relates to their biocontrol activity. *Streptomyces* sp. 201 produces a bioactive compound with antifungal and antibacterial activity, which was identified as 2-methylheptyl isonicotinate (Borodoloi et al. 2002). Dominant soil-borne phytopathogens belonging to the genera *Fusarium* and *Rhizoctonia* were suppressed in their growth after 2-methylheptyl isonicotinate applications, and seed inoculations of crucifer host plants with the substance resulted in resistance to fusarial wilt of crucifers. Both culture filtrate and spore suspension of the streptomycete exhibited protective activity, indicating that *Streptomyces* sp. 201 may be a promising biocontrol agent owing to the production of 2-methylheptyl isonicotinate. Lee et al. (2005) found that the aminoglycoside antibiotic paromomycin inhibited the in vitro growth of severe oomycete plant pathogens from the genera *Phytophthora* and *Pythium*, and showed potent in vivo activity against red pepper and tomato late blight. The paromomycin producer *Streptomyces* sp. AMG-P1 also exhibited high activity against late blight, indicating that it is a promising candidate for biological control (Lee et al. 2005).

6.9 Plant Endophytes as Sources for Antagonistic Secondary Metabolites

The soil-colonizing nature of streptomycetes would suggest that most of them must have evolved in close association with not only other soil microbes, insects and plasmodia, but also with plant root systems. A substantial part of the streptomycete populations in the rhizosphere are indeed capable of colonizing plant roots (Sardi et al. 1992; Coombs and Franco 2003). The interest in plant endophytes as biological control agents has recently increased. Such bacteria have been isolated from inside economically important plants and from plants traditionally used for medicinal purposes.

From a collection of endophytic actinomycete strains from surface-sterilized banana roots, Cao et al. (2005) analysed the most frequently isolated *Streptomyces* strain further, and identified it as *S. griseorubiginosus*. The antagonistic effect of *S. griseorubiginosus* against *Fusarium oxysporum* f. sp. *cubense* was not caused by antibiosis, but by the effective production of iron-chelating siderophores. *Fusarium* wilt disease symptoms were reduced and the mean fresh weight of the banana plants increased after *S. griseorubiginosus* application, suggesting that the bacterium is suitable for biocontrol approaches. A new microbial metabolite, designated as fistupyrone, was indicated as a novel inhibitor of the infection of Chinese cabbage by *Alternaria* leaf spot disease (Igarashi et al. 2000). Here, however, enhanced plant resistance against the pathogen was suggested to cause the disease suppression as no in vitro fungicidal activity against *Alternaria* was observed.

S. aureofaciens CMUAc130 is an inhabitant of the root tissue of *Zingiber officinale*. The strain is an antagonist of several plant pathogenic fungi, as tested in in vitro antibiosis tests with the bacterium and the culture filtrate (Taechowisan et al. 2005). The major active components in the culture filtrate of *S. aureofaciens* CMUAc130 were characterized as 5,7-dimethoxy-4-*p*-methoxylphenylcoumarin (Fig. 6.2) and 5,7-dimethoxy-4-phenylcoumarin, antifungal metabolites that were active against the fungi tested.

Castillo et al. (2002) detected novel peptide antibiotics (munumbicins A, B, C and D) from a streptomycete isolate growing endophytically inside the medicinal plant snakevine (*Kennedia nigriscans*). The munumbicins were major components of the *Streptomyces* sp. culture broth extract. Their biological activities showed a surprisingly wide spectrum of target organisms, ranging from *Staphylococcus aureus* to the malarial parasite *Plasmodium falciparum*. In a similar approach, Ezra et al. (2004) isolated a streptomycete strain that produces coronamycins, a complex of novel peptide antibiotics with activity against oomycete fungi and the human fungal pathogen *Cryptococcus neoformans*. In this report, the streptomycete isolate was an endophyte from an epiphytic vine, *Monstera* sp. As found by Castillo et al.

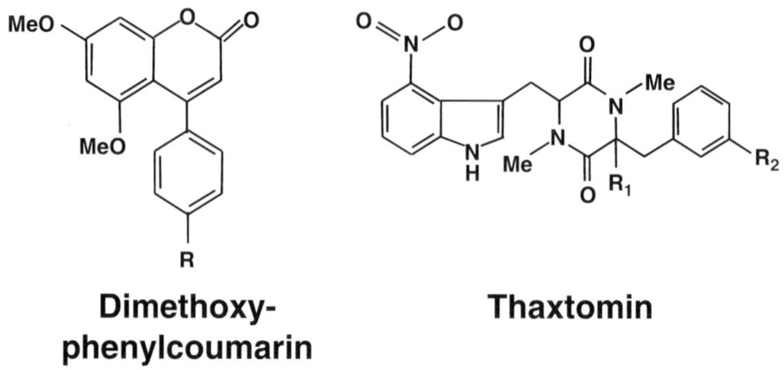

Fig. 6.2 Basic structures of dimethoxyphenylcoumarin and thaxtomin (*R* is a variable group)

(2002), other compounds with antifungal activities were also detected in the culture broth of the endophytic streptomycete isolate.

6.10 Induction of Plant Disease Resistance

We have recently studied the mechanisms of biocontrol against *Heterobasidion* root rot in Norway spruce seedlings by *Streptomyces* sp. GB 4-2. Although bacterial inoculation leads to enhanced mycelial growth of the phytopathogenic fungus *Heterobasidion*, promoted germination rate of fungal spores, faster extension of germ tubes and rapid colonization of outer cortical layers of the plant root by the fungus, later stages of disease development are suppressed by *Streptomyces* sp. GB 4-2. Colonization of inner tissues is namely hampered by the induction of cell wall appositions in the inner cortical cell layers and increased xylem formation in the vascular cylinder (our unpublished results).

Interestingly the infection of needles by grey mould is inhibited by bacterial preinoculation of the roots, as is the infection of *Arabidopsis thaliana* leaves with the phytopathogen *Alternaria brassicicola*. By using a set of *Arabidopsis* genes related to plant defence, Schrey and von Rad (unpublished results) observed that the response of *Arabidopsis thaliana* to *Streptomyces* sp. GB 4-2 includes the induction of genes involved in two major disease resistance pathways, systemic acquired resistance (SAR) and induced systemic resistance (ISR). In both SAR and ISR, prior treatment results in a stronger defence response against subsequent challenge by a pathogen, which is suggested by the negative influence of *Streptomyces* sp. GB 4-2 on *Alternaria brassicicola* infection. Until now, three dominant secondary metabolites of *Streptomyces* sp. GB 4-2 have been tested with spruce and *Arabidopsis* seedlings, but none of these proved to be the plant resistance inducing signal (D. Schulz, S. Schrey, and M. Tarkka, unpublished results).

6.11 Plant Pathogenic Members of *Streptomyces*

Among the hundreds of *Streptomyces* species described, only four species to date have been described as plant pathogens. These species, *S. scabies, S. acidiscabies, S. turgidiscabies* and *S. ipomoeae*, are agents of common scab disease in potato and other taproot crops (Loria et al. 1997). These diseases lead to reduction of root and shoot length, dramatic radial swelling of roots, tissue chlorosis and necrosis. The mechanisms of pathogenicity behind scab diseases are well documented, since plant disease symptoms are related to the production of a family of cyclic dipeptides, thaxtomins, by the streptomycete (Fig. 6.2). Two important findings have underlined the necessary role of thaxtomins in scab disease development. The application of purified thaxtomins was shown to lead to symptoms identical to those of the disease itself, i.e. cell hypertrophy and stunted growth (Lawrence et al. 1990; King et al. 1992). Secondly, when chemically mutagenized *S. scabies* strains were tested for their virulence, all of the mutants that produced lower levels of thaxtomin A relative to the parent strain showed reduced virulence in plant inoculation assays (Goyer et al. 1998).

The plant pathogenic *Streptomyces* species possess a conserved biosynthetic pathway for the phytotoxin thaxtomin. The importance of the thaxtomin synthesis cluster was confirmed by an elegant genetic analysis (Kers et al. 2005). A large pathogenicity island, conserved among the plant pathogenic *Streptomyces* species, was transferred from *S. turgidiscabies* to the non-pathogen *S. diastatochromogenes*. As a result the latter bacterium conferred a plant pathogenic phenotype.

6.12 Modulation of Plant Beneficial Symbioses by Streptomycetes

Interactions between symbiotic partners and rhizosphere streptomycetes have profound influences on plant root symbioses. Both negative and positive effects have been observed, ranging from a complete block of the growth of the microbial partner to promoted establishment and improved functioning of the symbiotic tissues (Wyss et al. 1992; Tokala et al. 2002; Schrey et al. 2005).

6.12.1 Root Nodule Symbiosis of Leguminous Plants

Nitrogen acquisition is facilitated in leguminous plants by an endophytic symbiosis with bacteria belonging to *Rhizobiales* (rhizobia). Several actinomycetes inhibit the growth of rhizobia, and their presence may cause an unsuccessful nodulation under field conditions (Patel 1974; Rangarajan et al. 1984). Recently, however, plant beneficial interactions between streptomycetes and rhizobia have been characterized.

Gregor et al. (2003) evaluated the utilization of actinomycetes as potential soybean coinoculants. The authors showed that wild-type strains of *Bradyrhizobium japonicum* were unable to form root nodules following coinoculation with the antagonistic *S. kanamyceticus*. In contrast, *B. japonicum* mutants with increased antibiotic resistance formed significantly more root nodules in the presence of the streptomycetes than in their absence (Gregor et al. 2003).

S. lydicus WYEC108 hosts a rare combination of plant beneficial characteristics. This strain suppresses root pathogenic fungi by mycoparatisism and by the secretion of antifungal metabolites (Crawford et al. 1993; Yuan and Crawford 1995). It also promotes plant growth, possibly owing to siderophore production. Tokala et al. (2002) showed that *S. lydicus* also colonizes the outer layers in pea root nodules and promotes root nodulation. Most importantly, the root nodule colonization by the streptomycete leads to a significantly increased rate of nitrogen fixation (Tokala et al. 2002).

6.12.2 Mycorrhizal Symbiosis

Roots of most terrestrial plants develop symbiotic structures (mycorrhiza) with soil-borne fungi. In these interactions, the fungal partner provides the plant with improved access to water and nutrients in the soil owing to more or less complex hyphal structures, which emanate from the root surface and extend far into the soil. The plant, in return, supplies carbohydrates for fungal growth and maintenance (Hampp and Schaeffer 1998; Smith and Read 1997). Owing to the leakage and the turnover of mycorrhizal structures, these solutes are also released into the mycorrhizosphere, where they can be accessed by other microorganisms. Some of the mycorrhiza-forming fungi have been shown to reduce bacterial viability (Green et al. 1999, Meyer and Linderman 1986). Owing to the transfer and exudation of plant-derived organic compounds to soil microsites not accessible to roots, the mycorrhizal fungi can, however, promote bacterial growth and survival (Frey-Klett et al. 1997; Hobbie 1992; Söderström 1992). There is also evidence that soil bacteria can enhance the formation of mycorrhizal structures, either by promoting growth (mycorrhization helper bacteria, MHB; Bending et al. 2002; Garbaye 1994) or by protecting them from pathogenic microorganisms (Pedersen et al. 1999; Schelkle and Peterson 1996). Plant-growth-promoting rhizobacteria (Kloepper et al. 1989) reported so far include species and strains which belong to the genera *Azotobacter, Pseudomonas, Burkholderia, Acetobacter, Herbaspirillum* and *Bacillus* (Glick 1995; Probanza et al. 1996).

Arbuscular mycorrhizal (AM) symbiosis is the most common form of a symbiotic relationship between plants and microbes (Smith and Read 1997). Most of the tested actinomycete strains until now, including the *S. griseoviridis* biocontrol agent, suppressed the germination of spores from AM fungi in coinoculation assays, and halted symbiosis development (Meyer and Linderman 1986; Ames 1989; Wyss et al. 1992). In contrast, substances that stimulate AM fungal spore germination are produced by several streptomycete strains (Mugnier and Mosse 1987; Tylka et al. 1991). The active substances are released to the gas phase, but their identity is so far unknown (Tylka et al. 1991). Specific streptomycete strains stimulate AM development. For example, coinoculation with *S. coelicolor* significantly increased the intensity of mycorrhizal root colonization and arbuscule formation by *Glomus intraradices* in sorghum plants (Abdel-Fattah and Modamedin 2000).

6.12.3 Ectomycorrhiza Helper Streptomycetes

When 12 actinomycete isolates were tested for their effects on mycelial growth of ectomycorrhizal fungi (Richter et al. 1989), the bacterial isolates inhibited, promoted or showed no significant effects on hyphal extension in dual culture. Three fungal species were tested in dual culture, *Laccaria bicolor, L. laccata* and *Thelephora terrestris*, of which the slowest growing fungus, *T. terrestris*, was most sensitive to both growth-promoting and antagonistic actinomycetes (Richter et al. 1989). Some antagonistic actinomycetes are also producers of plant-growth-promoting substances, helping plants to withstand adverse conditions and attacks by pathogens (Igarashi et al. 2002).

In recent studies, interactions between actinomycetes and rhizosphere fungi have been investigated in more detail (Maier et al. 2004; Schrey et al. 2005; Riedlinger et al. 2006). Maier et al. (2004) collected Gram-positive bacteria from the rhizosphere from a spruce stand rich with the widespread mycorrhiza-forming fungus, fly agaric (*Amanita muscaria*). Using an axenic culture system, these authors reported that a range of the bacteria distinctly and highly reproducible promoted growth of hyphae of *A. muscaria*. One of these strains was shown to additionally inhibit growth of pathogenic fungi such as *Armillaria obscura* (wide host range) and *Heterobasidion annosum* (causes wood decay in conifers). Taxonomic characterization of the effective bacterial isolates by their morphological appearance, by the analysis of diaminopimelic acid, cell wall sugars and DNA sequencing (16S ribosomal DNA) identified them as actinomycetes (Maier et al. 2004).

6.13 Interaction of Streptomycetes with Rhizosphere Fungi

Out of a collection of actinomycetes originating from the rhizosphere of a spruce stand (Maier 2003), the isolate *Streptomyces* sp. nov. 505 (AcH 505) significantly promoted the mycelial growth and mycorrhization rate of *A. muscaria* in the presence of spruce seedlings, while suppressing the mycelial extension of the plant pathogenic fungi, *Armillariella obscura* and *Heterobasidion annosum* (Maier et al. 2004; Hampp and Maier 2004). In contrast to AcH 505, the second MHB isolated, *S. annulatus* 1003 (AcH 1003), did not affect the growth of the plant pathogenic fungi tested.

In order to test the MHB function of AcH 505 and AcH 1003, these bacteria were grown in a seminatural perlite-moss culture system in the presence of the fungus of interest (Schrey et al. 2005). With use of suppression subtractive hybridization (Diatchenko et al. 1996), alterations in fungal (*A. muscaria*) gene expression in response to the interaction with MHB was investigated. In the actively growing hyphal front of 9-week-old dual cultures where AcH 505 most strongly promoted mycelial growth, a series of *A. muscaria* genes were differentially transcribed.

6.14 Effective Bacterial Compounds

In order to screen for bacterial compounds, responsible for the effects induced, the supernatants of batch fermentations of strain AcH 505 were cultured either alone or in the presence of *A. muscaria*. Culture filtrates were chromatographed on Amberlite XAD-16 columns, subjected to different purification steps, and finally analysed by reversed-phase high-performance liquid chromatography (Riedlinger et al. 2006). The major peaks could be identified as auxofuran and as the antibiotics WS-5995 B and WS-5995 C (Fig. 6.3) (Keller et al. 2006).

6.14.1 Auxofuran

Auxofuran was the dominant fungal growth-promoting substance excreted by AcH 505. In the case of the fermentation of the strain AcH 505, a nearly constant level of about 1.5 mg/l auxofuran was measured over a fermentation period of 21 days. Co-cultivation of both organisms stimulated the auxofuran production by the streptomycete to a continuously increasing amount, reaching a maximal value of 6 mg/l

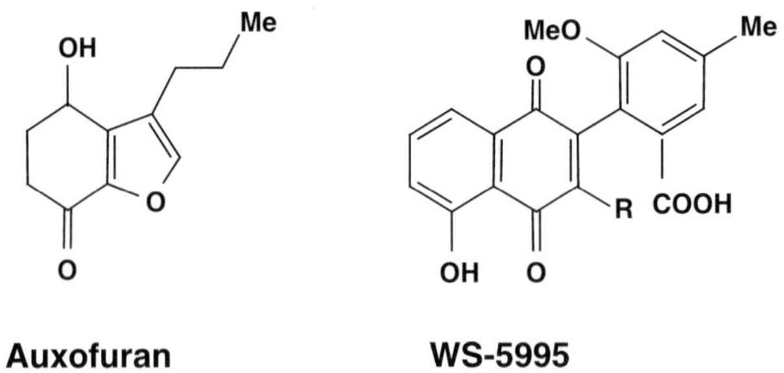

Fig. 6.3 Structures of auxofuran and WS-5995 (*R* is H in WS-5995 B and OH in WS-5995 C)

after incubation for 21 days. As expected, no production of auxofuran was observed in a single culture of *A. muscaria*.

Auxofuran stimulated *A. muscaria* hyphal growth most effectively at a concentration of 15 µM, and the fungus responded significantly to concentrations in the nanomolar range. It is possible that auxofuran and its synthetic dehydroxylated derivative, 7-dehydroxyauxofuran, could display specificity in their effects as growth-stimulating substances, because of differences in their structure and solubility. In order to investigate this possibility, fungal mycelia were grown on solid medium supplemented with one of the compounds. As shown for auxofuran, the strongest positive effect towards the growth of *A. muscaria* on solid media was observed with 15 µM 7-dehydroxyauxofuran. Up to 4 weeks of culture *A. muscaria* responded to both substances, although to a significantly greater extent to 7-dehydroxyauxofuran (Riedlinger et al. 2006). At 6 weeks of growth, a further stimulation of growth was detected for 7-dehydroxyauxofuran, whereas the growth-promoting effect of auxofuran disappeared, indicating that 7-dehydroxyauxofuran shows not only a stronger but also a more persistent stimulatory effect towards fly agaric than auxofuran.

Although AcH 505 promotes mycelial extension of *A. muscaria*, it sharply reduces the hyphal biomass to colony area ratio owing to a reduction in mycelial density, indicating that AcH 505 does not promote the accumulation of biomass in *A. muscaria*, but instead enhances the spread of the fungal mycelium (Schrey et al. 2007). Moreover, it reduces the thickness of the fungal hyphae (Maier, 2003). We recently analysed the structural background of this hyphal thinning: bacterial inoculation leads to a changed organization of the fungal actin cytoskeleton (Schrey et al. 2007).

6.14.2 WS-5995 B and WS-5995 C

WS-5995 B and WS-5995 C differ in that WS-5995 C contains a hydroxyl group which is not present in WS-5995 B. A great variety of Gram-positive and Gram-negative bacteria and fungi were tested using the agar plate diffusion assay. Only WS-5995 B exhibited a growth inhibition of Gram-positive bacteria and *Haemophilus influenzae*, whereas other Gram-negative bacteria, such as *Escherichia coli* K12, *Pseudomonas fluorescens* DSM 50090 and *Proteus mirabilis* ATCC 35501, yeasts, such as *Saccharomyces cerevisiae* ATCC 9080 and *Candida albicans* Tü 164, and filamentous fungi, such as *Botrytis cinerea* Tü 157, *Aspergillus viridi nutans* CBS 12756, *Penicillium notatum* Tü 136 and *Paecilomyces variotii* Tü 137, were not sensitive against WS-5995 B and C (Riedlinger et al. 2006). The minimal inhibitory concentration of WS-5995 B was determined in a microtiter plate assay as 33 µM for *Arthrobacter aurescens, Bacillus subtilis* and *Staphylococcus aureus*. With regard to the growth inhibition of *A. muscaria*, WS-5995 B was more effective than WS-5995 C.

6.14.3 *Effects of Auxofuran and WS-5995 B on Fungal Gene Expression*

In order to screen for altered gene expression in the fungus *A. muscaria*, three genes from an AcH 505-induced *A. muscaria* complementary DNA library (Schrey et al. 2005) were selected, i.e. acetoacylcoenzyme A synthetase (*Aacs*), cyclophilin 40 (*Cyp40*) and γ-aminobutyric acid (GABA) permease (*Uga4*), as these genes were previously shown to be related to growth promotion in *A. muscaria* hyphae (Schrey et al. 2005). Within 3 h, the fungal cells responded to both compounds. The level of *Aacs* expression was increased approximately twofold by auxofuran and threefold by WS-5995 B, that of *Cyp 40* threefold with WS-5995 B, and the level of *Uga4* expression increased threefold with WS-5995 B. As *Uga4* expression levels correlated with GABA concentrations in the budding yeast *Saccharomyces cerevisiae* (Andre et al. 1993), the amounts of GABA contained in *A. muscaria* hyphae were determined in parallel. In line with the increased level of *Uga4* expression, GABA concentration increased after the addition of WS-5995 B, but not with auxofuran (Riedlinger et al. 2006). Overall, these observations demonstrate the ability of *A. muscaria* to rapidly respond to the stimulatory and suppressive substances, and to specifically alter its physiological functioning in response to the compounds excreted by strain AcH 505.

6.15 Cooperation Between WS 5995 B/WS 5995 C and Auxofuran: Select and Amplify

The effects of the antibiotics WS-5995 B and WS 5995 C in combination with the growth promoter auxofuran against target fungi represent a novel mode of cooperative action between secondary metabolites. The fungal strains tested thus far, *A. muscaria, Hebeloma cylindrosporum* and *Heterobasidion annosum*, have responded similarly to auxofuran: the strongest promotion of mycelial growth occurred at 1–15 μM auxofuran, while a significant promotion was still observable in the nanomolar range. In contrast, the fungi showed different responses towards WS-5995 B/WS-5995 C. Most importantly, the fungi which are suppressed in their growth during coculture with AcH 505 were more sensitive to WS-5995 B than the ones promoted through this streptomycete, indicating that the resistance towards WS-5995 B/WS-5995 C serves as a selector, leading to the growth-promotion phenotype only in resistant organisms. A similar interaction between secondary metabolites can be suggested from the root nodule assays reported by Gregor et al. (2003), mentioned before.

Because of their selectivity, it has been suggested that bacteria that promote mycorrhizal fungi but suppress pathogenic fungi could become an alternative to soil fumigation, and they could be simultaneously used to improve symbiosis and to prevent disease development (Duponnois et al. 1993). However, when considering such applications, our data on the phytopathogen *Heterobasidion annosum* and

AcH 505 are of concern (Lehr et al. 2007). While 11 of the 12 *Heterobasidion annosum* strains that were tested against with AcH 505 and WS-5995 B were inhibited, one *Heterobasidion annosum* isolate was not. More importantly, the colonization of plant roots by this fungal strain was actually promoted by the bacterium, by a mechanism that was based on the suppression of plant defence response. This suggests that some MHB behave as helpers of both symbiotic and pathogenic fungi. The plant defence inhibiting signal(s) are currently unknown.

In conclusion, cooperative action of streptomycete secondary metabolites may thus be even more complex than the synergistic interactions observed between different antibiotic substances or contingent action between siderophores (Challis and Hopwood 2003).

6.16 Conclusions

According to recent literature, the secondary metabolite production in streptomycetes allows for ecological adaptation. This has been highlighted in the reports concerning the interactions of *Streptomyces* spp. with each other, with other microbes and with plants. One of the most striking characteristics of secondary metabolite production is the ability of these bacteria to simultaneously produce several synergistically or cooperatively acting substances. Further work has to be done to understand the selective advantage of multiple secondary metabolite production in natural surroundings. This knowledge would be instrumental to fully utilize these bacterial species in biocontrol or symbiosis-promotion approaches.

Acknowledgements As far as our own results are presented, we gratefully acknowledge financial support by the DFG-Graduiertenkolleg "Infection biology", which is the basis for the very efficient scientific collaboration with the microbiology group at the University of Tübingen (H.-P. Fiedler, J. Riedlinger, D. Schulz).

References

Abdel-Fattah GM, Mohamedin AH (2000) Interactions between a vesicular-arbuscular mycorrhizal fungus (*Glomus intraradices*) and *Streptomyces coelicolor* and their effects on sorghum plants grown in soil amended with chitin of brawn scales. Biol Fertil Soils 32:401–409

Ames BN (1989) Mycorrhiza development in onion in response to chitin-decomposing actinomycetes. New Phytol 112:423–427

Andre B, Hein C, Grenson M, Jauniaux JC (1993) Cloning and expression of the *UGA4* gene coding for the inducible GABA-specific transport protein of *Saccharomyces cerevisiae*. Mol Gen Genet 237:17–25

Bending GD, Poole EJ, Whipps JM, Read DJ (2002) Characterisation of bacteria from *Pinus sylvestris-Suillus luteus* mycorrhizas and their effects on root-fungus interactions and plant growth. FEMS Microbiol Ecol 39:219–227

Berdy J (1974) Recent developments of antibiotic research and classification of antibiotics according to chemical structure. Adv Appl Microbiol 18:309–406

Berdy J (1995) Are actinomycetes exhausted as a source of secondary metabolites? In: Debabov V, Dudnik Y, Danlienko V (eds) Proceedings of the 9th international symposium on the biology of actinomycetes. All-Russia Scientific Research Institute for Genetics and Selection of Industrial Microorganisms, Moscow, pp 13–34

Berdy J (2005) Bioactive microbial metabolites: a personal view. J Antibiot 58:1–26

Bibb MJ (2005) Regulation of secondary metabolism in streptomycetes. Curr Opin Microbiol 8:208–215

Bordoloi GN, Kumari B, Guha A, Thakur D, Bordoloi M, Roy MK, Bora TC (2002) Potential of a novel antibiotic, 2-methylheptyl isonicotinate, as a biocontrol agent against fusarial wilt of crucifers. Pest Manag Sci 58:297–302

Cao L, Qiu Z, You J, Tan H, Zhou S (2005) Isolation and characterization of endophytic streptomycete antagonists of *Fusarium* wilt pathogen from surface-sterilized banana roots. FEMS Microbiol Lett 247:147–152

Castillo UF, Strobel GA, Ford EJ, Hess WM, Porter H, Jensen JB, Albert H, Robison R, Condron MA, Teplow DB, Stevens D, Yaver D (2002) Munumbicins, wide-spectrum antibiotics produced by *Streptomyces* NRRL 30562, endophytic on *Kennedia nigriscans*. Microbiology 148:2675–2685

Challis GL, Hopwood DA (2003) Synergy and contingency as driving forces for the evolution of multiple secondary metabolite production by *Streptomyces* species. Proc Natl Acad Sci USA 100(Suppl 2):14555–14561

Cho KW, Lee HS, Rho JR, Kim TS, Mo SJ, Shin J (2001) New lactone-containing metabolites from a marine-derived bacterium of the genus *Streptomyces*. J Nat Prod 64:664–667

Cocito C (1969) Metabolism of macromolecules in bacteria treated with virginiamycin. J Gen Microbiol 57:179–194

Cocito C, Di Giambattista M, Nyssen E, Vannuffel P (1997) Inhibition of protein synthesis by streptogramins and related antibiotics. J Antimicrob Chemother 39:7–13

Coombs JT, Franco CM (2003) Isolation and identification of actinobacteria from surface-sterilized wheat roots. Appl Environ Microbiol 69:5603–5608

Crawford DL, Lynch JM, Whipps JM, Ousley MA (1993) Isolation and characterization of actinomycete antagonists of a fungal root pathogen. Appl Environ Microbiol 59:3899–3905

Curl EA, Truelove B (1986) The rhizosphere. Springer, Berlin

Davelos AL, Kinkel LL, Samac DA (2004) Spatial variation in frequency and intensity of antibiotic interactions among *Streptomycetes* from prairie soil. Appl Environ Microbiol 70:1051–1058

Diatchenko L, Lau YF, Campbell AP, Chenchik A, Moqadam F, Huang B, Lukyanov S, Lukyanov K, Gurskaya N, Sverdlov ED, Siebert PD (1996) Suppression subtractive hybridization: a method for generating differentially regulated or tissue-specific cDNA probes and libraries. Proc Nat Acad Sci USA 93:6025–6030

Doumbou CL, Hamby Salove MK, Crawford DL, Beaulieu C (2001) Actinomycetes, promising tools to control plant diseases and to promote plant growth. Phytoprotection 82:85–102

Duponnois R, Garbaye J, Bouchard D, Churin JL (1993) The fungus-specificity of mycorrhization helper bacteria (MHBs) used as an alternative to soil fumigation for ectomycorrhizal inoculation of bare-root Douglas-fir planting stocks with *Laccaria laccata*. Plant Soil 157:257-262

Egan S, Wiener P, Kallifidas D, Wellington EM (1998) Transfer of streptomycin biosynthesis gene clusters within streptomycetes isolated from soil. Appl Environ Microbiol 64:5061–5063

Egan S, Wiener P, Kallifidas D, Wellington EM (2001) Phylogeny of *Streptomyces* species and evidence for horizontal transfer of entire and partial antibiotic gene clusters. Antonie Van Leeuwenhoek 79:127–133

Emmert EA, Handelsman J (1999) Biocontrol of plant disease: a (gram-) positive perspective. FEMS Microbiol Lett 171:1–9

Ezra D, Castillo UF, Strobel GA, Hess WM, Porter H, Jensen JB, Condron MA, Teplow DB, Sears J, Maranta M, Hunter M, Weber B, Yaver D (2004) Coronamycins, peptide antibiotics produced by a verticillate *Streptomyces* sp. (MSU-2110) endophytic on *Monstera* sp. Microbiology 150:785–793

Fiedler HP, Krastel P, Muller J, Gebhardt K, Zeeck A (2001) Enterobactin: the characteristic catecholate siderophore of *Enterobacteriaceae* is produced by *Streptomyces* species. FEMS Microbiol Lett 196:147–151

Firn RD, Jones CG (2000) The evolution of secondary metabolism—a unifying model. Mol Microbiol 37:989–994

Fravel DR (1988) Role of antibiosis in the biocontrol of plant diseases. Annu Rev Phytopathol 26:75–91

Frey-Klett P, Pierrat JC, Garbaye J (1997) Location and survival of mycorrhiza helper *Pseudomonas fluorescens* during establishment of ectomycorrhizal symbiosis between *Laccaria bicolor* and Douglas Fir. Appl Environ Microbiol 63:139–144

Frey-Klett P, Chavatte M, Clausse M-L, Courrier S, Le Roux C, Raaijmakers J, Martinotti MG, Pierrat J-C, Garbaye J (2005) Ectomycorrhizal symbiosis affects functional diversity of rhizosphere fluorescent pseudomonads. New Phytol 165:317–328

Garbaye J (1994) Helper bacteria: a new dimension to the mycorrhizal symbiosis. New Phytol 128:197–210

Garcia I, Job D, Matringe M (2000) Inhibition of *p*-hydroxyphenylpyruvate dioxygenase by the diketonitrile of isoxaflutole: a case of half-site reactivity. Biochemistry 39:7501–7507

Glick BR (1995) The enhancement of plant growth by free-living bacteria. Can J Microbiol 41:109–117

Goodfellow M, Williams ST (1983) Ecology of actinomycetes. Annu Rev Microbiol 37:189–216

Goyer C, Vachon J, Beaulieu C (1998) Pathogenicity of *Streptomyces scabies* mutants altered in thaxtomin A production. Phytopathology 88:442–445

Green H, Larsen J, Olsson PA, Jensen DF, Jakobsen I (1999) Suppression of the biocontrol agent *Trichoderma harzianum* by mycelium of the mycorrhizal fungus *Glomus intraradices* in root-free soil. Appl Environ Microbiol 65:1428–1434

Gregor AK, Klubek B, Varsa EC (2003) Identification and use of actinomycetes for enhanced nodulation of soybean co-inoculated with *Bradyrhizobium japonicum*. Can J Microbiol 49:483–491

Hampp R, Maier A (2004) Interaction between soil bacteria and ectomycorrhiza-forming fungi. In: Varma A, Abbott LK, Werner D, Hampp R (eds) Plant surface microbiology. Springer, Berlin, pp 197–210

Hampp R, Schaeffer C (1998) Mycorrhiza—carbohydrate and energy metabolism. In: Varma A, Hock B (eds) Mycorrhiza—structure, function, molecular biology and biotechnology. Springer, Berlin, pp 273–303

Hiltner L (1904) Über neuere Erfahrungen und Probleme auf dem Gebiet der Bodenbakteriologie und unter besonderer Berücksichtigung der Gründüngung und Brache. Arb Dtsch Landwirtsch Ges 98:59–78

Hobbie SE (1992) Effects of plant species on nutrient cycling. Trends Ecol Evolution 7:336–339

Huddleston AS, Cresswell N, Neves MC, Beringer JE, Baumberg S, Thomas DI, Wellington EM (1997) Molecular detection of streptomycin-producing streptomycetes in Brazilian soils. Appl Environ Microbiol 63:1288–1297

Igarashi Y, Ogawa M, Sato Y, Saito N, Yoshida R, Kunoh H, Onaka H, Furumai T (2000) Fistupyrone, a novel inhibitor of the infection of Chinese cabbage by *Alternaria brassicicola*, from *Streptomyces* sp. TP-A0569. J Antibiot 53:1117–1122

Igarashi Y, Iida T, Yoshida R, Furumai T (2002) Pteridic acids A and B, novel plant growth promoters with auxin-like activity from *Streptomyces hygroscopicus* TP-A0451. J Antibiot 55:764–767

Jensen SE, Paradkar AS (1999) Biosynthesis and molecular genetics of clavulanic acid. Antonie Van Leeuwenhoek 75:125–133

Jensen PR, Mincer TJ, Williams PG, Fenical W (2005) Marine actinomycete diversity and natural product discovery. Antonie Van Leeuwenhoek 87:43–48

Katsifas, EA, Koraki, TG, Karagouni, AD (2000) Determination of metabolic activity of streptomycetes in soil microcosms. J Appl Microbiol 89:178–184

Keller S, Schneider K, Sussmuth RD (2006) Structure elucidation of auxofuran, a metabolite involved in stimulating growth of fly agaric, produced by the mycorrhiza helper bacterium *Streptomyces* AcH 505. J Antibiot 59:801–803

Kers JA, Cameron KD, Joshi MV, Bukhalid RA, Morello JE, Wach MJ, Gibson DM, Loria R (2005) A large, mobile pathogenicity island confers plant pathogenicity on *Streptomyces* species. Mol Microbiol 55:1025–1033

King RR, Lawrence CH, Calhoun LA (1992) Chemistry of phytotoxins associated with *Streptomyces scabies*, the causal organism of potato common scab. J Agric Food Chem 40:834–837

Kloepper JW, Lifshitz R, Zablotowicz RM (1989) Free-living bacterial inocula for enhancing crop productivity. Trends Biotechnol 7:39–43

Lawrence CH, Clark MC, King RR (1990) Induction of common scab symptoms in aseptically cultured potato tubers by the vivotoxin, thaxtomin. Phytopathology 80:606–608

Lee HB, Kim Y, Kim JC, Choi GJ, Park SH, Kim CJ, Jung HS (2005) Activity of some aminoglycoside antibiotics against true fungi, *Phytophthora* and *Pythium* species. J Appl Microbiol 99:836–843

Lehr NA, Schrey SD, Bauer R, Hampp R, Tarkka MT (2007) Suppression of plant defence response by a mycorrhiza helper bacterium. New Phytol 174, 892–903

Liras P (1999) Biosynthesis and molecular genetics of cephamycins. Cephamycins produced by actinomycetes. Antonie Van Leeuwenhoek 75:109–124

Loria R, Bukhalid RA, Fry BA, King RR (1997) Plant pathogenicity in the genus *Streptomyces*. Plant Dis 81:836–846

Maier A (2003) Einfluss bakterieller Stoffwechselprodukte auf Wachstum und Proteom des Ektomykorrhizapilzes *Amanita muscaria*. PhD thesis, University of Tübingen

Maier A, Riedlinger J, Fiedler H-P, Hampp R (2004) Actinomycetales bacteria from a spruce stand: characterization and effects on growth of root symbiotic, and plant parasitic soil fungi in dual culture. Mycol Prog 3:129–136

Mayfield CI, Williams ST, Ruddick SM, Hatfield HL (1972) Studies on the ecology of actinomycetes in soil. IV. Observations on the form and growth of *Streptomycetes* in soil. Soil Biol Biochem 4:79–91

McDaniel R, Welch M, Hutchinson R (2005) Genetic approaches to polyketide antibiotics. Chem Rev 105:543–558

Meyer JR, Linderman RG (1986) Selective influence of rhizosphere or rhizoplane bacteria and actinomycetes by mycorrhizas formed by *Glomus fasciculatum*. Soil Biol Biochem 18:191–196

Mugnier J, Mosse B (1987) Spore germination and viability of vesicular arbuscular mycorrhizal fungus *Glomus mossae*.Trans Br Mycol Soc 88:411–413

Nunes LR et al (2005) Transcriptome analysis of *Paracoccidioides brasiliensis* cells undergoing mycelium-to-yeast transition. Eukaryot Cell 4:2115–2128

Patel JJ (1974) Antagonism of actinomycetes against rhizobia. Plant Soil 41:395–402

Paulitz TC, Belanger RR (2001) Biocontrol in greenhouse systems. Annu Rev Phytopathol 39:103–133

Pedersen EA, Reddy MS, Chakravarty P (1999) Effect of three species of bacteria on damping-off, root rot development, and ectomycorrhizal colonization of lodgepole pine and white seedlings. Eur J For Pathol 29:123–134

Probanza A, Lucas JA, Guiterrez-Mañero JF (1996) The influence of native rhizobacteria on European alder (*Alnus glutinosa* (L.) Gaertn.) growth. I. Characterization of growth promoting and growth inhibiting bacterial strains. Plant Soil 182:59–66

Rangarajan M, David Ravindran A, Hariharan K (1984) Occurrence of a lysogenic *Streptomyces* sp. on the nodule surface of black gram (*Vigna mungo* (L.) Hepper). Appl Environ Microbiol 48:232–233

Richter DL, Zuellig TR, Bagley ST, Bruhn JN (1989) Effects of red pine (*Pinus resinosa* Ait.) mycorrhizoplane-associated actinomycetes on *in vitro* growth of ectomycorrhizal fungi. Plant Soil 115:109–116

Riedlinger J, Schrey SD, Tarkka MT, Hampp R, Kapur M, Fiedler H-P (2006) Auxofuran, a novel metabolite stimulating growth of fly agaric, produced by the mycorrhiza helper bacterium Streptomyces AcH 505. Appl Environ Microbiol 72:3550–3557

Sanchez-Lopez JM, Martinez Insua M, Perez Baz J, Fernandez Puentes JL, Canedo Hernandez LM (2003) New cytotoxic indolic metabolites from a marine *Streptomyces*. J Nat Prod 66:863–864

Sanglier JJ, Haag H, Huck TA, Fehr T (1996) Review of actinomycetes compounds 1990-1995. Expert Opin Invest Drugs 5:207–223

Sardi P, Saracchi M, Quaroni S, Petrolini B, Borgonovi GE, Merli S (1992) Isolation of endophytic *Streptomyces* strains from surface-sterilized roots. Appl Environ Microbiol. 58:2691–2693

Schelkle M, Peterson RL (1996) Supression of common root pathogens by helper bacteria and ectomycorrhizal fungi *in vitro*. Mycorrhiza 6:481–485

Schrey SD, Schellhammer M, Ecke M, Hampp R, Tarkka MT (2005) Mycorrhiza helper bacterium *Streptomyces* AcH 505 induces differential gene expression in the ectomycorrhizal fungus *Amanita muscaria*. New Phytol 168:205–216

Schrey SD, Salo V, Raudaskoski M, Hampp R, Nehls U, Tarkka MT (2007) Interaction with mycorrhiza helper bacterium *Streptomyces* sp. AcH 505 modifies organisation of actin cytoskeleton in the ectomycorrhizal fungus *Amanita muscaria* (fly agaric). Curr Genet 52:77–85

Siddiqui ZA, Mahmood I (1999) Role of bacteria in the management of plant parasitic nematodes: a review. Bioresour Technol 69:167–179

Smalla K, Wieland G, Buchner A, Zock A, Parzy J, Kaiser S, Roskot N, Heuer H, Berg G (2001) Bulk and rhizosphere soil bacterial communities studied by denaturing gradient gel electrophoresis: plant-dependent enrichment and seasonal shifts revealed. Appl Environ Microbiol 67:4742–4751

Smith SE, Read DJ (1997) Mycorrhizal symbiosis, 2nd edn. Academic, Cambridge, pp 1–605

Söderström B (1992) Ecological potential of ectomycorrhizal mycelium. In: Read DJ, Lewis DH, Fitter AH, Alexander IJ (eds) Mycorrhizas in ecosystems. Cambridge University Press, Cambridge, pp 77–83

Sola-Landa A, Moura RS, Martin JF (2003) The two-component PhoR-PhoP system controls both primary metabolism and secondary metabolite biosynthesis in *Streptomyces lividans*. Proc Natl Acad Sci USA 100:6133–6138

Stone MJ, Williams DH (1992) On the evolution of functional secondary metabolites (natural products). Mol Microbiol 6:29–34

Taechowisan T, Lu C, Shen Y, Lumyong S (2005) Secondary metabolites from endophytic *Streptomyces aureofaciens* CMUAc130 and their antifungal activity. Microbiology 151:1691–1695

Tokala RK, Strap JL, Jung CM, Crawford DL, Salove MH, Deobald LA, Bailey JF, Morra MJ (2002) Novel plant-microbe rhizosphere interaction involving *Streptomyces* lydicus WYEC108 and the pea plant (*Pisum sativum*). Appl Environ Microbiol 68:2161–2171

Tylka GL, Hussey RS, Roncadori RW (1991) Axenic germination of vesicular arbuscular mycorrhital fungi: effects of selected *Streptomyces* species. Phytopathology 81:754–759

Weissman KJ, Leadlay PF (2005) Combinatorial biosynthesis of reduced polyketides. Nat Rev Microbiol 3:925–936

Weist S, Süssmuth RD (2005) Mutational biosynthesis–a tool for the generation of structural diversity in the biosynthesis of antibiotics. Appl Microbiol Biotechnol 68:141–150

Weller DM, Raaijmakers JM, Gardener BB, Thomashow LS (2002) Microbial populations responsible for specific soil suppressiveness to plant pathogens. Annu Rev Phytopathol 40:309–348

Whipps JM (2001) Microbial interactions and biocontrol in the rhizosphere. J Exp Bot 52:487–511

Whipps JM, Lynch JM (1986) The influence of the rhizosphere on crop productivity. Adv Microb Ecol 9:187–244

Wiener P (2000) Antibiotic production in a spatially structured environment. Ecol Lett 3:122–133

Wiener P, Egan S, Huddleston AS, Wellington EM (1998) Evidence for transfer of antibiotic-resistance genes in soil populations of streptomycetes. Mol Ecol 7:1205–1216

Wyss P, Boller T, Wiemken A (1992) Testing the effect of biological control agents on the formation of vesicular arbuscular mycorrhiza. Plant Soil 147:159–162

Yuan WM, Crawford DL (1995) Characterization of *Streptomyces* lydicus WYEC108 as a potential biocontrol agent against fungal root and seed rots. Appl Environ Microbiol 61:3119–3128

Part III
Fungal Metabolites

Chapter 7
The Effect of Fungal Secondary Metabolites on Bacterial and Fungal Pathogens

N. Mathivanan(✉), V.R. Prabavathy, and V.R. Vijayanandraj

7.1 Introduction

Fungi are an extremely diverse group of organisms, with about 230,000 species distributed widely essentially in every ecosystem. Among them, only limited species are considered to be effective biocontrol agents. The fungal antagonists restrict the growth of plant pathogens by the three suggested mechanisms: antibiosis, competition and parasitism. Besides, they also induce the defense responses in host plants, termed "induced systemic resistance" (van Loon et al. 1998). Among the abovementioned mechanisms, antibiosis is considered the most important, in which the antagonists produce an array of secondary metabolites such as antibiotics and toxin, which contribute to the antagonistic activity of fungal biocontrol agents against plant pathogens. Antagonistic strains belonging to the *Trichoderma* and *Fusarium* genera were able to produce various secondary metabolites which can play a role in the mechanism of their biological activity (http://www.item.ba.cnr.it/biopesti.htm). Production of antimicrobial secondary metabolites has been reported in many fungal biocontrol agents (Gottlieb and Shaw 1970; Fries 1973; Hutchinson 1973; Sivasithamparam and Ghisalberti 1998; Vyas and Mathur 2002). In this review, we highlight the secondary metabolites of selected fungal biocontrol agents and their involvement in the control of plant pathogens.

7.2 Secondary Metabolites of *Trichoderma*

Many *Trichoderma* species have been used as biocontrol agents against various plant pathogens. The biocontrol mechanisms exercised by *Trichoderma* could be attributed to competition for nutrients, release of toxic metabolites and extracellular

Dr. N. Mathivanan
Biocontrol and Microbial Metabolites Lab, Centre for Advanced Studies in Botany,
University of Madras, Guindy Campus, Chennai 600 025, India
e-mail: prabamathi@yahoo.com

P. Karlovsky (ed.), *Secondary Metabolites in Soil Ecology. Soil Biology 14.*

hydrolytic enzymes (Elad 2000; Mathivanan et al. 2004). The production of secondary metabolites by different species of *Trichoderma* is well documented. It has been reported that *Trichoderma* spp. produce a wide range of volatile and non-volatile antibiotic substances (Weindling and Emerson 1936; Sivasithamparam and Ghisalberti 1998; Vyas and Mathur 2002) and two such compounds, namely, trichodermin and viridin, produced by *Trichoderma* sp. inhibited pathogenic fungal growth at very low concentrations (Weindling and Emerson 1936; Weindling 1941). The volatile and nonvolatile substances produced by *Trichoderma* spp. effectively inhibited the growth of *Rhizoctonia solani* (Roy 1977), *Sclerotium rolfsii* (Upadhyay and Mukhopadhyay 1983) and *Thanatephorus cucumeris* (Dubey and Patel 2001).

A crystalline organic metabolite isolated from *Trichoderma* which inhibited *R. solani* at high dilutions was identified as gliotoxin (Weindling, 1941). Brain and McGowan (1945) described the production of viridin, a highly fungistatic antibiotic by *Trichoderma viride*. Godtfredsen and Vangedal (1965) reported the production of trichodermin, a sesquiterpenoid metabolite by *Trichoderma*. According to Dennis and Webster (1971), *Trichoderma polysporum* and *T. viride* also produced trichodermin and *Trichoderma hamatum* produced peptide antibiotics. They further demonstrated the fungi-toxic action of *Trichoderma* metabolites on pathogenic *Pythium*. Trichodermin-4 is an antibiotic produced by *Trichoderma lignorum* that was used to control plant diseases (Fedorinchik et al. 1975).

Isolates of *T. hamatum* produced toxic water-soluble metabolites and two of them were identified as isonitrile acids (Brewer and Taylor 1981). Papavizas et al. (1982) used several UV-induced mutants of *T. harzianum* for the production of secondary metabolites. They obtained two unidentified metabolites from *T. harzianum*: one is thermolabile and the other is thermostable. Stipanovic and Howell (1982) isolated a new toxic metabolite, gliovirin from *Gliocladium virens* (synonym *Trichoderma virens*) that was later found to be active against *Pythium ultimum* (Howell and Stipanovic 1983). Antifungal pyrones isolated from the culture filtrates of *T. harzianum* and *Trichoderma koningii* inhibited the growth of many fungal pathogens, including *Bipolaris sorokiniana, Fusarium oxysporum, Gaeumannomyces graminis* var. *tritici, Phytophthora cinnamomi, Pythium middletonii* and *R. solani* (Claydon et al. 1987; Simon et al. 1988). A phenol-like compound isolated from *T. harzianum* inhibited the uredospore germination of the rust pathogen of groundnut, *Puccinia arachidis* (Govindasamy and Balasubramanian 1989).

Seven *Trichoderma* spp. were evaluated for antagonistic activity against *F. oxysporum, Fusarium equiseti, Fusarium solani, Sclerotinia sclerotiorum, Sclerotinia minor, Rhizoctonia* sp. and *S. rolfsii* and the presence of diffusible metabolites in the medium was demonstrated in almost 80% of the pathogen–antagonist interactions (Monaco et al. 1994). A strain of *T. harzianum* isolated from wheat roots produced five different metabolites. Among them, three new octaketide-derived compounds exhibited antifungal activity against *G. graminis* var. *tritici*, the causal agent of take-all disease of wheat (Ghisalberti and Rowland 1993). DiPietro et al. (1993) obtained gliotoxin from the culture filtrate of *G. virens*, which inhibited the spore germination of *Botrytis cinerea*. Isolates of *T. viride* and *T. harzianum* inhibited the growth of *Fusarium moniliforme* and *Aspergillus flavus* by producing inhibitory volatile compounds (Calistru et al. 1997). Cotton seedling disease incited by *R. solani* has been suppressed by *T. viride* owing to mycoparasitism and antibiotic production (Howell et al. 2000). The volatile secondary metabolites produced by *Trichoderma pseudokoningii, T. viride* and *Trichoderma aureoviride* affected the mycelial growth and protein synthesis in two isolates of *Serpula lacrymans* in varying degrees (Humphris et al. 2002). But the production of nonvolatile metabolites appears to be one of the mechanisms involved in the biological control of the tomato root pathogen *Pyrenochaeta lycopersici* by four different *T. harzianum* isolates in addition to the high secretion of chitinases (Perez et al. 2002).

Mukherjee and Raghu (1997) studied the effect of temperature on metabolites production by *Trichoderma* sp. They observed that *Trichoderma* produced high concentration of fungi-toxic metabolites in broth culture at high temperatures. However, *Trichoderma* sp. was not effective in suppressing *S. rolfsii* at temperatures above 30 °C. Mischke et al. (1997) measured the efficacy of metabolites produced by *Trichoderma* spp. based on growth inhibition of *R. solani*. Further, they observed that the aqueous extracts from light-grown germlings of *T. virens* inhibited *R. solani* more than the extracts from germlings grown in the dark. In addition, they demonstrated that the extracts from *T. virens* grown under low pH showed increased inhibitory activity. Endophytic *Trichoderma* sp.-DIS 172ai produced metabolites that inhibited the growth of *Moniliophthora roreri* (Bailey et al. 2006). Vinale et al. (2006) isolated secondary metabolites from two commercialized strains of *T. harzianum*, T22 and T39, for the first time. Three major bioactive compounds were produced by strain T22, of which one is a new azaphilone that showed remarkable antifungal activity against *R. solani, P. ultimum* and *G. graminis* var. *tritici* under in vitro condition.

Peptaibols, the linear polypeptides produced by *Trichoderma* spp., showed interesting physicochemical and biological properties, including antibacterial, antifungal and occasionally antiviral activities. Further, these peptaibols can also induce plant resistance. Many species of *Trichoderma*, viz.,, *T. asperellum, T. harzianum, T. koningii, T. virens* and *T. viride*, were reported to produce these peptaibols (Iida et al. 1995; Wada et al. 1995; Huang et al. 1996; Landreau et al. 2002; Chutrakul and Peberdy, 2005; Szekeres et al. 2005; Wei et al. 2005; Xiao-Yan et al. 2006). The biosynthesis and biological properties of peptaibols were reviewed in detail recently (Szekeres et al. 2005). Wiest et al. (2002) demonstrated the production of peptaibols antibiotic in *T. virens* and their role in biocontrol activity. Three new groups of peptaibols, trichodecenins, trichorovins and trichocellins, have been isolated from conidia of *T. viride*. The structures of trichodecenin-I and trichodecenin-II were established by positive-ion fast-atom bombardment, collision-induced dissociation mass spectrometry and two-dimensional NMR spectroscopy. Trichodecenin-I and trichodecenin-II have a (*Z*)-4-decenoyl group, six amino acid residues and a leucinol moiety in the molecules. Xiao-Yan et al. (2006) purified peptaibol-type metabolites from *T. koningii* SMF2 by gel filtration and high-performance liquid chromatography which showed antimicrobial activity against a wide range of Gram-positive bacterial and fungal phytopathogens. Three bioactive metabolites were identified as trichokonin VI, VII and VIII by liquid chromatography–electrospray ionization tandem mass spectrometry. All three trichokonins were stable and exhibited antimicrobial activity over a wide range of pH and temperature. Interestingly, these trichokonins were insensitive to proteolytic enzymes and did not lose their bioactivity even after autoclaving.

The expressed sequence tag (EST) database developed by Liu and Yang (2005) using a directional complementary DNA library constructed from the mycelial DNA of *T. harzianum* gave useful information on *Trichoderma* gene sequences to elucidate the integrated biocontrol mechanism. They subjected 3,298 clones to single-pass sequencing from the 5~ end of the vector, and identified sequence similarity against gene sequences in the databases. Of the 3,298 clones, 2,174 exhibited similarity to known genes and 451 similarity to known ESTs, while 673 represented novel gene sequences. Analysis of the identified clones indicated sequence similarity to a broad diversity of genes encoding proteins such as enzymes, structural proteins and regulatory factors. According to them, a significant proportion of genes identified in the mycelium were involved in processes related to mycoparasitism and production of fungicidal metabolites.

Reithner et al. (2005) studied the signal transduction pathways and analyzed the *tga1* gene encoding a G α-subunit of *Trichoderma atroviride* P1. A *Δtga1* mutant showed continuous sporulation and elevated internal steady-state cyclic AMP levels. Deletion of the *tga1* gene resulted in complete loss of mycoparasitic overgrowth and lysis of *R. solani, B. cinerea* and *S. sclerotiorum* during direct confrontation, although the formation of infection structures was unaffected. The reduced mycoparasitic ability was due to decreased chitinase activity and reduced *nag1* and *ech42* gene transcription. Furthermore, production of 6-pentyl-α-pyrone and metabolites with sesquiterpene structure was reduced in the Δ*tga1* mutant. Despite these deficiencies, the mutant showed enhanced growth inhibition in the host fungus by overproducing other low molecular weight antifungal metabolites, suggesting opposite roles of *tga1* in regulating the biosynthesis of different antifungal substances in *T. atroviride*.

Mukherjee et al. (2006) for the first time cloned the secondary metabolism related genes from *T. virens*. They identified six genes similar to those involved in secondary metabolism in other fungi, which include four cytochrome P450 genes, one *O*-methyl transferase and one terpene cylase by a transcriptional comparison of a wild-type and a secondary metabolite deficient *T. virens* mutant. Of the six, four genes (three cytochrome P450s and the cyclase) were located as a cluster. Three genes, viz., the P450 genes, the *O*-methyl transferase and the terpene cyclase, were underexpressed in the mutant, which lacks the major secondary metabolites viridin and viridiol. The gene-expression pattern and associated secondary metabolite profile were similar to the other secondary metabolic pathways in related fungi, indicating that the cluster is associated with the production of a terpene, possibly viridin.

Several other metabolites, viz., trichocaranes (Macias et al. 2000), demethylsorbicillin, oxosorbicillinol (Abe et al. 2000), trichodenones, harzialactone A and B, (*R*)-mevalonolactone (Amagata et al. 1998), 6-*n*-pentyl pyrone, isonitrile acid

(Graeme-Cook and Faull 1991; Brewer and Taylor 1981), trichoviridin, 3-(3-isocyano-6-oxabicyclo[3,1,0]hex-2-en-5-yl)acrylic acid and 3-(3-isocyanocyclopent-2-enylidene)propionic acid (Brewer et al. 1982) have already been reported to be produced by *Trichoderma* spp.

7.3 Secondary Metabolites of *Fusarium*

In general, *Fusarium* spp. are common plant pathogens causing diseases associated with roots such as wilts and rots. Furthermore, they are also considered as deleterious fungi because of their ability to produce mycotoxins, the secondary metabolites with adverse health effect. However, several species of *Fusarium*, viz., *F. chlamydosporum, F. decemcellulare, F. heterosporum, F. longipes, F. semitectum* var. *majus, F. solani* and nonpathogenic *F. oxysporum*, have been reported as biocontrol agents against various phytopathogenic fungi (Kapooria and Sinha 1969; Hornok and Walcz 1983; Gill and Chahal 1988; Rao and Thakur 1988; Amorim et al. 1993; Navi and Singh 1993; Mathivanan 2000; Mathivanan and Murugesan 2000). These biocontrol agents produced a number of toxic substances that inhibit the growth of pathogenic microorganisms (Baker et al. 1990; Sawai et al. 1981; Diekmann 1970; Robinson and Garrett 1969). Garrett and Robinson (1969) isolated nonanoic acid from *F. oxysporum*, which inhibited the spore germination of *Cunninghamella elegans*. Two antifungal substances isolated from the culture filtrate of *F. solani* effectively inhibited the growth of the Japanese apple canker pathogen *Valsa ceratosperma* and also *Helminthosporium oryzae* and *Stereum purpureum* (Sawai et al. 1981). Goodman and Burpee (1991) observed that the hyphal growth of *Sclerotinia homoeocarpa* was inhibited by the metabolites of *F. heterosporum*.

An antifungal metabolite of *F. solani* inhibited the uredospore germination of *P. arachidis* (Jayapal Gowdu 1986) and decreased the rust disease development in groundnut (*Arachis hypogaea*). *F. chlamydosporum* was isolated from pustules of groundnut rust, *P. arachidis*, and it was also found to be nonpathogenic to groundnut plants (Mathivanan 1995). The application of conidia of *F. chlamydosporum* reduced the pustule number in both detached and intact groundnut leaves. The antagonist rapidly colonized the rust pustules and as a result the uredopsores of *P. arachidis* greatly lost its ability to germinate, which indicated the possible production of toxic metabolites by *F. chalydosporum*. Further, an unidentified antifungal metabolite of *para*-disubstituted, aromatic in nature having carbonyl and methylene groups, with a molecular weight of 257 was isolated from the culture filtrate. This metabolite completely inhibited the uredospore germination at 30 μg/ml (Mathivanan and Murugesan 1999). Two α-pyrones, fusapyrone and deoxyfusapyrone, isolated from rice cultures of *F. semitectum* (Evedente et al. 1994, 1999) showed considerable antifungal activity against several plant pathogenic fungi (Altomare et al. 2000, 2004). Among the two compounds, fusapyrone was consistently more active than deoxyfusapyrone. Further, these two pyrones were highly active against *Alternaria alternata, Ascochyta rabiei, A. flavus, B. cinerea, Cladosporium cucumerinum,*

Phoma tracheiphila and *Penicillium verrucosum*, but they less active against *Fusarium* spp. (Altomare et al. 2000).

7.4 Secondary Metabolites of *Trichothecium roseum*

Trichothecium roseum Link. is a common soil fungus, which is synonymous with *Cephalothecium roseum* Corda. Antagonism between *T. roseum*, and certain plant pathogenic fungi was reported by Freeman and Morrison (1949), Pohjakallio and Makkonen (1957), Makkonen and Pohjakallio (1960), Kashyap (1978), Rod (1984), Jindal and Thind (1990), Singh (1991), Gouramani (1995), Lacicowa and Pieta (1996), Mandal et al. (1999) and Vanneste et al. (2002). Al-Heeti and Sinclair (1985) reported that 18 isolates of *T. roseum* grown separately on modified Czapeck Dox broth, and their culture filtrate diluted to 5% in water, inhibited sporogenesis of *Phytophthora megasperma* f. sp. *glycinea*. *T. roseum* produces water-soluble, heat-resistant metabolites, which showed toxic effects on the mycelial growth and conidial germination of *Pestalotia funerea* (Urbasch 1985). Further, Urbasch (1985) demonstrated the penetration of conidia of *P. funerea* by simple or lobed appressoria of *T. roseum*. The presence of granulated and vacuolated cytoplasm in cortical and medullary cells suggested that *T. roseum* may produce toxic metabolites in infected tissues, in addition to cell wall degrading enzymes (Huang and Kokko 1993). Trichothecin produced by *T. roseum* is an ester of isocrotonic acid and an α,β-unsaturated ketonic alcohol, trichothecolone ($C_{15}H_{20}O_4$). Trichothecin crystallizes in light petroleum as long fibrous needles, and has a melting point of 118 °C. It is readily soluble in chloroform, ethanol, acetone and hexane. Application of trichothecin to cottonseeds and crop plants prevented wilt diseases (Askarova and Ioffe 1962).

Bawden and Freeman (1952) were the first investigators to report the antiviral effects of trichothecene compounds on plant viral infections. These workers discovered two heat-stable substances present in the culture filtrate of *T. roseum* inhibited viral infection of bean and tobacco. One of these compounds was identified as trichothecin. Trichothecin is more effective in managing viral infection in bean plants than in tobacco, whereas the reverse is true for trichothecolone and acetyltrichothecolone. Trichothecin and its derivatives inhibited infection when applied 1 day after the plants had been inoculated with the viruses (*Tobacco necrosis virus, Tobacco mosaic virus* and *Tomato bushy stunt virus*).

7.5 Antifungal Metabolites from Other Fungi

Forrer (1977) demonstrated that the teliospore formation in *P. graminis* was delayed by the metabolites of *Aphanocladium album*. Toxins of *Scytalidium uredinicola* inhibited the uredospore germination of *Endocronartium harknessii* (Fairbairn et al.

1983). Jayapal Gowdu (1986) isolated many secondary metabolites from the culture filtrates of *Acremonium obclavatum* and *Myrothecium verrucaria*. These metabolites showed antifungal activity against groundnut rust by inhibiting uredospore germination. Leinhos and Buchenauer (1992) isolated several antifungal compounds from the culture filtrates of *Verticillium chlamydosporium*. These compounds reduced development of cereal rusts by inhibiting the growth of *Puccinia coronata* on oat, *Puccinia recondita* on wheat and *Puccinia sorghi* on corn. Culture filtraes of *Penicillium brevicompactum, Penicillium expansum* and *Penicillium pinophilum* effectively inhibited the mycelial growth of *R. solani*, suggesting the production of antifungal metabolites. Three purified compounds, mycophenolic acid, patulin and 3-*O*-methylfunicone, which were extracted from the culture filtrate of *Penicillium* strains inhibited *R. solani* in vitro. Further, their production was detected in dual cultures of the same *Penicillium* strains with *R. solani* in sterilized soil (Nicoletti et al. 2004).

Machida et al. (2001) isolated two 3(2*H*)-benzofuranones and three chromanes from the culture liquid of a mycoparasitic fungus, *Coniothyrium minitans*. Later McQuilken et al. (2003) isolated four closely related antifungal metabolites from *C. minitans*. Among the four, a major metabolite identified as macrosphelide A inhibited the growth of *Sclerotinia sclerotiorum* and *Sclerotium cepivorum* by 50% at 46.6 and 2.9 μg/ml, respectively.

7.6 Conclusions

Several fungal biocontrol agents are reported to produce various secondary metabolites, *Trichoderma* is considered the most important as many of its species produce a variety of metabolites. These metabolites are toxic to plant pathogens at very low concentrations. The secondary metabolites of *Fusarium* and *Trichothecium* also show antifungal, antibacterial and antiviral activities against various plant pathogens. A large amount of literature has documented the potential of fungal secondary metabolites in controlling plant pathogens. However, most of the published information is restricted to the laboratory or greenhouse experiments and the use of these fungal metabolites in agriculture for the management of plant diseases has not been realized. In this scenario, it is worth testing these metabolites under field conditions in order to promote them as fungicides, if they show promising results.

References

Abe N, Yamamoto K, Hirota A (2000) Novel fungal metabolites, demethylsorbicillin and oxosorbicillinol, isolated from *Trichoderma* sp. USF-2690. Biosci Biotechnol Biochem 64:620–622

Al-Heeti MB, Sinclair JB (1985) Inhibition of sporogenesis of *Phytophthora megasperma* f sp. *glycinea* by 18 isolates of *Trichothecium roseum*. Phytopthology 75:1327

Altomare C, Perrone G, Zonno MC, Evidente A, Pengue R, Fanti F, Polonelli L (2000) Biological characterization of fusapyrone and deoxyfusapyrone, two bioactive secondary metabolites of *Fusarium semitectum*. J Nat Prod 63:1131–1135

Altomare C, Pengue R, Favilla M, Evidente A, Visconti A (2004). Structure-activity relationships of derivatives of fusapyrone, an antifungal metabolite of *Fusarium semitectum*. J Agric Food Chem 52:2997–3001

Amagata T, Minoura K, Numata A (1998) Cytotoxic metabolites produced by a fungal strain from a *Sargassum* algae. J Antibiot 51:432–434

Amorim EPR, Pio-Ribeiro G, Menezes M Coelho RSB (1993) The pathogenicity and hyperparasitic action of *Fusarium decemcellulare* on *Puccinia psidii* in guava (*Psidium guajava*). Fitopatol Bras 18:226–229

Askarova SA, Ioffe RI (1962). On the possibility of the use of the fungicidal preparation trichotecin in the control of cotton wilt. Antibiotiki 7:929–930

Bailey BA, Bae H, Strem MD, Roberts DP, Thomas SE, Crozier J, Samuels GJ, Choi IY, Holmes KA (2006) Fungal and plant gene expression during the colonization of cacao seedlings by endophytic isolates of four *Trichoderma* species. Planta 224:1449–1464

Baker RA, Tatum JH, Nemec S (1990) Antimicrobial activity of napthoquinones from Fusaria. Mycopathologia 111:9–16

Bawden FC, Freeman GG (1952). The nature and behaviour of inhibitors of plant viruses produced by *Trichothecium roseum* Link. J Gen Microbiol 7:154–168

Brain PW, McGowan, JC (1945) Viridin: a highly fungistatic substance produced by *Trichoderma viride*. Nature 156:144–145

Brewer D, Taylor A (1981). *Trichoderma hamatum* isolated from pasture soil. Mycopathologia 76:167–173

Brewer D, Feicht A, Taylor A, Keeping JW, Taha AA, Thaller V (1982) Production of experimental quantities of isocyanide metabolites of *Trichoderma hamatum*. Can. J Microbiol 28:1252–1260

Calistru C, McLean M, Berjak P (1997) In vitro studies on the potential for biological control of *Aspergillus flavus* and *Fusarium moniliforme* by *Trichoderma* species. Mycopathologia 137:115–124

Chutrakul C, Peberdy JF (2005) Isolation and characterisation of a partial peptide synthetase gene from *Trichoderma asperellum*. FEMS Microbiol Lett 252:257–265

Claydon N, Allan M, Hanson JR, Avent AG (1987) Antifungal alkyl pyrones of *Trichoderma harzianum*. Trans Br Mycol Soc 88:505–513

Dennis L, Webster J (1971) Antagonistic properties of species group of *Trichoderma* III hyphal interactions. Trans Br Mycol Soc 57:363–369

Diekmann H (1970) Stoffwechselprodukte von Mikroorganismen. Arch Mikrobiol 73:65–76

DiPietro A, Lorito M, Hayes CK, Broadway RM, Harman GE (1993) Endochitinase from Gliocladium virens: Isolation, characterization and synergistic antifungal activity in combination with gliotoxin. Phytopathology 83:308–313

Dubey SC, Patel B (2001). Evaluation of fungal antagonists against *Thanatephorus cucumeris* causing web blight of urd and mung bean. Indian Phytopathol 54:206–209

Elad Y (2000) Biological control of foliar pathogens by means of *Trichoderma harzianum* and potential modes of action. Crop Prot 19:709–714

Evidente A, Conti L, Altomare C, Bottalico A, Sindona G, Segre AL, Logrieco A (1994) Fusapyrone and deoxyfusapyrone, two antifungal alpha-pyrones from *Fusarium semitectum*. Nat Toxins 2:4–13

Evidente A, Amalfitano C, Pengue R Altomare C (1999) High performance liquid chromatography for the analysis of fusapyrone and deoxyfusapyrone, two antifungal alpha-pyrones from *Fusarium semitectum*. Nat Toxins 7:133–137

Fairbairn N, Pickard MA, Hiratuska Y (1983) Inhibition of *Endocronartium harknesii* spore germination by metabolites of *Scytalidium uredinicola* and the influence of growth medium on inhibitor production. Can J Bot 61:2147–2152

Fedorinchik NS, Tarunina, TA, Tyutyunnikov MG, Kudryatseva KI (1975) Trichodermin-4, a new biological preparation for plant disease control. Plant Prot 3:67–72

Forrer HR (1977) Der Einfluss von Stoffwechsel Produkten des Mycoparasiten *Aphanocladium album* auf die Teleutosporenbildung von Rostpilzen. Phytopathol Z 88:306–311

Freeman GG, Morrision RI (1949) The isolation and chemical properties of trichothecin, an antifungal substance from *Trichothecium roseum* Link. Biochem. 44:1–5

Fries N (1973) Effects of volatile organic compounds on the growth and development of fungi. Trans Br Mycol Soc 60:1–21

Garrett MK, Robinson PM (1969) A stable inhibitor of spore germination produced by fungi. Arch Mikrobiol 67: 370–377

Ghisalberti EL, Rowland CY (1993). Antifungal metabolites from *Trichoderma harzianum*. J Nat Prod 56:1799–1804

Gill KS, Chahal SS (1988). Growth inhibition of *Claviceps fusiformis* with culture filtrate of *Fusarium chlamydosporum*. Plant Dis Res 3:64–65

Godtfredsen WO, Vangedal S (1965) Trichodermin, a new sesquiterpene antibiotic. Acta Chem Scand 19:1088–1102

Goodman DM, Burpee, LL (1991) Biological control of dollar spot disease of creeping bentgrass. Phytopathology 81:1438–1446

Gottlieb D, Shaw PD (1970) Mechanism of action of antifungal antibiotics. Annu Rev Phytopathology 8:371–402

Gouramani GD (1995) Biological and chemical control of rice blast disease (*Pyricularia oryzae*) in north Greece. Cah Mediterr 15:61–68

Govindasamy V, Balasubramanian R (1989) Biological control of groundnut rust, *Puccinia arachidis* by *Trichoderma harzianum*. Z Pflanzenkr Pflanzenschutz 96:337–345

Graeme-Cook KA, Faull JL (1991) Effect of ultraviolet-induced mutants of *Trichoderma harzianum* with altered antibiotic production on selected pathogens in vitro. Can J Microbiol 37:659–664

Hornok L, Walcz I (1983) *Fusarium heterosporum*, a highly specialized hyperparasite of *Claviceps purpurea*. Trans Br Mycol Soc 80: 377–380

Howell CR, Hanson LE, Stipanovic RD, Puckhaber LE (2000) Induction of terpenoid synthesis in cotton roots and control of *Rhizoctonia solani* by seed treatment with *Trichoderma virens*. Phytopathology 90:248–252

Howell CR, Stipanovic, RD (1983) Gliovirin, a new antibiotic from *Gliocladium* virens and its role in the biological control of *Pythium ultimum*. Can J Microbiol 29:321–324

Huang HC, Kokko EG (1993) *Trichothecium roseum*, a mycoparasite of *Sclerotinia sclerotiorum*. *Can J Bot* 71:1631–1638

Huang Q, Tezuka Y, Hatanaka Y, Kikuchi T, Nishi A, Tubaki K (1996) Studies on metabolites of mycoparasitic fungi. V. Ion-spray ionization mass spectrometric analysis of trichokonin-II, a peptaibol mixture obtained from the culture broth of *Trichoderma koningii*. Chem Pharm Bull (Tokyo) 44:590–593

Humphris SN, Bruce A, Buultjens E, Wheatley RE (2002) The effects of volatile microbial secondary metabolites on protein synthesis in *Serpula lacrymans*. FEMS Microbiol Lett 210:215–219

Hutchinson SA (1973) Biological activities of volatile fungal metabolites. Annu Rev Phytopathol 11:223–246

Iida A, Sanekata M, Wada S, Fujita T, Tanaka H, Enoki A, Fuse G, Kanai M, Asami K (1995) Fungal metabolites. XVIII. New membrane-modifying peptides, trichorozins I-IV, from the fungus *Trichoderma harzianum*. Chem Pharm Bull (Tokyo) 43:392–397

Jayapal Gowdu B (1986) Studies on some biological aspects of fungal associates on groundnut rust, Puccinia arachidis Speg. PhD thesis, University of Madras, Chennai

Jindal KK, Thind BS (1990) Microflora of cowpea seeds and its significance in the biological control of seed borne infection of *Xanthomonas campestris* pv. *vignicola*. Seed Sci Technol 18:393–403

Kapooria RG, Sinha S (1969) Phylloplane mycoflora of pearl millet and its influence on the development of *Puccinia pennisetti*. Trans Br Mycol Soc 53:153–155

Kashyap U (1978) Biological control of some fungal diseases of tomatoes by phyllosphere microorganisms. Indian J Mycol Plantpathol 8:37

Lacicowa B, Pieta D (1996) The efficiency of microbiological dressing of pea seeds (*Pisum sativum* L.) against pathogenic soil-borne fungi. Rocz Nauk Roln Ser E 25:15–21

Landreau A, Pouchus YF, Sallenave-Namont C, Biard JF, Boumard MC, Robiou du PT, Mondeguer F, Goulard C, Verbist JF (2002) Combined use of LC/MS and a biological test for rapid identification of marine mycotoxins produced by *Trichoderma koningii*. J Microbiol Methods 48:181–194

Leinhos GME, Buchenauer H (1992) Inhibition of rust diseases of cereals by metabolic products of *Verticillium chlamydosporium*. J Phytopathol 136:177–193

Liu PG, Yang Q (2005) Identification of genes with a biocontrol function in *Trichoderma harzianum* mycelium using the expressed sequence tag approach. Res Microbiol 156:416–423

Machida, KM, Trifonov LS, Ayer WA, Lu ZX, Laroche A, Huang HC, Cheng KJ (2001) 3(2H)-Benzofuranones and chromanes from liquid cultures of the mycoparasitic fungus *Coniothyrium minitans*. Phytochemistry 58:173–177

Macias FA, Varela RM, Simonet AM, Cutler HG, Cutler SJ, Eden MA, Hill RA (2000) Bioactive carotanes from *Trichoderma virens*. J Nat Prod 63:1197–2000.

Makkonen R, Pohjakallio O (1960) On the parasitic attacking the sclerotia of some fungi pathogenic to higher plants and on the resistance of these studies to their parasites. Acta Agric Scand 10:105–126

Mandal S, Srivatava KD, Rashmi A, Singh DV, Mandal S, Aggarwal R (1999). Mycoparasitic action of some fungi on spot blotch pathogen (*Drechslera sorokiniana*) of wheat. Indian Phytopathol 52:39–43

Mathivanan N (1995) Studies on extracellular chitinase and secondary metabolites produced by *Fusarium chlamydosporum*, an antagonist to *Puccinia arachidis*, the rust pathogen of groundnut. PhD thesis, University of Madras, Chennai

Mathivanan N (2000) Chitinase and β-1,3 glucanase of *Fusarium solani*: effect of crude enzymes on *Puccinia arachidis*. J Mycol Plant Pathol 30:327–330

Mathivanan N, Murugesan K (1999) Isolation and purification of an antifungal metabolite from *Fusarium chlamydosporum*, a mycoparasite to *Puccinia arachidis*, the rust pathogen of groundnut. Indian J Exp Biol 37:98–101

Mathivanan N, Murugesan K (2000) *Fusarium chlamydosporum*, a potent biocontrol agent to groundnut rust, *Puccinia arachidis*. Z Pflanzenkr Pflanzenschutz 107:225–234

Mathivanan N, Prbavathy VR, Murugesan K (2004) Biocontrol potential of microorganisms—an overview: focus on *Trichoderma* as biofungicide for the management of plant diseases. In: Mayee CD, Manoharachary C, Tilak KVBR, Mukadam KS Deshpande J (eds) Biotechnological approaches for the integrated management of crop diseases. Daya, Delhi, pp 88–109

McQuilken MP, Gemmell J, Hill RA, Whipps JM (2003) Production of macrosphelide A by the mycoparasite *Coniothyrium minitans*. FEMS Microbiol Lett 219:27–31

Mischke S (1997) A quantitative bioassay for extracellular metabolites that antagonize growth of filamentous fungi, and its use with biocontrol fungi. Mycopathologia. 137:45–52

Monaco C, Perello A, Rollan MC (1994). In vitro tests of the antagonistic behavior of *Trichoderma* spp. against pathogenic species of the horticultural region of La Plata, Argentina. Microbiologia 10:423–428

Mukherjee M, Horwitz BA, Sherkhane PD, Hadar R, Mukherjee PK (2006) A secondary metabolite biosynthesis cluster in *Trichoderma virens*: evidence from analysis of genes underexpressed in a mutant defective in morphogenesis and antibiotic production. Curr Genet 50:193–202

Mukherjee PK, Raghu K (1997) Effect of temperature on antagonistic and biocontrol potential of *Trichoderma* sp. on *Sclerotium rolfsii*. Mycopathologia. 139:151–155

Navi SS, Singh SD (1993) *Fusarium longipes*—a mycoparasite of *Sclerospora graminicola* on pearl millet. Indian Phytopathol 46:365–368

Nicoletti R, De Stefano M, De Stefano S, Trincone A, Marziano F (2004) Antagonism against *Rhizoctonia solani* and fungitoxic metabolite production by some *Penicillium* isolates. Mycopathologia 158:465–474

Papavizas GC, Lewis JA, Abd-El Moity TH (1982) Evaluation of new biotypes of *Trichoderma harzianum* for tolerence to benomyl and enhanced biocontrol capabilities. Phytopathology 72:126–132

Perez LM, Besoain X, Reyes M, Pardo G, Montealegre J (2002) The expression of extracellular fungal cell wall hydrolytic enzymes in different *Trichoderma harzianum* isolates correlates with their ability to control *Pyrenochaeta lycopersici*. Biol Res 35:401–410

Pohjakallio O, Makkonen R (1957) On the resistance of the sclerotia of some phytopathological fungi against their parasites. Acta Chem Fenn 30:222

Rao VP, Thakur RP (1988) *Fusarium semitectum* var. *majus*—a potential biocontrol agent of ergot (*Claviceps fusiformis*) of pearlmillet. Indian Phytopath 41:567–574

Reithner B, Brunner K, Schuhmacher R, Peissl I, Seidl V, Krska R, Zeilinger S (2005) The G protein alpha subunit Tga1 of *Trichoderma atroviride* is involved in chitinase formation and differential production of antifungal metabolites. Fungal Genet Biol 42:749–760

Robinson PM, Garrett MK (1969) Identification of volatile sporostatic factors from cultures of *Fusarium oxysporum*. Trans Br Mycol Soc 52:293–299

Rod J (1984) Antagonistic effects of some fungi on fungal pathogens causing storage rot of onion (*Allium cepa* L.) Ceska Mykol 38:235–239

Roy AK (1977) Parasitic activity of *Trichoderma viride* on sheath blight fungus of rice. J Plant Dis Prot 84:675–683

Sawai K, Okuno, T, Terada Y, Harada Y, Sawamura K, Sasaki H, Takao S (1981) Isolation and properties of two antifungal substances from *Fusarium solani*. Agric Biol Chem 45:1223–1228

Simon A, Dunlop RW, Ghissalberti EL, Sivasithamparam (1988) *Trichoderma koningii* produces a pyrone compound with antibiotic properties. Soil Biol Biochem 20:263–264

Singh D (1991) Biocontrol of *Sclerotina sclerotiorum* (Lib.) de Bary by *Trichoderma harzianum*. Trop Pest Manag 37:374–378

Sivasithamparam K, Ghisalberti EL (1998) In: Kubicek CP, Harman GE (eds) *Trichoderma* and *Gliocladium*, vol 1. Taylor and Francis, London, pp 139–191

Stipanovic RD, Howell CR (1982) The structure of gliovirin, a new antibiotic from *Gliocladium virens*. J Antibiot 35:1326–1330

Szekeres A, Leitgeb B, Kredics L, Antal Z, Hatvani L, Manczinger L, Vagvolgyi C (2005) Peptaibols and related peptaibiotics of *Trichoderma*—a review. Acta Microbiol Immunol Hung 52:137–168

Upadhyay JP, Mukhopadhyay AN (1983) Effect of non-volatile and volatile antibiotics of *Trichoderma harzianum* on growth of *Sclerotium rolfsii*. Indian J Mycol Plant Pathol 13:232–233

Urbasch I (1985) Antagonistische wirkung von *Trichothecium roseum* (Pers.) Link ex Gray auf *Pestalotia funereal*. Desm Phytopathol Z 113:343–347

van Loon LC, Bakker PAHM, Pieterse CMJ (1998) Systemic resistance induced by rhizosphere bacteria. Annu Rev Phytopathol 36:453–483

Vanneste JL, Hill RA, Kay SJ, Farrel RL, Holland PT (2002) Biological control of sapstain fungi with natural products and biological control agents: a review of the work carried out in New Zealand. Mycol Res 106:228–232

Vinale F, Marra R, Scala F, Ghisalberti EL, Lorito M, Sivasithamparam K (2006) Major secondary metabolites produced by two commercial *Trichoderma* strains active against different phytopathogens. Lett Appl Microbiol 43:143–148

Vyas RK, Mathur K (2002) *Trichoderma* spp. in cumin rhizosphere and their potential in suppression of wilt. Indian Phytopathol 55:455–457

Wada S, Iida A, Akimoto N, Kanai M, Toyama N, Fujita T (1995) Fungal metabolites. XIX. Structural elucidation of channel-forming peptides, trichorovins-I-XIV, from the fungus *Trichoderma viride*. Chem Pharm Bull (Tokyo) 43:910–915

Wei X, Yang F, Straney DC (2005) Multiple non-ribosomal peptide synthetase genes determine peptaibol synthesis in *Trichoderma virens*. Can J Microbiol 51:423–429

Weindling R (1941) Experimental consideration of the mould toxins of *Gliocladium* and *Trichoderma*. Phytopathology 31:991–1003

Weindling R, Emerson OH (1936) The isolation of a toxic substance from the culture of a *Trichoderma*. Phytopathology 26:1068–1070

Wiest A, Crzegorski D, Xu BW, Goulard C, Rebuffat S, Ebbole DJ, Bodo B, Kenerley C (2002) Identification of peptaibols from *Trichoderma virens* and cloning of a peptaibol synthetase. J Biol Chem 277:20862–20868

Xiao-Yan S, Qing-Tao S, Shu-Tao X, Xiu-Lan C, Cai-Yun S, Yu-Zhong Z (2006) Broad-spectrum antimicrobial activity and high stability of trichokonins from *Trichoderma koningii* SMF2 against plant pathogens. FEMS Microbiol Lett 260:119–125.

Chapter 8
Biological Significance of Truffle Secondary Metabolites

Richard Splivallo

8.1 Introduction

Fungal primary and secondary metabolites have an important impact on our society. Best known as mycotoxins, phytotoxins, antibiotics and natural aromas; they represent industries worth billions of dollars. Fungi are also of major importance in terms of biomass: they rank first with an estimated dry weight of 450 kg/ha, which represents 91% of the total soil biomass (microflora and microfauna) (Müller and Loeffler 1976). Yet our knowledge of the ecological significance of fungal metabolites is limited. Despite the pioneer work of Dick and Hutchinson (1966) and Hutchinson (1973) on the effect of volatile fungal metabolites on fungi and plants, this argument seems to have raised little interest in the scientific community. Since then, most studies have focused on parasitic interactions with plants (phytopathogens), while much less attention has been given to the ecological role of the metabolites of symbiotic fungi. An important group of the latter is represented by mycorrhizal fungi. Mycorrhizas are one of the oldest associations between plants and fungi. Dating back to the early colonization of the terrestrial environment (Brundrett 2002), they are classified as endomycorrhizas (arbuscular, ericoid, orchid mycorrhizas) or ectomycorrhizas depending on their ability to penetrate the host-plant root. Truffles fall in the last category of the ectomycorrhizal fungi. Best known for the complex aroma of their hypogeous fruitbodies, truffles were already known to the Greeks and the Romans, but only reached their luxury standing in the last 20 years owing to decreasing production (Fauconnet and Delher 1998; Hall and Yun 2001) and an ever-increasing demand. Despite their high commercial value, very little is known about their biology. Indeed, the unique features of mycorrhizal fungi, from their formation to signal exchange with the surrounding environment (the rhizosphere), are still poorly understood. In addition to the compounds involved in nutritional exchanges between the host plant and the fungus, various micromolecules and macromolecules are secreted into the rhizosphere. These exudates and volatile

Richard Splivallo
University of Torino, Department of Plant Biology, IPP-CNR, 25 Viale Mattioli, 10125 Torino, Italy
e-mail: ricsi17@hotmail.com

P. Karlovsky (ed.), *Secondary Metabolites in Soil Ecology. Soil Biology 14.*

organic compounds (VOCs) play an active role in the regulation of symbiosis and interactions with other organisms, including nonhost plants.

More than 200 VOCs and many nonvolatile compounds have been identified from various truffle species. The aim of this chapter is to discuss the ecological significance of these metabolites (VOCs and/or exudates) associated with three levels of differentiation: fruitbody, free-living mycelium and mycorrhizas. Furthermore, the possible role of these metabolites in the interaction with the host plants and nonhost plants (the so-called burnt, a zone with scarce herbaceous cover) shall be discussed.

8.2 Truffles: Life Cycle and Distribution

Ectomycorrhizal symbiosis has evolved repeatedly over the last 130 million to 180 million years (LePage et al. 1997). In boreal and temperate forests, 95% of the short roots of plants form ectomycorrhizae (Martin et al. 2001), with 5,000–6,000 species of basidiomycetes or ascomycetes—including truffles (Buscot et al. 2000; Martin et al. 2001). Ectomycorrhizae positively impact plant growth in nature (Read 1991) owing to improved nutrient uptake and protection against pathogens (Borowicz 2001; Buscot et al. 2000).

Truffles are hypogeous ascomycete fungi belonging to the genus *Tuber*, the family *Tuberaceae* and the order *Pezizales* (O'Donnell et al. 1997; Trappe 1979). Their mycorrhizal status was established worldwide in the 1960s (Harley and Smith 1983; Trappe 1962). Truffles live in symbiosis with plant roots, generally forming ectomycorrhizas. In contrast to the high degree of promiscuity exhibited by arbuscular mycorrhizal (AM) fungi towards their hosts, ectomycorrhizal fungi are rather host-specific. Indeed, truffles tend to associate with angiosperms and gymnosperms, predominantly with oaks, hazels, some species of pines, but also some species of shrubs like *Cistus*. For a complete list of the host plants of European truffle species, refer to Ceruti et al. (2003). Recently truffle mycelium has also been identified within orchid roots—even though it does not form ectomycorrhizas (Selosse et al. 2004).

The present information about truffle's life cycle is very patchy. On the basis of observations both in nature and in the laboratory, as well as possible similarities with the life cycle of other ascomycetes fungi, Lanfranco et al. (1995) proposed a model for the life cycle of truffles which can be divided into three phases: (1) a reproductive phase (fruitbody), (2) a vegetative phase (free-living mycelium—saprotrophic phase actually only observed in the laboratory) and (3) a symbiotic phase (mycorrhizas) (Fig. 8.1). Indeed difficulties arise from the impossibility to follow the full life cycle in the laboratory. Even though Fassi and Fontana (1969) reported production in pots of fruitbodies of *Tuber maculatum* in association with *Pinus strobus* (Fig. 8.2), this achievement has not been repeated since then for any truffle species! Nevertheless more insight has recently been gained into the life cycle of truffles by Paolocci et al. (2006). In an elegant experiment the authors

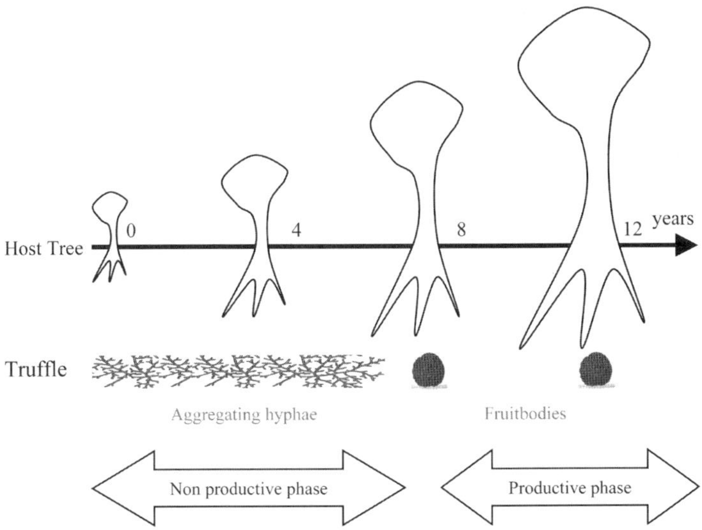

Fig. 8.1 A model of truffle's biological cycle. Associated with a young tree (nonproductive phase, 0 to approximately 7 years old), no fruitbodies are produced. During that period the fungi probably oscillates between the symbiotic and vegetative phases. Once the host has reached a certain maturity, the truffle can enter its reproductive phase (productive phase, starting approximately 8 years). The hyphae aggregate to eventually produce fruitbodies. During that period, the three phases (symbiotic, vegetative and reproductive) might succeed or coexist with each other

applied polymorphic microsatellites to compare the allelic configuration at different stages of *T. magnatum*'s life cycle (asci and surrounding mycelium in fruit bodies; ectomycorrhizal root tips). Their results suggest that *T. magnatum* outcrosses and that its life cycle is predominantly haploid. Nevertheless if outcrossing occurs, the proportion of ascocarps that do so is for the moment unknown as is how well these observations apply to other truffle species.

If fruitbodies are generally not obtainable in the laboratory, methods for obtaining mycorrhizas in 3–4 months are rather well established (Miozzi et al. 2005; Sisti et al. 1998; Zambonelli and Branzanti 1989). Success has been reported starting either with fresh fruitbodies or with mycelium grown in pure culture. The latter method (mycelium) has recently been adapted by Zeppa et al. (2004) in order to study the VOCs emitted by the mycelium/plant system before, during and after the formation of the ectomycorrhizas.

In vitro mycorrhization with *T. borchii, T. brumale* or *T. albidum* has been described for *Tilia platyphyllos, Cistus incanus, Alnus cordata, Castanea sativa, Populus alba* and *Corylus sativa* (Giomaro et al. 2002; Miozzi et al. 2005; Sisti et al. 1998; Zambonelli and Branzanti 1989, 1990).

More than 60 truffle species have been described so far (Trappe 1979), of which 20 are present in Europe (Gandeboeuf 1997). The fruitbodies and spores of two species of commercial interest are illustrated in Fig. 8.3. For a historical review and

Fig. 8.2 *Tuber maculatum* and *Pinus strobus*. Association between the host and the truffle with visible fruitbodies (*left*). Fruitbodies of *T. maculatum* (*right*). (Reproduced from Fassi and Fontana 1969)

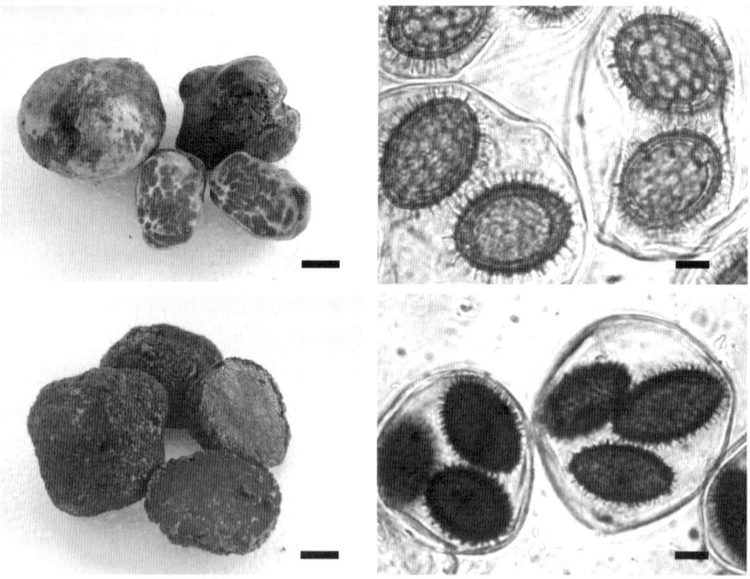

Fig. 8.3 Fruitbodies and spores of two truffle species. Fruiting bodies and spores of the white truffle *T. borchii* (top) and the black truffle *T. melanosporum* (bottom) (scale fruitbodies 0.50 cm, spores 10 μm)

identification of European truffle species, refer to Ceruti et al. (2003). The most famous and expensive ones—because of their intense and complex organoleptic properties—are *T. magnatum*, otherwise referred to as Alba's white truffle (and actually found in Italy and the Carpathian Basin), and the black truffle *T. melanosporum*, also referred to as the Périgord truffle and found mostly in Spain, France and Italy. Recent population-genetic studies suggest that *T. melanosporum* recolonized western Europe from southern Italy after the last glaciation period, and that the colonization pattern is closely related to the route followed by oak—its major host (Murat et al. 2004). However, the natural habitat of other truffle species is not limited to western Europe, but extends in the Northern Hemisphere and the Southern Hemisphere, spreading from northern Africa to Sweden towards the north (Weden et al. 2004) and the Carpathian Basin towards the east (Bratek et al. 1999). Lastly, truffles are also found in northern America (Amaranthus et al. 1999), Australia and New Zealand (where *T. melanosporum* has been recently introduced) and Asia (Yang Mei 1999).

Our knowledge of the truffle distribution is limited by the fact that fruitbodies are hypogeous and require trained dogs or pigs to locate them. Thus, the present distribution map reflects the zones where fruitbodies are collected. Mycorrhizas or nongerminated spores could tell another story, and reveal a much larger distribution than the one known today. This is exemplified by the lack of correlation between the presence of mycorrhizas and fruitbodies of *T. magnatum* observed in a truffle field in northern Italy (Murat et al. 2005).

8.3 Field Observations and Open Questions

In the Northern Hemisphere, most truffle species tend to form mature fruitbodies in the winter. In the case of *T. melanosporum*, small truffles 2 mm in diameter and reddish in color appear in June/July (Sourzat 1997). The fruitbody swells to reach its "mature" size in September/October. In the next 2 months, the peridium and gleba become darker owing to the melanization, indicating spore formation. Whether the fruitbody always remains connected with the mycorrhizas through mycelium is still unclear.

Interestingly Barry et al. (1994) suggested that the fruitbody of *T. melanosporum* and that of *T. aestivum* could absorb nutrients and take up water through de novo formed mycelial tufts at the surface of the peridium. If such a nutrition mechanism is generalized among truffles and whether it is sufficient to satisfy the full nutritional requirements of the fruitbody have nevertheless not been established yet.

It is believed that spores remain dormant—sometimes for many years—until a potential host plant gets in their vicinity. Whether spore germination in truffles involves some signaling from the host or from the fungi to the host is still not known. The mycelium from the germinated spore then comes into contact with the plant root and forms ectomycorrhizas within a few months, thus closing the life cycle (Fig. 8.1). It is not known what growth free mycelium can achieve in soil, and how long it can survive without a host. Nonetheless, mycellia of diverse truffle species grown in the laboratory on agar and supplemented with glucose or sucrose as a carbon source display an extremely slow growth (Ceccaroli et al. 2001; Iotti et al. 2002; Saltarelli et al. 1998), suggesting similar behavior in nature.

From the planting of a young mycorrhized tree in the wild, fruitbody formation is a rather long process, and seems to be related to the age, and also the species of the host plant. For example, a young oak tree mycorrhized with *T. melanosporum* will generally not induce any fruitbody formation before it has reached 7–15 years (Fig. 8.1). Some associations might actually never do so owing to the harsh competition of truffles and other microorganisms in the soil. However, the trigger for the fruitbody production is still totally obscure. It is not clear how the age of the tree or maybe its size influences fruitbody formation. In various plantations of hazels and oaks of the same age, all mycorrhized with *T. melanosporum*, fruitbodies have been observed 2–4 years earlier under hazels than under oaks (P. Sourzat, personal communication), suggesting that some change in metabolism due to aging in the host could somehow trigger fruitbody formation.

Having briefly described the life cycle of truffles, we shall now focus on the molecules that could act as signals in the complex interaction of truffle with its environment. Before discussing their potential involvement in the interaction with host and nonhost plants, let us have a brief look at what they are.

8.4 An Overview of Truffle Metabolites

8.4.1 VOCs from Fruitbodies

VOCs emitted from truffle fruitbodies have been widely studied, mainly though for species of commercial interest such as *T. melanosporum, T. magnatum, T. borchii, T. uncinatum* and *T. aestivum*. Most of those studies focused on aroma description (Claus et al. 1981; Flament et al. 1990; Ney and Freytag 1980; Splivallo et al. 2007a; Talou et al. 1987a, b, 1989a–d), influence of storage conditions on shelf life and aroma evolution (Bellesia et al. 1996, 2001, 2002; Falasconi et al. 2005; Pelusio et al. 1995) and only recently the possible ecological role of some VOCs of *T. borchii* has been discussed (Zeppa et al. 2004).

While certain VOCs such as 1-octen-3-ol, 2-methyl-1-butanol, 3-methyl-1-butanol and dimethyl sulfide are generally common to all truffle species, other VOCs only present in trace amounts might vary in intensity and structure depending on the truffle species. Aditionally, a strong variability in the VOC blend of truffles of the same species has been demonstrated (Mauriello et al. 2004; Splivallo et al. 2007a) and can be attributed to factors such as the fruitbody's maturity (Zeppa et al. 2004), its origin (Diaz.et al. 2003) and also to associated microorganisms which might feed on the fruitbody (Buzzini et al. 2005).

To date, more than 200 VOCs have been reported from various truffle species, and that number is likely to continue growing as VOC extraction techniques such

Table 8.1 Selected fruitbody volatile organic compounds (VOCs) of some truffle species Nonexhaustive list of VOCs reported in Splivallo et al. (2007a) for the following truffle species: Tuber melanosporum (MEL), T. borchii (BORC), T. indicum (IND), T. aestivum (AEST), T. magnatum (MAGN). ND not determined, – the VOCs have not been detected so far to the best of our knowledge

Molecule	Aroma description	MEL	BORC	IND	AEST	MAGN
Fatty acid derived VOCs						
1-Octen-3-ol	Fungal	x	x	x	x	–
3-Octanone	Fungal, sweet	x	x	x	x	–
3-Methyl-1-butanol	Whiskey	x	x	x	x	–
2-Methyl-1-butanal	Green, malty	x	x	x	x	x
3-Methyl-1-butanal	Sweet, malty	x	x	x	x	x
Hexanal	Cut grass	x	x	x	x	x
Terpenoids						
trans-Ocimene	Warm, herbaceous	–	x	–	–	–
Aromatic compounds						
2-Phenylethanol	Rose	x	x	x	x	–
1-Methoxy-3-methylbenzene	ND	x	x	–	x	–
Benzaldehyde	Almond	x	–	x	x	–
Anisole (methoxybenzene)	Anise	x	–	–	–	–
Sulfur compounds						
Dimethyl sulfide	Garlic	x	–	–	x	x
Dimethyl disulfide	Rubber	x	–	–	x	x
Dimethyl trisulfide	Fecal	x	–	–	x	x
2-Methyl-4,5-dihydrothiophene	ND	–	x	–	–	–

as solid-phase microextraction and related techniques become more and more sensitive (Diaz et al. 2003; Mauriello et al. 2004; Splivallo et al. 2007a), permitting detections limits at the parts per billion level. The VOCs identified so far are simple hydrocarbons that contain functional groups such as alcohols, aldehydes, esters, ketones, aromatic groups and sulfur compounds. Some frequently reported VOCs of truffles are listed in Table 8.1, with some structures being given in Fig. 8.4.

The aim of the following section is not to give an exhaustive list of truffle metabolites, but instead to focus on the most characteristic ones (sulfur compounds, fatty acid derived VOCs) and on the classes with a major ecological importance (terpenoids for signaling, or phenolics for phytotoxicity).

8.4.1.1 Fatty Acid Derived VOCs

Most of the linear chain hydrocarbons, alcohols, aldehydes and ketones are derived from fatty acid metabolism. Among those, 1-octen-3-ol and 3-octanone have been reported for most truffle species (Table 8.1). They are responsible for the strong

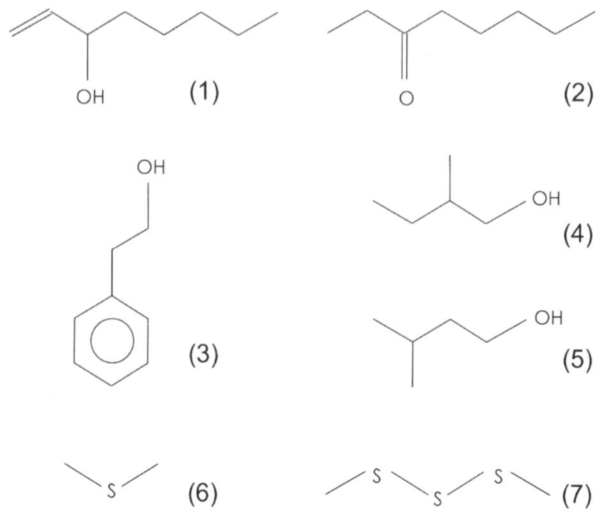

Fig. 8.4 Structure of some truffle volatile organic compounds (*VOCs*): *1* 1-octen-3-ol, *2* 3-octanone, *3* 2-phenylethanol, *4* 2-methyl-1-butanol, *5* 3-methyl-1-butanol, *6* dimethyl sulfide, *7* dimethyl trisulfide

fungal smell typical of *T. borchii*, but are also common to most other fungi (Abraham and Berger 1994; Chiron and Michelot 2005; Venkateshwarlu et al. 1999; Wnouk et al. 1983), some plants and have even been reported in fish (Ingvarsdóttir et al. 2002). 1-Octen-3-ol has recently been identified as a fungal hormone able to inhibit mycelial growth and trigger sporulation in *Penicillium paneum* (Chitarra et al. 2004). In *Pleurotus pulmonarius* mycelium grown in liquid culture, Assaf et al. (1997) confirmed that 1-octen-3-ol was directly derived from linoleic acid breakdown by a lipoxygenase. Refer to Combet et al. (2006) for a detailed review of the properties and biosynthesis of eight-carbon volatiles in fungi. 2-Methyl-1-butanol, 3-methyl-1-butanol and their respective aldehydes, all derived from fatty acid catabolism, are also well represented among truffle species (Table 8.1). They have been reported, along with dimethyl sulfide, as the major contributors to the final aroma of *T. melanosporum*. 2-Methyl-1-butanol and 3-methyl-1-butanol seem to be widespread among higher fungi (Abraham and Berger 1994: Chiron and Michelot 2005) and molds (Meruva et al. 2004), and might have phytotoxic properties (Pacioni 1991). Their production has also been reported for yeasts isolated from fruitbodies of *T. melanosporum* and *T. magnatum* (Buzzini et al. 2005), confirming the hypothesis that the VOC blends of truffle fruitbodies could be produced by more that one organism (in this case ectomycorrhizal fungi and yeasts).

Other linear-chain C6, C7, C8 and C9 aldehydes and alcohols seem also to be common among VOCs of different truffle species. For *Arabidopsis thaliana*, C6 aldehydes are known to activate defense genes and induce resistance against fungal pathogens such as *Botrytis cinerea* (Kishimoto et al. 2005). Whether the C6 compounds from truffle fruitbodies serve a similar self-defense role or might induce resistance in neighboring plants is not known.

8.4.1.2 Terpenoids

Terpenoids have only been identified recently in fruitbodies of *T. borchii* (Zeppa et al. 2004) and *T. brumale* (Mauriello et al. 2004). Unlike in many flowers or fruits (Aharoni et al. 2004), they represent a minor part of fruitbody VOCs in terms of concentration, but might be of major ecological importance. Zeppa et al. (2004) identified four monoterpenes and seven sesquiterpenes in *T. borchii*'s fruitbodies at different maturation stages, which could be involved in defense against microbes, interactions with insects and signaling with the host plant. One of these, aromadendrene, was only found in very immature fruitbodies of *T. borchii*, rendering that molecule a good marker of fruitbody maturity. Furthermore three major genes of the isoprenoid pathway, upregulated in mature fruitbodies, have recently been cloned and characterized in *T. borchii* (Guidi et al. 2006).

8.4.1.3 Aromatic Compounds

VOCs containing aromatic rings have been reported in all *Tuber* species; however, none seem to be common to all of them—maybe due to different VOC extraction techniques used in the different studies. They might somehow contribute to the so-called burnt area (area with scarce herbaceous cover) observed with some truffle species (Sect. 8.6) as simple phenolics are known for their phytotoxicity (Gallet and Pellissier 1997).

8.4.1.4 Sulfur-Containing Compounds

Sulfur-containing compounds seem to be characteristic of most truffle species (thiols, thioesters, sulfides, thioalcohols and thiophenones), but are generally present in trace amounts in fresh fruitbodies. One sulfur-containing compound, dimethyl trisulfide, has also been identified in pure mycelial cultures of *T. borchii* (Tirillini et al. 2000), while different yeast strains isolated from truffle fruitbodies also have the capacity to produce them (Buzzini et al. 2005). Most sulfur-containing compounds have very low olfactory detection limits and are thus major contributors to the final aroma of truffle fruitbodies. They derive from the catabolism of L-methionine, their major precursor (Berger et al. 1999; Spinnler et al. 2001). In *Geotrichum candidum* L-methionine is first converted to 4-methylthio-2-oxobutyric acid, which is then transformed into methanethiol, a key precursor of most sulfur VOCs (Arfi et al. 2003; Bonnarme et al. 2001a, b; Spinnler et al. 2001).

From an ecological point of view, sulfur-containing compounds might act as fumigants against microbes in decomposing roots of cabbage (Bending and Lincoln

1999) and as repellents against amphipods in marine algae (Schnitzler et al. 1998). In fruitbodies of *T. magnatum*, the concentration of sulfur-containing compounds has been reported to decrease within 2 weeks of storage at room temperature (Bellesia et al. 1996), while an increase was observed in *T. borchii* upon aging (Bellesia et al. 1996, 2001).

8.4.2 Fruitbody Non-VOCs

Nonvolatile metabolites from fruitbodies have been investigated for *T. aestivum* (Mannina et al. 2004) and *T. indicum* (Jin-Ming 2004). The authors identified sugars, polyols, amino acids, organic acids, fatty acids, sterols and lipids, among which were two sphingolipids, highly bioactive molecules known to be involved in regulation of cell growth, differentiation and apoptosis. De Angelis et al. (1996) also identified quinonoid and polyphenolic biopolymers as the major constituents of *T. melanosporum*'s melanin, and suggested a polyketidic origin.

8.4.3 Mycelial VOCs

VOCs produced by only one species (*T. borchii*) have been investigated so far. When grown either on potato dextrose agar or in liquid cultures, *T. borchii* mycelium (strain ATCC 96540) produced eight VOCs, including aromatic compounds, alcohols and a ketone, most of which have also been described from the fruitbodies of various truffle species (Splivallo et al. 2007a). Nevertheless cultural conditions strongly influence the production of volatile compounds, as exemplified by Tirillini et al. (2000), who identified 29 VOCs from submerged cultures of *T. borchii* mycelium (modified Melin-Norkans medium). Under those conditions, most of the VOCs had not been reported in truffle fruitbodies, with the exception of butan-2-one and dimethyl trisulfide described, respectively, in *T. melanosporum* (Bellesia et al. 1998a) and *T. magnatum* (Bellesia et al. 1998b). Mycelium of other truffle species has not been investigated so far, mainly owing to their poor growth.

Having considered the major group of metabolites reported in the literature, we should say a word of caution regarding their ecological significance. On one hand, they were generally identified under rather unnatural conditions (sterile system, or in the case of the fruitbody's VOCs with the fruitbodies generally washed free of soil, thus certainly under a high-stress condition). The occurrence of these metabolites should be checked in situ to understand their possible ecological role. Besides, it is likely that we see only the tip of the iceberg as far as truffle metabolite diversity is concerned. Indeed VOCs at an early stage of fruitbodies (when they are only a few millimeters in diameter) have never been investigated. Neither have VOCs been reported for advanced stages of decomposition (overmaturity). Ecologically sound studies should thus focus on these aspects in situ, and unfortunately require the assistance of a very collaborative truffle hunter!

8.5 Metabolites Involved in Truffle and Host Plant Interaction

Signaling in the rhizosphere between plants and microorganisms is regulated by molecules which permit host–symbiont recognition and induce morphological changes in each partner. Such early signaling events have been extensively studied for *Rhizobium* and legumes, and led to the understanding that some flavonoids secreted by the plant trigger the production of the nodulation (nod) factor in the bacteria, which in turn induces morphological changes in the rooting system of the plant (Dénarié and Cullimore 1993; Heidstra and Bisseling 1996). A similar molecular dialogue has been observed in the case of the early interaction between AM fungi and their host. Indeed Akiyama et al. (2005) have recently identified a sesquiterpene lactone (5-deoxystrigol) from *Lotus japonicus* root exudates inducing branching in *Gigaspora margarita* hyphae. For a recent review on signaling between AM fungi and plants, refer to Harrison (2005).

Much less is known on the recognition events between ectomycorrhizal fungi and plants than on the AM fungi–plant interaction (reviewed in Martin et al. 2001). In the case of truffles, VOCs produced during ectomycorrhizas formation of *T. borchii* with *Tilia americana* have been recently investigated (Gioacchini et al. 2002; Menotta et al. 2004a). Twenty-nine VOCs specific to the premycorrhizal stage (where the host and the symbiont are separated by a few centimeters) have been identified as hydrocarbons, alcohols, ketones, a brominated cholesterol derivative and terpenoids, including the sesquiterpene germacrene D, as well as dehydroaromadendrene, ß-cubebene and longicyclene—which might be involved in chemotropism of hyphae towards the roots of the host (Menotta et al. 2004a). Molecular changes in mycelium during early interaction between *T. borchii* and *Tilia americana* were also investigated by Menotta et al. (2004b). Suppressive subtractive hybridization and reverse northern blots allowed the identification of differentially expressed genes in the mycelium and involved in cellular detoxification, secretion and apical growth, or general metabolism.

Nutrient availability may act as a further regulatory signal: for example, a phospholipase A, strongly upregulated by nitrogen starvation in *T. borchii* mycelium (Soragni et al. 2001), was shown to be expressed mostly during the early steps of the fungus–plant interactions (Miozzi et al. 2005). This provides confirmation of the hypothesis that mycorrhization is a response to nutrient stress. However, the road to the identification of molecular messengers involved in interaction between truffles and hosts is still long. As ectomycorrhizal fungi present much higher host specificity than AM fungi, it seems reasonable to argue that the structures of the signal molecules might be characteristic of each specific association of ectomycorrhiza and plant. The metabolites released by truffles in the rhizosphere might not only be involved in host recognition, but might also serve other functions such as defense or competition with other organisms, as exemplified in the following section.

8.6 Interaction with Nonhost Plants

The burnt (or *brûlé* in French) is the only phenomenon where the presence of truffles mycorrhizas/mycelium is obvious. It is a zone around or near the host tree where vegetal cover is scarce (Fig. 8.5). The phenomenon, generally observed with trees mycorrhized with *T. melanosporum, T. aestivum* and *T. indicum*, is not seen with *T. magnatum*. Its occurrence with other truffle species is more controversial.

During the life cycle of the fungus, the burnt becomes apparent when the host tree is 5–10 years old, and its appearance precedes the formation of the first fungal fruitbodies by a few years. The burnt tends to form as a more or less circular zone (of a few meters in diameter) around the trunk of the host tree, and moves over years, spreading in diameter and/or moving away from the tree (sometimes as far as 15–20 m). Additionally, most herbaceous plants inside the burnt are smaller than their counterparts outside it, but some plants such as *Festuca ovina* (Mamoun and Olivier 1997) and other Graminaceae (such as *Bromus inermis, B. erectus*) seem to be less affected (Montacchini and Caramiello Lomagno 1977; Sourzat 1997).The burnt phenomenon has been known for a long time (Cicarello 1564); however, its causes still remain unclear. Explanations for the formation of the burnt have been proposed by Delmas (1983), who hypothesized that truffle mycorrhizas may compete for nutrient or water, by Plattner and Hall (1995), who suggested that *T. melanosporum* hyphae could penetrate the roots of the herbaceous plants, maybe acting as parasites, and in a series of other publications highlighting the phytotoxic effect of truffle fruitbody's metabolites (Fasolo-Bonfante et al. 1971; Lanza et al. 2004; Pacioni 1991; Papa 1980; Splivallo et al. 2007b).

In nature, the distinction between mycorrhizal and saprobic behavior is not always an easy one as the organisms involved might switch between one and the other depending on changing biotic and abiotic factors (Fitter 1991; Hibbett et al. 2000). Tibbett and Sanders (2002) reported a case of necrotrophy for an ectomycorrhizal fungus, demonstrating that *Hebeloma syrjense* P. Krast, colonizing willow roots, was able with its extraradical mycelium to find nutrient patches within the soil (dead seeds, fruits, pollen), and absorb them after digestion with exoenzymes. In the case of *T. melanosporum*, such necrotrophic and parasitic behavior (as mentioned in the preceding section) has not been clearly demonstrated. Could such a dualistic behavior—mutualistic symbiont with the host and endophytic with nonhosts—be driven by mycoeterotrophic behavior of some plants interconnected through the mycelial network of truffle? This behavior has been observed in the case of achlorophyllous orchids able to get their nutrients through the mycelial network interconnecting them with photosynthetic plants (Bidartondo et al. 2004; Girlanda et al. 2006). Could the high energy requirement necessary to produce

Fig. 8.5 The burnt. *Top*: An 8-year-old hazel mycorrhized with *T. uncinatum*—winter period (Murisengo, northern Italy). *Bottom*: A 10-year-old hazel mycorrhized with *T. melansporum*—summer period (Cravanza, northern Italy). The burnt is clearly visible in both pictures as a circular zone with scarce vegetal cover surrounding the host tree

fruitbodies induce saprobic behavior of the truffle on the herbaceous plants? In both cases why is the phenomenon (burnt) not observed with all truffle species? This question remains unanswered for the moment. Consequently further evidence must be obtained to support this dualistic behavioral theory. Its final contribution to the burnt should be quantified with other possible factors such as the competition for nutrients among the nonhost plants and truffle or the production of phytotoxic metabolites by the truffles, so far only observed in laboratory experiments.

8.6.1 Phytotoxic Metabolites in Soil from Truffle Fields?

The presence of toxic substances in "burnt" soil is supported by the retarded germination observed by Papa (1978–1979) when treating *Lepidium sativum* with aqueous extracts of burnt and non burnt soils from a *T. melanosporum* truffle field. The author however noted that if differences were obvious with 2-day-old seedlings, they had almost completely disappeared on the third day, implying either a degradation of metabolites or low starting concentrations. More recently, Lanza et al. (2004) reported reduction in primary root length of *Vicia faba* planted in soil collected from a burnt zone produced by *T. aestivum*. The authors also tested long-term toxicity (genotoxicity) using a *Vicia faba* root micronucleus test, and reported a significant increase in the number of micronuclated cells for burnt soil compared with the control. We similarly observed a reduction in root length (approximately 14%) for two consecutive years with cucumber planted in burnt soil compared with non burnt soil from a truffle field of hazels mycorrhized with *T. melanosporum* (Fig. 8.6). In contrast, no differences between burnt and non burnt soil were observed for cucumber germination (R. Splivallo, unpublished data).

One should keep in mind that neither cucumber, nor *Vicia faba* nor *Lepidium sativum* is generally found in truffle fields. Consequently, if the reduction in root length can be considered a good indicator for the presence of some inhibiting/stimulating metabolites inside/outside the burnt, herbaceous species associated with

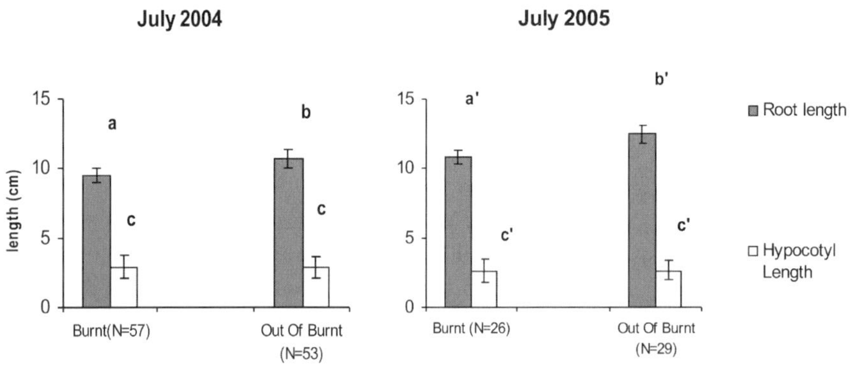

Fig. 8.6 Average root and hypocotyl length of cucumber germinated in burnt and non burnt soil. Soils samples were collected from a truffle field in Cravanzana, northern Italy. In 2004, three soil samples were taken from two different burnt zones and three soil samples were taken from just outside (approximately 2 m) the burnt zones. In 2005, eight soil samples were taken from eight different burnt zones and eight samples were taken from outside the burnt zones. For each year, the soil samples were sieved (5 mm), then pooled in two categories: burnt and out of burnt. Cucumber seeds (10 per pot) were germinated in 250 g soil wetted with 80 ml H_2O for 8 days, after which root and hypocotyl length were recorded. A significant reduction in root length was observed each year for seedlings grown in burnt compared with out of burnt soil. Hypocotyl length was not affected. The results are presented with the standard deviation (*bar*). Different letters indicate significant differences $P < 0.05$ (Kruskal–Wallis) test. *N* total number of seedlings

truffle fields should be used for an ecologically sound argument, especially as plant responses to secondary metabolites are species- and dose-dependent. Lastly, the soil collected from the field might contain very low concentrations of active secondary metabolites owing to bacterial degradation and/or owing to the physical separation from or disruption of the producing organisms (possibly mycorrhizas, mycelium) at the time of the collection. Therefore, laboratory assays might underestimate the real or long-time effect.

Neither the metabolites responsible for the effects described above nor their source has been identified yet. Truffles might potentially produce them, at one or different stages of their life cycle; there are two grounds for this. First, truffles are clearly associated with the burnt, and thus appear as obvious candidates. Second, production of phytotoxic substances by truffles has been documented in laboratory experiments.

8.6.2 Phytotoxic Metabolites from the Fruitbody

Some authors focused on the phytotoxicity of *T. melanosporum* and *T. aestivum* fruitbodies to try to explain the burnt (Fasolo-Bonfante et al. 1971; Lanza et al. 2004; Pacioni 1991; Papa 1980). Aqueous extracts of *T. melanosporum* have been tested by Montacchini and Caramiello Lomango (1977) on a series of seeds collected from truffle fields and including Graminacea, Caryophyllaceae, Lamiaceae, Scrophulariaceae, Plantaginaceae and Asteraceae. Germination bioassays with different extract concentrations always led to a reduced number of germinated seeds and reduction of root length for germinated seeds compared with the control. Similar results were obtained also with *T. melanosporum* extracts by Fasolo-Bonfante et al. (1971) and Papa (1978–1979), and for *T. aestivum* by Lanza et al. (2004), who furthermore highlighted the genotoxicity of the fruitbody using the the *Vicia faba* micronucleus test. The metabolites responsible for the phytotoxicity in the above-mentioned experiments are not fully known. On one hand, Papa (1980) reported isolating a strongly phytotoxic brown substance form *T. melanosporum* fruitbodies, however without characterizing its molecular structure. On the other hand, Pacioni (1991) tested the effect of ten VOCs characteristic of *T. melanosporum*, and reported a significant root shortening of wheat induced by 2-methylbutanol, 3-methylbutanol and 3-methylbutanal already at a concentration of 7.5 ppm of each single VOC. Similarly, lentil roots were significantly reduced at 5.0 ppm for 3-methylbutanol, 7.5 ppm for 3-methylbutanal and 10.0 ppm for 2-methylbutanol. The other seven VOCs tested, namely, dimethyl sulfide, 2-butanone, 2-butanol, 2-methylpropanol, 2-methylpropanal, 2-methylbutanal and methylanisole did not show any significant effects on wheat or lentils at concentrations of 10 and 25 ppm. We similarly tested the effect of three truffle species on cucumber. We chose *T. uncinatum* and *T. indicum* for the burnt associated with those species, while *T. borchii* was used as a negative control because it was thought it did not produce any burnt. In a first set of experiments, the fruitbodies were cut into small pieces and incorporated into the sand where cucumber was germinated, thus allowing slow diffusion of VOCs and exudates into the sand. Root length reduction was the strongest with *T. borchii*, followed by *T. uncinatum*, while no significant difference was observed for *T. indicum* (Fig. 8.7). The trend was reversed for hypocotyl length (stimulation instead of reduction; Fig. 8.7). In order to control the effect of the fruitbody's VOCs on cucumber (and not the exudates), a second set of experiments was carried out placing the fruitbodies in a small open plastic container at the sand surface, thus only allowing free diffusion of VOCs. Exactly the same trend as in the first set of experiments was observed (with the fruitbody in the sand) (Fig. 8.7), suggesting that the VOC blends released by *T. borchii* and *T. uncinatum* are responsible for the observed root shortening and hypocotyl elongation.

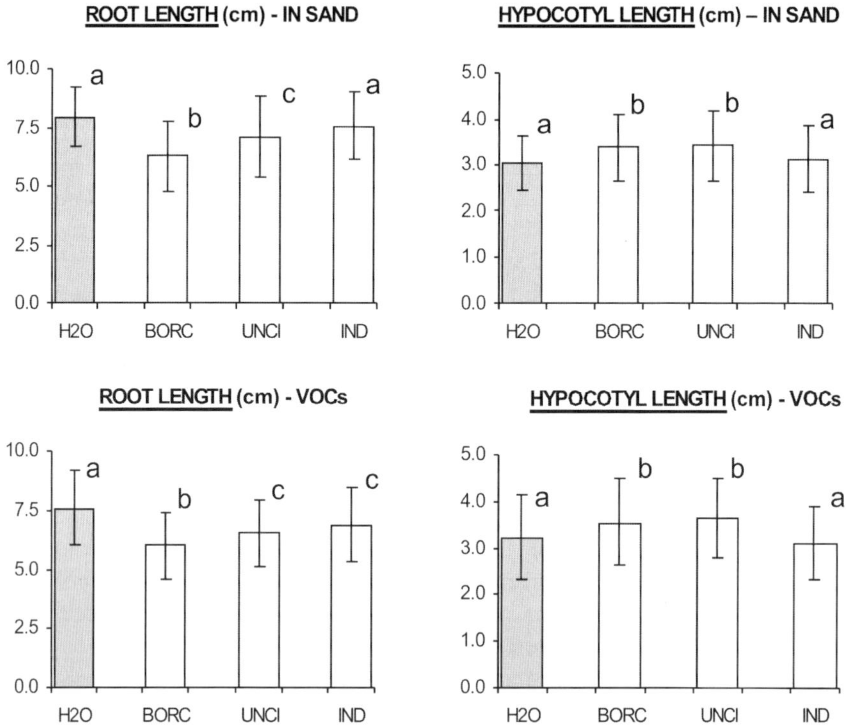

Fig. 8.7 Effect of truffle metabolites on cucumber growth. Cucumber was germinated for 8 days (ten seeds per pot with 350 g sand and 80 ml H_2O, and 1.0 g was taken from five fruitbodies of *T. borchii* (*BORC*), seven fruitbodies of *T. uncinatum* (*UNCI*) and five fruitbodies of *T.indicum* (*IND*) or without a fruitbody as the control (*H2O*). The fruitbodies (frozen at –80 °C for long-term conservation) were either chopped into small pieces and mixed into the sand or placed in a small plastic container on the surface of the sand to allow solely diffusion of VOCs. Pots were sealed to prevent loss of VOCs. All bioassays were repeated eight times, so the total number of seedlings per treatment was always more than 75. In both sets of experiments (in sand and in pots) root length was significantly reduced for all truffle species (with the exception of *IND* in sand), while hypocotyl length significantly increased in the cases of *BORC* and *UNCI*. The results are presented with the standard deviations (*bar*). Different letters represent significantly different results $P < 0.05$ (Kruskal–Wallis test)

As far as the burnt is concerned, the results described above suggest that the fruitbody is not the major cause of the burnt, as on one hand, the strongest reduction in root length was observed with *T. borchii*, which probably does not produce a burnt and, on the other hand, no reduction in root length was observed with *T. indicum*, which produces one. This is further supported by the fact that in laboratory bioassays fruitbody volatiles from various truffle species inhibited the development of both host (*Cistus incanus*) and nonhost (*Arabidopsis thaliana*) plants, suggesting that truffle fruitbody volatiles might not be involved in premycorrhizal signaling, but simply serve as defense molecules against plants (Splivallo et al. 2007b).

Further evidence from the truffle life cycle supports the fact that the fruitbody is not the initiating agent of the burnt. Indeed the burnt appears a few months to a few years before the first fruitbodies. Furthermore fruitbodies have sometimes been found well outside the burnt. Nevertheless, fruitbody metabolites might somehow enhance the phytotoxicity of the burnt, and it cannot be excluded that in truffle grounds metabolites produced by the fruitbodies are also synthesized by the mycelium and/or the mycorrhizas. Indeed, this has recently been demonstrated for 1-octen-3-ol, a volatile that strongly inhibited the development of *Arabidopsis thaliana* in laboratory bioassays (Splivallo et al. 2007b), and that has been reported from truffle fruitbodies, mycelial pure cultures (Splivallo et al. 2007a) and the symbiotic association *T. borchii*/*Tilia americana* (Menotta et al. 2004b).

8.6.3 Phytotoxic Metabolites from the Mycelium

Germination inhibition of *Sinapis alba* treated with culture broth of *T. melanosporum* mycelium was reported by Fasolo-Bonfante et al. (1971); however in the 1970s, species identification was only based on morphological observations. The experiment awaits a fresh confirmation using mycelium whose identity was confirmed by currently available molecular techniques (Douet et al. 2004).

Truffle is nonetheless not the only organism capable of producing phytotoxic metabolites. Other microorganisms, including fungi and bacteria, are indeed associated with truffle fields as we shall see now.

8.6.4 Possible Contribution to the Burnt by Other Organisms

Many plants and fungi are known to produce phytotoxic substances. In a large screening experiment for isolating new bioactive metabolites, Schulz et al. (2002) reported that 18% of endophytes fungi isolated from various soils had an antialgal/herbicidal effect, or in other words were able to produce some phytotoxic secondary metabolites in vitro. Phytotoxic substances also seem to be a common competitive weapon used by invasive plants (Barney et al. 2005).

Regarding truffles, other microorganisms are characteristic of the burnt zone, and might contribute to its toxicity. Luppi Mosca and Fontana (1977), using plate isolation techniques to quantify and identify the saprotrophyc mycoflora in and outside the burnt (*T. melanosporum*), concluded that not only was the burnt area much "richer" in saprotrophyc flora, but that some species such as *Penicilium diversum, Penicilium restrictum* and *Acremonium breve* were strongly stimulated inside the burnt. Similar plate isolation techniques have been applied to yeast populations in truffle fields of *T. aestivum* by Zacchi et al. (2003), who identified one strain of *Cryptococcus albidus* specific to the truffle field. Bacterial population living inside the fruitbody could also produce phytotoxic metabolites. Barbieri et al. (2005) identified many bacteria living inside the fruitbody of *T. borchii*. Even though the burnt is not associated with that truffle species; it is possible that other bacterial strains associated with burnt-"producing" truffles emit some phytotoxic substances.

Lastly, AM fungi have also been known to inhibit herbaceous plant growth, especially in the interaction with a nonhost plant, probably mediated by some chemicals (Francis and Read 1995). Unfortunately, the presence of AM fungi in truffle fields (and differences in community composition inside and outside the burnt) has not been studied yet. As a consequence, whether the truffle is responsible alone or with some other organisms for the inhibition of herbaceous plants inside the burnt remains to be determined (Fig. 8.8).

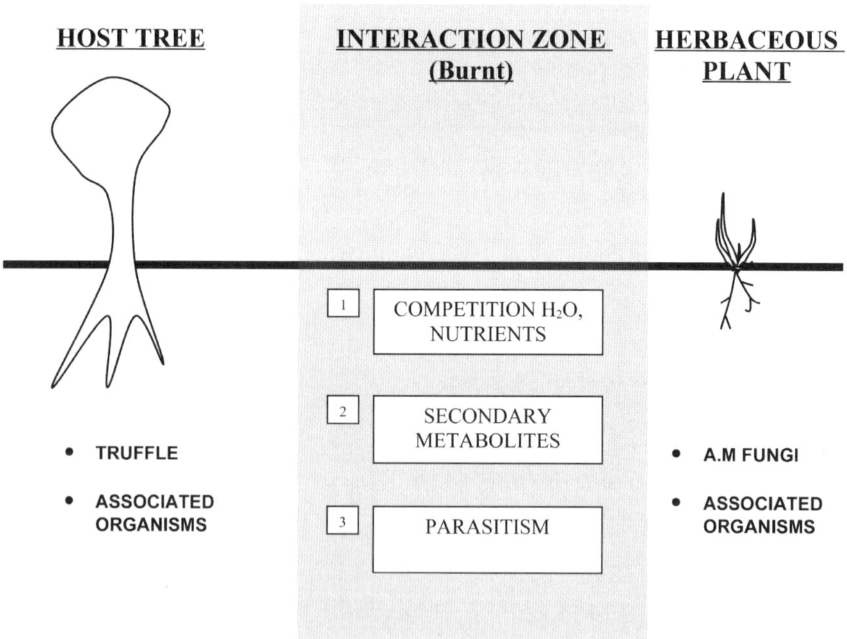

Fig. 8.8 The interactions inside the burnt. Factors possibly involved in the formation of the burnt are as follows: *1* competition for nutrients and water; *2* interactions involving phytotoxic secondary metabolites; *3* parasitism from the truffle on the nonhost herbaceous plants. *A.M* arbuscular mycorrhizal

8.6.5 Ecological Significance of Truffle Metabolites

The results obtained so far in laboratory experiments confirm that truffle fruitbodies can indeed produce some phytotoxic substances (Fig. 8.7) that might also be present in soil (Fig. 8.6). Nevertheless, those results do not reproduce the strong phytotoxic effect observed in the field (Fig. 8.5). The reasons for this can be various. First, most test plants used for the bioassays are not generally found in truffle fields, and can have different responses to phytotoxic metabolites from those of the plants typically found in truffle fields (in this case they could be less sensitive). Second, only rather short-term effects have been tested so far in laboratory experiments (days to weeks), while phytotoxicity could be induced in nature on a much longer time scale. Last but not least, the metabolites (volatiles or exudates) might be present in nature at concentrations different from those in the laboratory or might act in synergy with other unknown metabolites not present in laboratory experiments.

At this stage the ecological significance of truffle metabolites and specifically their contribution to the burnt are not known, as all the data obtained so far are once again not from field experiments but are rather from laboratory experiments. Indeed, production of secondary metabolites from fungi is known to be drastically influenced by biotic and abiotic factors (Bode et al. 2002) and consequently the metabolite production pattern by mycelium or fruitbody might drastically vary between the laboratory conditions and the field. For these reasons, investigation of the ecological roles of secondary metabolites should be possibly done *in vivo*, or in the case of the metabolites identified in laboratory experiments, their occurrence and biological role (in synergy with other metabolites present in the field) should be investigated in nature.

8.7 Conclusions

Our knowledge of the interaction of truffles with host plants is still at an early stage. Early signals between host and truffle have only been investigated recently. The genes involved in such interactions are also under investigation. The molecular and/or environmental signals triggering fruitbody formation are still completely obscure owing to the long and complex life cycle of truffles.

If truffles are a good model in which to study the interactions between ectomycorrhizal ascomycetes fungi and their host, they also offer an interesting perspective to understand the interactions with nonhost plants. Indeed bioassays with nonhost plants evidenced that truffle metabolites interact with root elongation. Experiments with soil collected from truffle fields also suggest the presence of secondary metabolites interacting with plants roots. Nevertheless, the occurrence of these metabolites and their origin in nature is not yet known. Finally, as the burnt is not reproducible in the laboratory, further investigation should be done in vivo in order to quantify the contribution of secondary metabolites to the scarcity of the vegetal cover observed inside it, and to shed a little more light on these delicious, yet mysterious fungi!

Acknowledgements The author is grateful to P. Karlovsky (University of Göttingen, Germany) for his suggestion to write this chapter and his support, to Caroline Gutjahr (University of Turin, Italy) for stimulating discussions, to J.H. Craddock and the Azienda Nasio (Communità Montana Alta Langa) for their assistance and permission to do the field work in Cravanzana, and to P. Bonfante (University of Turin, Italy), for her comments on and corrections to the manuscript. Experimental work was supported by grants to P. Bonfante (Biodiversity Project (CNR), Cebiovem, the Italian Ministry of Education, University and Research). R.S. was supported by a fellowship from the Fondation Pour des Bourses d'Études Italo-Suisses.

References

Abraham BG, Berger RG (1994) Higher fungi for generating aroma components through novel biotechnologies. J Agric Food Chem 42:2344–2348

Aharoni A, Giri AP,Verstappen FWA, Bertea CM, Sevenier R, Zhongkui S, Jongsma MA, Schwab W, Bouwmeestera HJ (2004) Gain and loss of fruit flavor compounds produced by wild and cultivated strawberry species. Plant Cell 16:3110–3131

Akiyama K, Matsuzaki K, Hayashi H (2005) Plant sesquiterpenes induce hyphal branching in arbuscular mycorrhizal fungi. Nature 435:824–827

Amaranthus M, Luoma D, Eberhart JL, Trappe JM (1999). Truffle dominance and diversity in natural vegetative communities. In: Actes du Vème congrès international, science et culture de la truffe. Fédération Française des Trufficulteurs, Aix-en-Provence, pp 4.183–4.187

Arfi K, Tâche R, Spinnler HE, Bonnarme P (2003) Dual influence of the carbon source and L-methionine on the synthesis of sulfur compounds in the cheese-ripening yeast *Geotrichum candidum*. Appl Microbiol Biotechnol 61:359–365

Assaf S, Hadar Y, Dosoretz CG (1997) 1-Octen-3-ol and 13-hydroperoxylinoleate are products of distinct pathways in the oxidative breakdown of linoleic acid by *Pleurotus pulmonarius*. Enzyme Microb Technol 21:484–490

Barbieri E, Bertini L, Rossi I, Ceccaroli P, Saltarelli R, Guidi C, Zambonelli A, Stocchi V (2005) New evidence for bacterial diversity in the ascoma of the ectomycorrhizal fungus *Tuber borchii* Vittad. FEMS Microbiol Lett 247:23–35

Barney JN, Hay AG, Weston LA (2005) Isolation and characterization of allelopathic volatiles from mugwort (*Artemisia vulgaris*). J Chem Ecol 31(2):247–265

Barry D, Staunton S, Callot G (1994) Mode of absorption of water and nutrients by ascocarps of *Tuber melanosporum* and *Tuber aestivum*: a radioactive tracer technique. Can J Bot 72:317–322

Bellesia F, Pinetti A, Bianchi A, Tirillini B (1996) I composti solforati dell'aroma del tartufo: loro evoluzione durante la conzervatione. Atti Soc Nat Mat Modena 127:177–187

Bellesia F, Pinetti A, Bianchi A, Tirillini B (1998a) The volatile organic compounds of black truffle (*Tuber melanosporum* Vitt.) from middle Italy. Flavour Fragr J 13:56–58.

Bellesia F, Pinetti A, Bianchi A, Tirillini B (1998b) Volatile compounds of white truffle (*Tuber magnatum* Pico.) from middle Italy. Flavour Fragr J 11:239–243

Bellesia F, Pinetti A, Tirillini B, Bianchi A (2001) Temperature-dependant evolution of volatiles organic compounds in *Tuber borchii* from Italy. Flavour Fragr J 16:1–6

Bellesia F, Pinetti A, Tirillini B, Paolocci F, Rubina A. Arcioni S, Bianchi A (2002) The headspace volatiles of the Asian truffle *Tuber indicum* Cooke et Mass. J Essent Oil Res 14:3–5

Bending GD, Lincoln SD (1999) Characterization of volatile sulphur-containing compounds produced during decomposition of *Brassica juncea* tissues in soil. Soil Biol Biochem 31:695–703

Berger C, Khan JA, Molimard P, Martin N, Spinnler HE (1999) Production of sulfur flavors by ten strains of *Geotrichum candidum*. Appl Environ Microbiol 65:5510–5514

Bidartondo MI, Burghardt B, Gebauer G, Bruns TD, Read DJ (2004) Changing partners in the dark: isotopic and molecular evidence of ectomycorrhizal liaisons between forest orchids and trees. Proc R Soc Lond Ser B 271:1799–1806

Bode HB, Bethe B, Hofs R, Zeeck A (2002) Big effects from small changes: possible ways to explore nature's chemical diversity. Chembiochem 3:619–627

Bonnarme P, Arfi K, Dury C, Helinck S, Yvon M, Spinnler HE (2001a) Sulfur compounds production by *Geotrichum candidum* from L-methionine: importance of the transaminase step. FEMS Microbiol Lett 205:247–252

Bonnarme P, Lapadatescu C, Yvon M, Spinnler HE (2001b) L-Methionine: degradation potentiality of cheese-ripening microorganisms. J Dairy Res 68:663–674

Borowicz VA (2001) Do arbuscular mycorrhizal fungi alter plant-pathogen relations? Ecology 82(11):3057–3068

Bratek Z, Albert L, Gagi I, Pálfy B, Takács T, Rudnóy S, Alász K (1999) New and rare hypogenus fungi of Carpathian Basin. In: Actes du Vème congrès international, science et culture de la truffe. Fédération Française des Trufficulteurs, Aix-en-Provence, pp 2.55–2.56

Brundrett MC (2002) Coevolution of roots and mycorrhizas of land plants. New Phytol 154(2):275–304

Buscot F, Munch JC, Charcosset JY, Gardes M, Nehls U, Hampp R (2000) Recent advances in exploring physiology and biodiversity of ectomycorrhizas highlight the functioning of these symbioses in the ecosystem. FEMS Microbiol Rev 24:601–614

Buzzini P, Gasparetti C, Turchetti B, Cramarossa MR, Vaughan-Martini A, Martini A, Pagnoni UM, Forti L (2005) Production of volatile organic compounds (VOCs) by yeasts isolated from the ascocarps of black (*Tuber melanosporum* Vitt.) and white (*Tuber magnatum* Pico) truffles. Arch Microbiol 184:187–193

Ceccaroli P, Saltarelli R, Cesari P, Zambonelli A, Stocchi V (2001) Effects of different carbohydrates sources on the growth of *Tuber borchii* Vitta. Mycelium strains in pure culture. Mol Cell Biochem 218:65–70

Ceruti A, Fontana A, Nosenzo C (2003) Le specie europee del genere *Tuber*. Una revisione storica. Monografie XXXVII. Museo Regionale di Scienze Naturali, Turin

Chiron N, Michelot D (2005) Mushrooms odors, chemistry and role in the biotic interactions–a review (in French). Cryptogam Mycol 26(4):299–364

Chitarra GS, Abee T, Rombouts FM, Posthumus MA, Dijksterhuis J (2004) Germination of *Penicillium paneum* conidia is regulated by 1-octen-3-ol, a volatile self-inhibitor. Appl Environ Microbiol 70:2823–2829

Ciccarello A (1564) Opusculum de tuberibus. Padua

Claus R, Hoppe HO, Karg H (1981) The secret of truffle: a steroidal pheromone? Experientia 37:1178–1179

Combet E, Henderson J, Eastwood DC, Burton KS (2006) Eight-carbon volatiles in mushrooms and fungi: properties, analysis, and biosynthesis. Mycoscience 47:317–326

De Angelis F, Arcadi A, Marinelli F, Paci M, Botti D, Pacioni G, Miranda M (1996) Partial structures of truffle melanins. Phytochemistry 43(5):1103–1106

Delmas J (1983) La truffe et sa culture, 2nd edn. INRA, Paris

Dénarié J, Cullimore J (1993) Lipo-oligosaccharide nudulation factors: a new class of signalling molecules mediating recognition and morphogenesis. Cell 74:951–954

Diaz P, Ibanez E, Senorans FJ, Reglero G (2003) Truffle aroma characterization by headspace solid-phase microextarction. J Chromatogr A 1017:207–214

Dick CM, Hutchinson SA (1966) Biological activity of volatile fungal metabolites. Nature 211:868

Douet JP, Castroviejo M, Mabru D, Chevalier G, Dupre C, Bergougnoux F, Ricard JM, Medina B (2004) Rapid molecular typing of *Tuber melanosporum*, *T. brumale* and *T. indicum* for tree seedlings and canned truffles. Anal Bioanal Chem 379(4):668–673

Falasconi M, Pardo M, Sberveglieri G, Battistutta F, Piloni M, Zironi R (2005) Study of white truffle aging with SPME-GC-MS and the Pico2-electronic nose. Sens Actuators B 106:88–94

Fasolo-Bonfante P, Fontana A, Montacchini F (1971) Studi sull'ecologia del *Tuber melanosporum*. Domonstrazione di un effetto fitotossico. Allionia 17:47–54

Fassi B, Fontana A (1969) Sintesi micorrhizica tra *Pinus strobus* e *Tuber maculatum*. Allionia 15:115–120

Fauconnet C, Delher G (1998) Influence des facteurs climatiques sur la production des truffes en Quercy. Trufficult Fr 24(3):19–21

Fitter AH (1991) Costs and benefits of mycorrhizas—implications for functioning under natural conditions. Experientia 47:350–354

Flament I, Chevalier G, Debonneville C (1990) Analysis of the volatile flavor constituents of Périgord black truffle (Tuber melanosporum Vitt.). Riv Ital EPPOS 9:280–299

Francis R, Read DJ (1995) Mutualism and antagonism in the mycorrhizal symbiosis, with special reference to impacts on plant community structure. Can J Bot 73:S1301–S1309

Gallet C, Pellissier F (1997) Phenolic compounds in natural solutions of a coniferous forest. J Chem Ecol 23(10):2401–2412

Gandeboeuf D (1997) Caractérisation et identification moléculaire de différentes espèces de genre *Tuber*. PhD thesis, Université Blaise Pascal, Clermont-Ferrand

Gioacchini AM, Menotta M, Polidori E, Giomaro G, Stocchi V (2002) Solid-phase microextraction gas chromatography/ion trap mass spectrometry and multistage mass spectrometry experiments in the characterization of germacrene D. J Mass Spectrom 37:1229–1235

Giomaro G, Sisti D, Zambonelli A, Amicucci A, Cecchini M, Comandini O, Stocchi V (2002) Comparative study and molecular characterization of ectomycorrhizas in *Tilia americana* and *Quercus pubescens* with *Tuber brumale*. FEMS Microbiol Lett 216(1):9–14

Girlanda M, Selosse MA, Cafasso D, Brilli F, Delfine S, Fabbian R, Ghignone S, Pinelli P, Segreto R, Loreto F, Cozzolino S, Perotto S (2006) Inefficient photosythesis in the mediterrean orchid *Limodorum abortivum* is mirrored by specific association to ectomycorrhizal *Russulaceae*. Mol Ecol 15(2):491–504

Guidi C, Zeppa S, Annibalini G, Pierleoni R, Guescini M, Buffalini M, Zambonelli A, Stocchi V (2006) The isoprenoid pathway in the ectomycorrhizal fungus *Tuber borchii* Vittad.: cloning and characterization of the *tbhmgr*, *tbfpps* and *tbsqs* genes. Curr Genet 50:393–404

Hall I, Yun W (2001) Truffles and other edible mycorrhizal mushrooms—some new crops for the Southern Hemisphere. In: Hall I, Yun W, Danell E, Zambonelli A (eds) Edible mycorrhizal mushrooms and their cultivation. Proceedings of the 2nd international conference on edible mycorrhizal mushrooms, New Zealand, pp 1–7

Harley FRSJL, Smith SE (1983) Mycorrhizal symbiosis. Academic. London

Harrison MJ (2005) Signaling in the arbuscular mycorrhizal symbiosis. Annu Rev Microbiol 59:19–42

Heidstra R, Bisseling T (1996) Nod factor induced hosts responses and mechanisms of Nod factor perception. New Phytol 133:25–43

Hibbett DS, Gilbert LB, Donoghue MJ (2000) Evolutionary instability of ectomycorrhizal symbioses in basidiomycetes. Nature 407:506–510

Hutchinson SA (1973) Biological activities of volatile fungal metabolites. Annu Rev Phytopathol 11:223–246

Ingvarsdóttir A, Birkett MA, Duce I, Genna RL, Mordue W, Pickett JA, Wadhams LJ, Mordue LAJ (2002) Semiochemical strategies for sea louse control: host location cues. Pest Manag Sci 58:537–545

Iotti M, Amicucci A, Stocchi V, Zambonelli A (2002) Morphological and molecular characterization of mycelia of some *Tuber* species in pure culture. New Phytol 155:499–505

Jin-Ming G, Wei-Ming Z, She-Qi Z, Xing Z, An-Ling Z, Hui C, Yue-Ying S, Ming T (2004) Sphingolipids from the edible fungus *Tuber indicum*. Eur J Lipid Sci Technol 106:815–821

Kishimoto K, Matsui K, Ozawa R, Takabayashi J (2005) Volatile C6-aldehyde and allo-ocimene activate defense genes and induce resistance against *Botrytis cinerea* in *Arabidopsis thaliana*. Plant Cell Physiol 46(7):1093–1102

Lanfranco L, Arlorio M, Matteucci A, Bonfante (1995) Truffles: their life cycle and molecular characterization. In: Stocchi V, Bonfante P, Nuti M (eds) Biotechnology of ectomycorrhizae. Molecular approaches. Plenum, New York, pp 139–150

Lanza B, Owezarek M, De Marco A, Raglione M (2004) Evaluation of phytotoxicity and genotoxicity of substances produced by *Tuber asetivum* and distributed in the soil using *Vicia faba* root micronucleus test. Fresenius Environ Bull 13:1410–1414

LePage BA, Currah RS, Stockey RA, Rothwell GW (1997) Fossil ectomycorrhizae from the middle Eocene. Am J Bot 84:410–412

Luppi Mosca AM, Fontana A (1977) Studi sull'ecologia del *Tuber melanosporum*. Analisi micologiche di terreni tartufieri dell'italia centrale. Allionia 22:105–113

Mamoun M, Olivier JM (1997) Mycorrhizal inoculation of cloned hazels by Tuber melanosporum: effect of soil disinfestation and co-culture with *Festuca ovina*. Plant Soil 188:221–226

Mannina L, Cristinzio M, Sobolev AP, Ragni P, Serge A (2004) High-field nuclear magnetic resonance (NMR) study of truffles (*Tuber aestivum vittadini*). J Agric Food Chem 52:7988–7996

Martin F, Dupleissis S, Ditengou F, Lagrange H, Voiblet C, Lapeyrie F (2001) Developmental cross talking in the ectomycorrhizal symbiosis: signals and communication genes. New Phytol 151:145–154

Mauriello G, Marino R, D'Auria M, Cerone G, Rana GL (2004) Determination of volatile organic compounds from truffles via SPME-GC-MS. J Chromatogr Sci 42:299–305

Menotta M, Gioacchini AM, Amicucci A, Buffalini M, Sisti D, Stocchi V (2004a) Headspace solid-phase microextraction with gas chromatography and mass spectrometry in the investigation of volatile organic compounds in an ectomycorrhizae synthesis system. Rapid Commun Mass Spectrom 18:206–210

Menotta M, Amicucci A, Sisti D, Gioacchini AM, Stocchi V (2004b) Differential gene expression during pre-symbiotic interaction between *Tuber borchii* Vittad. and *Tilia Americana* L. Curr Genet 46:158–165

Meruva NK, Penn JM, Farthing DE (2004) Rapid identification of microbial VOCs from tobacco molds using closed-loop stripping and gas chromatography/time-of-flight mass spectrometry. J Ind Microbiol Biotechnol 31482–488

Miozzi L, Balestrini R, Bolchi A, Novero M, Ottonello S, Bonfante P (2005) Phospholipase A(2) up-regulation during mycorrhiza formation in *Tuber borchii*. New Phytol 167(1):229–238

Montacchini F, Caramiello Lomango R (1977) Studi sull'ecologia del *Tuber melanosporum*. Azione inibitrice su specie erbacee della flora spontanea. Allionia 22:81–85

Müller E, Loeffler W (1976) Mycology. An outline for science and medical students. Translated by B Kendrick and F Bärlocher. Thieme, Stuttgart, p 15

Murat C, Diez J, Luis P, Delaruelle C, Dupre C, Chevalier G, Bonfante P, Martin F (2004) Polymorphism at the ribosomal DNA ITS and its relation to postglacial re-colonization routes of the Perigord truffle *Tuber melanosporum*. New Phytol 164:401–411

Murat C, Vizzini A, Bonfante P, Mello A (2005) Morphological and molecular typing of the below-ground fungal community in a natural Tuber magnatum truffle-ground. FEMS Microbiol Lett 245(2):307–313

Ney KH, Freytag WG (1980) Trüffel-aroma. Gordian 9:214

O'Donnell K, Cigelnik E, Weber NS, Trappe JM (1997) Philogenetic relationships among ascomycetous truffles and the true and false morels inferred from 18S and 28S ribosomal DNA sequence analysis. Mycologia 89(1):48–65

Papa G (1978–1979) Studi sull'ecologial del *Tuber melanosporum*. Analisi spettrofotometriche di estratti di terreni tartufigeni ed azione inibente la germinazione. Allionia 23:95–102

Papa G (1980) Purification attempts of the plant inhibitory principle of Tuber melanosporum Vitt Phytopathol Mediterr 19:177

Pacioni G (1991) Effects of *Tuber* metabolites on the rhizospheric environment. Mycol Res 95:1355–1358

Paolocci F, Rubini A, Riccioni C, Arcioni S (2006) Reevaluation of the life cycle of Tuber magnatum. Appl Environ Microbiol 72:2390–2393

Pelusio F, Nillsson T, Montanarella L, Tilio R, Larsen B, Facchetti S, Madsen JØ (1995) Headspace solid-phase microextraction analysis of volatile organic sulfur compounds in black and white truffle aroma. J Agric Food Chem 34:2138–2143

Plattner I, Hall IR (1995) Parasitism of non-host plants by the mycorrhizal fungus *Tuber melanosporum*. Mycol Res 99(11):1367–1370

Read DJ (1991) Mycorrhizas in ecosystems. Experientia 47:376–390

Saltarelli R, Ceccaroli P, Vallorani L, Zambonelli A, Citterio B, Malatesta M, Stocchi V (1998) Biochemical and morphological modifications during the growth of *Tuber borchii* mycelium. Mycol Res 102(4):403–409

Schulz B, Boyle C, Draeger S, Rommert AK, Krohn K (2002) Endophytic fungi: a source of novel biologically active secondary metabolites. Mycol Res 106(9): 996–1004

Schnitzler I, Boland W, Hay ME (1998) Organic sulfur compounds from *Dictyopteris* spp. deter feeding by an herbivorous amphipod *(Ampithoe longimana)* but not by an herbivorous sea urchin *(Arbacia punctulata)* J Chem Ecol 24(10):1715–1732

Selosse MA, Faccio A, Scappaticci G, Bonfante P (2004) Chlorophyllous and achlorophyllous specimens of *Epipactis microphylla* (Neottieae, Orchidaceae) are associated with ectomycorrhizal septomycetes, including truffles. Microbiol Ecol 47(4):416–442

Sisti D, Zambonelli A, Giomaro G, Rossi I, Ceccaroli P, Citterio B, Benedetti PA, Stocchi V (1998) *In vitro* mycorrhizal synthesis of micropropagated *Tilia platyphyllos* Scop. plantlets with *Tuber borchii* Vittad. mycelium in pure culture. Acta Hortic 457:379–387

Soragni E, Bolchi A, Balestrini R, Gambaretto C, Percudani R, Bonfante P, Ottonello S (2001) A nutrient-regulated, dual localization phospholipase A2 in the symbiotic fungus *Tuber borchii*. EMBO J 20(18):5079–5090

Sourzat P (1997) Guide pratique de trufficulture. Station d'experimentation sur la truffe. Lycee Professionnel Agricole de Cahors, Le Montat

Spinnler HE, Berger C, Lapadatescu C, Bonnarme P (2001) Production of sulfur compounds by several yeasts of technological interest for cheese ripening. Int Dairy J 11:245–252

Splivallo R, Bossi S, Maffei M, Bonfante P (2007a) Discrimination of truffle fruiting body versus mycelial aromas by stir bar sorptive extraction. Phytochemistry 68:2584–2598. doi:10.1016/j.phytochem.2007.03.030

Splivallo R, Novero M, Bertea CM, Bossi S, Bonfante P (2007b) Truffle volatiles inhibit growth and induce an oxidative burst in *Arabidopsis thaliana*. New Phytol 175:417–424

Talou T, Delmas M, Gaset A (1987a) Identification of the principal constituents of black truffle aroma. In: Frontiers of flavor. Proceedings of the 5th international flavor conference, Porto Karras, Chalkidiki, Greece, pp 367–371

Talou T, Delmas M, Gaset A (1987b) Principal constituents of black truffle (Tuber melanosporum) aroma. J Agric Food Chem 35:774–777

Talou T, Delmas M, Gaset A (1989a) Analysis of headspace volatiles from entire black truffle (Tuber melanosporum). J Sci Food Agric 48:57–62

Talou T, Delmas M, Gaset A (1989b) Direct capture of volatiles emitted from entire black Perigord truffle. J Essent Oil Res 1:281–286

Talou T, Delmas M, Gaset A (1989c) Black Perigord truffle: from aroma analysis to aromatizer formulation. In: Flavors and off-flavors, Proceedings of the 6th international flavor conference, Rethymnon, Crete, Greece, pp 715–728

Talou T, Delmas M, Gaset A (1989d) New trends in black truffle aroma analysis. ACS Symp Ser 388:202–212

Tibbett M, Sanders FE (2002) Ectomycorrhizal symbiosis can enhance plant nutrition through improved access to discrete organic nutrient patches of high resource quality. Ann Bot 89:783–789

Tirillini B, Verdelli G, Paolocci F, Ciccioli P, Frattoni M (2000) The volatile organic compounds from the mycelium of *Tuber borchii* Vitt. Phytochemistry 55:983–985

Trappe JM (1962) Fungus associates of ectotrophic mycorrhizae. Bot Rev 28:538–606

Trappe JM (1979) The orders, families and genera of hypogeous ascomycotina (truffles and their relatives). Mycotaxon 9:297–340

Venkateshwarlu G, Chandravadana MV, Tewari RP (1999) Volatile flavor components of some edible mushrooms (Basidiomycetes). Flavour Fragr J 14:191–194

Weden C, Danell E, Camacho FJ, Backlund A (2004) The population of the hypogeous fungus *Tuber aestivum* syn. *Tuber uncinatum* on the island of Gotland. Mycorrhiza 14(1):19–23

Wnouk S, Kinastowski S, Kaminski E (1983) Synthesis and analysis of 1-octen-3-ol, the main flavor component of mushrooms. Nahrung 27:479–486

Yang Mei C (1999) Truffles in southwest China. In: Actes du Vème congrès international, science et culture de la truffe. Fédération Française des Trufficulteurs, Aix-en-Provence, pp 4.248–4.249

Zacchi L, Vaughan-Martini A, Angelini P (2003) Yeast distribution in a truffle field ecosystem. Ann Microbiol 53(3):275–282

Zambonelli A, Branzanti MB (1989) Mycorrhizal synthesis of Tuber albidum Pico with Castanea sativa Mill. and Alnus *cordata* Loisel. Agric Ecosyst Environ 28:563–568

Zambonelli A, Branzanti MB (1990) Competizione fra Tuber albidum e alcuni basidiomiceti nella formazione di ectomycorrize su semenzali di *Pinus pinea*. In: Bencivenga M, Granetti B (eds) Atti del secondo congresso internazionale sul tartufo, Spoleto, pp 443–449

Zeppa S, Gioacchini AM, Guidi C, Guescini M, Pierleoni R, Zambonelli A, Stocchi V (2004) Determination of specific volatile organic compounds synthesized during *Tuber borchii* Fruit body development by solid-phase microextraction and gas chromatography/mass spectrometry. Rapid Commun Mass Spectrom 18:199–205

Chapter 9
Mycotoxins in the Soil Environment

Susanne Elmholt

9.1 Introduction

Hawksworth (1991) suggested the existence of 1.5 million fungal species, of which about 5% are known. As many fungi produce a plethora of secondary metabolites, only fantasy sets a limit to the number we may find in soil. Many secondary metabolites have been characterised chemically in line with improved analytical methods and increasing consensus that metabolite profiling is a valuable means of phenotype characterisation in taxonomical studies (Frisvad and Filtenborg 1989; Pitt and Samson 1990; Thrane 1989; Frisvad et al. 1998; Samson et al. 2000; Moss and Thrane 2004; Thrane et al. 2004). A review of the many secondary metabolites likely to be found in the soil environment is beyond the scope of this chapter especially because we lack detailed knowledge of the function and importance of most of them. Instead this chapter concentrates on mycotoxins that constitute a recognised safety hazard to humans and domestic animals.

The term 'mycotoxin' is confusing. Out of context, it is unclear if these toxins poison fungi or if they are produced by fungi to poison other organisms—and if so which. This was pointed out by Bennett (1987), who defined mycotoxins as 'natural products produced by fungi that evoke a toxic response when introduced in low concentrations to higher vertebrates and other animals by a natural route'. Mycotoxins are secondary metabolites, defined by Bennett and Bentley (1989) as 'metabolic intermediates or products, found as a differentiation product in restricted taxonomic groups, not essential to growth and life of the producing organism, and biosynthesised from one or more general metabolites by a wider variety of pathways than is available in general metabolism'.

Mycotoxins produced or deposited in the soil environment may affect all three aspects of soil quality (Doran et al. 1994), i.e. (1) biological productivity,

Susanne Elmholt
Department of Agroecology and Environment, Faculty of Agricultural Sciences, University of Aarhus, Blichers Allé 1, DK-8830 Tjele
e-mail: susanne.elmholt@agrsci.dk

P. Karlovsky (ed.), *Secondary Metabolites in Soil Ecology. Soil Biology 14.*

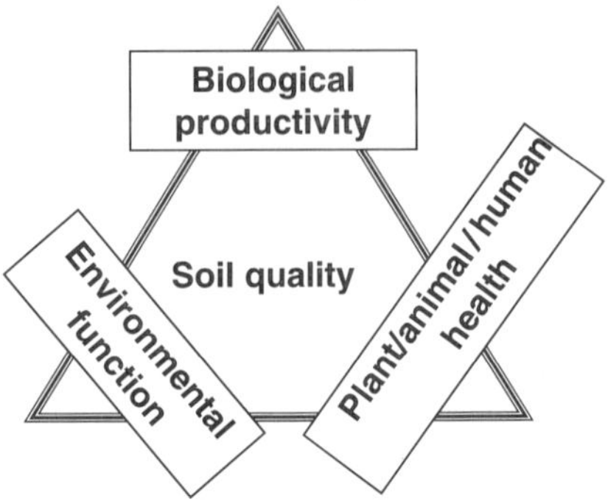

Fig. 9.1 Soil quality triangle. (After Doran et al. 1994)

(2) environmental function and (3) plant/animal/human health (Fig. 9.1), as will be exemplified in this chapter. Focus will be on soils and crops that are characteristic for arable soils in Europe and North America.

9.2 Agriculturally Relevant Mycotoxins

About 20 mycotoxins are found in foods and feeds at frequencies and levels to be of food safety concern (FAO 2004; Smith et al. 1994). The most important mycotoxins and their producers are briefly outlined below and will be addressed in this chapter (Fig. 9.2, Table 9.1). Mycotoxigenic fungi belong to genera where species look 'awfully like each other to non-specialists' (Moss and Thrane 2004)—and sometimes also to specialists! Misidentifications have therefore led to many erroneous reports on which species produce which mycotoxins (Frisvad 1989). Taxonomical methods of today combine morphological examination, molecular biology and secondary metabolite profiling and have led to major revisions of 'who does what' (Seifert and Lévesque 2004). Yet new taxonomical insight will inevitably add changes to Table 9.1.

9.2.1 Trichothecenes

The trichothecenes contain more than 150 sesquiterpenoids with a 12,13-epoxy group and well-described biosynthetic pathways (Tamm and Breitenstein 1980; Ueno 1983; Desjardins et al. 1993; Eriksen and Alexander 1998; Moss and Thrane 2004). This chapter will not address macrocyclic diesters and triesters but only the less complex non-macrocyclic alcohol and simple esters, which are subdivided according to absence (type A) or presence (type B) of a ketone at C-8. Most

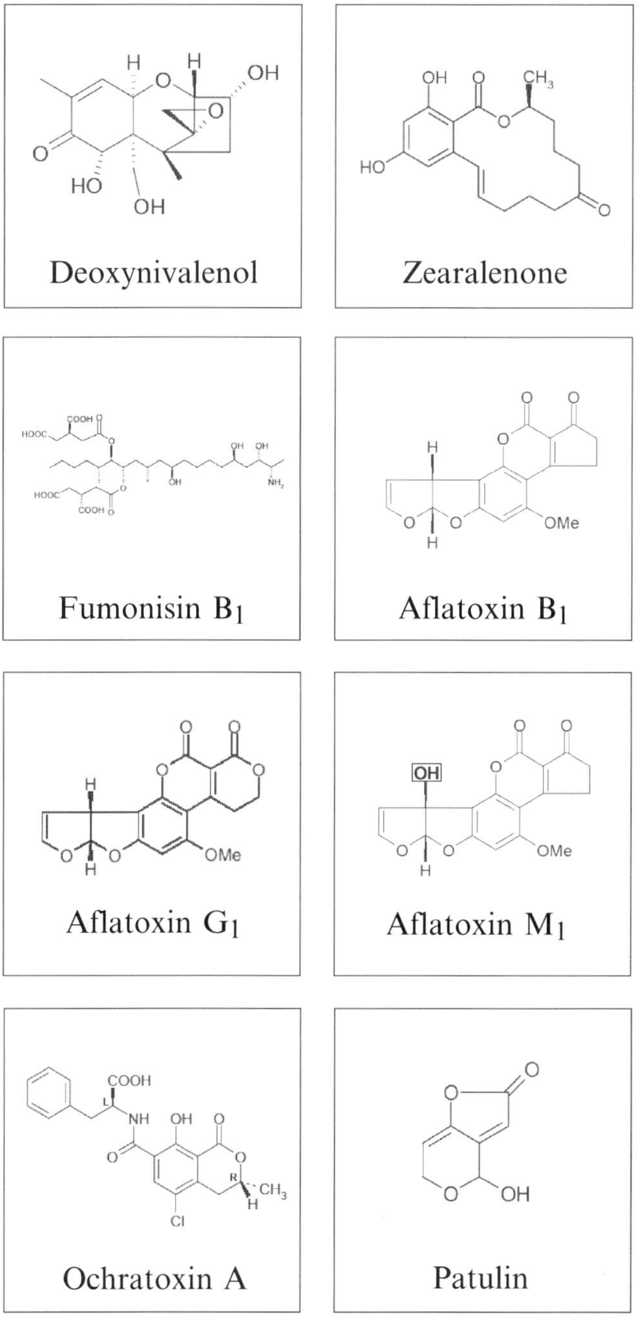

Fig. 9.2 Chemical structure of important mycotoxins

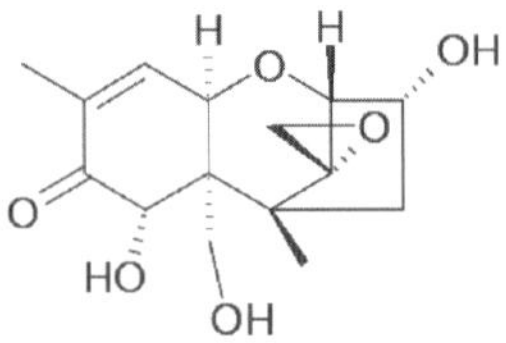

Deoxynivalenol

CAS no.: 51481-10-8

Water solubility: Slight

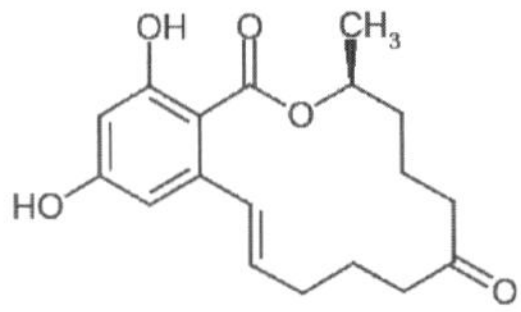

Zearalenone

CAS no.: 17924-92-4

Water solubility: Insoluble/ slight

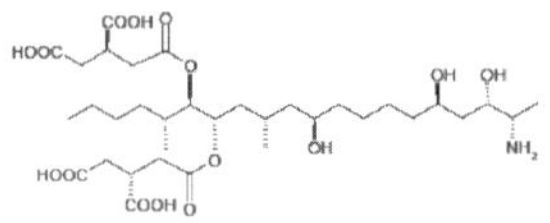

Fumonisin B_1

CAS no.: 116355-83-0

Water solubility: Soluble (>20 mg/ ml)

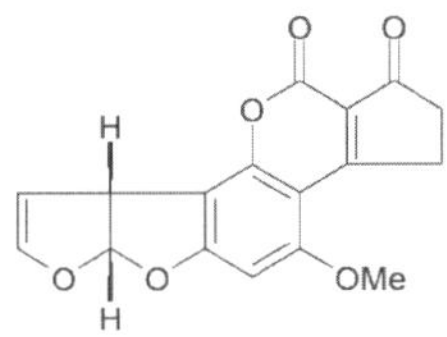

Aflatoxin B_1

CAS no.: 1162-65-8

Water solubility: slight

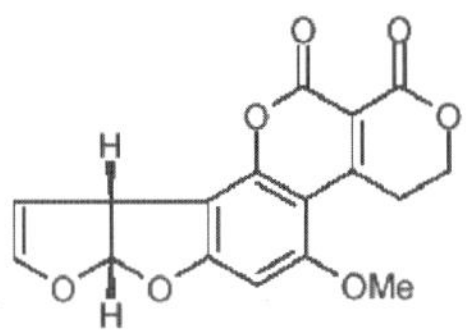

AflatoxinG_1

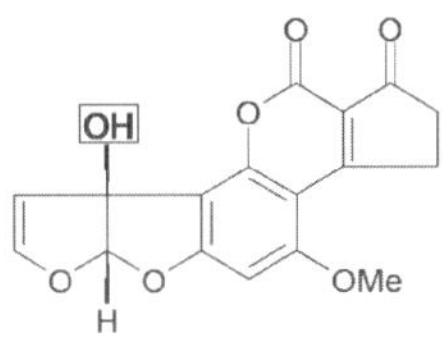

AflatoxinM_1

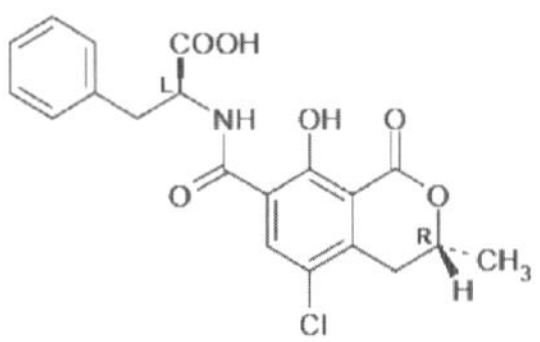

Ochratoxin A

CAS no.: 303-47-9

Water solubility: Slight

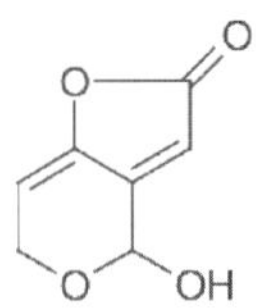

Patulin

CAS no.: 149-29-1

Water solubility: Soluble

Fig. 9.2 (continued)

Table 9.1 Agriculturally important mycotoxins and their producers

Trichothecenes, type A[a]	T-2 and HT-2 toxins	*Fusarium sporotrichioides*	Moss and Thrane (2004)
		F. sambucinum	
		F. tunidum	
		F. armeniacum	
		F. musarum	
		F. langsethiae	Thrane et al. (2004)
		F. poae (few strains)	
	Diacetoxyscipenol	*F. sporotrichioides*	Moss and Thrane (2004)
		F. poae	
		F. sambucinum	
		F. equiseti	
		F. tunidum	
		F. venenatum	
		F. musarum	
		F. langsethiae	Thrane et al. (2004)
Trichothecenes, type B[b]	Deoxynivalenol	*F. culmorum*	Moss and Thrane (2004)
		F. graminearum	
		F. pseudograminearum	
	Nivalenol	*F. culmorum*	Moss and Thrane (2004)
		F. graminearum	
		F. poae	
		F. equiseti	
		F. venenatum	
		F. crookwellense	
		F. kyushuense	
Zearalenone		*F. crookwellense*	Frisvad and Thrane (2000)
		F. culmorum	
		F. equiseti	
		F. graminearum	
		F. semitectum	
		F. pseudograminearum	Seifert and Lévesque (2004)
Fumonisins B1 and B2		*F. verticilloides*	Frisvad and Thrane (2000)
		F. proliferatum	
		F. nygamai	
	B	*Aspergillus flavus*, group I	Geiser et al. (2000)
Aflatoxin	B and G	*A. flavus*, group II	
		A. parasiticus	Seifert and Lévesque (2004)
		A. nomius	
Ochratoxin A[a]		*Penicillium verrucosum*	Larsen et al. (2001)
		Penicillium nordicum	

(continued)

Table 9.1 (continued)

	A. Ochratoxin	Frisvad and Thrane (2000)
	A. niger (few strains)	
	Petromyces alliaceus	
	Neopetromyces muricatus	
Patulin	*Penicillium expansum*	
	Penicillium carneum	Frisvad and Thrane (2000)
	Penicillium grandicola	
	Paecilomyces variotii	

[a]Other type A trichothecenes in agricultural commodities are monoacetoxyscirpenol and neosolaniol (Smith et al. 1994)

[b]Other type B trichothecenes in agricultural commodities are 3-acetyldeoxynivalenol, 15-acetyldeoxynivalenol and 4-acetylnivalenol (fusarenon-X) (Smith et al. 1994)

trichothecenes are only known from the laboratory, but about 12 have been reported from agricultural commodities (Smith et al. 1994), the most important being listed in Table 9.1. They are all regarded a health hazard, although deoxynivalenol (DON, Fig. 9.2) is less toxic than diacetoxyscirpenol, nivalenol (NIV) and T-2 toxin (Smith et al. 1994; WHO 1990). Trichothecenes are highly toxic at the subcellular, cellular and organic system level. They easily penetrate the cell lipid bilayers and inhibit protein synthesis (Marasas and Nelson 1987; Smith et al. 1994). All animals tested appear to be sensitive and symptoms include anemia, immunosuppression, haemorrhage, emesis (therefore the popular synonym vomitoxin for DON), and feed refusal in cattle, pigs and poultry (Marasas et al. 1984). Epidemiological data associate human disease outbreaks of alimentary toxic aleukia in the former Soviet Union in the 1940s with consumption of trichothecene-contaminated grain, but no firm causal relationship can be stated (Desjardins et al. 1993). More recently, disease outbreaks were related to trichothecene poisoning in India (Bhat et al. 1989). The non-macrocyclic trichothecenes are produced by species of *Fusarium* (Frisvad and Thrane 2000; Marasas et al. 1984). Taxonomical controversies have caused much confusion regarding which *Fusarium* species are able to produce a given trichothecene, but a recent overview was offered by Moss and Thrane (2004) (Table 9.1).

9.2.2 Zearalenone

Zearalenone (ZEA, Fig. 9.2) is a resorcyclic lactone with many—also naturally occurring—derivatives (Mirocha and Christensen 1974; Eriksen and Alexander 1998). Its acute toxicity is low, and ZEA and some of its derivatives are primarily known for their effects on reproduction in swine, poultry, rodents and possibly humans (Marasas et al. 1984; Smith et al. 1994; Eriksen and Alexander 1998; Creppy 2005). ZEA can be classified as an oestrogen in the sense that it produces oestrus, i.e. cornification of the vagina in adult mice (Mirocha and Christensen 1974). Compared with 17ß-estradiol, ZEA is 10–1,000 times weaker, depending on the route of administration and animal age. Young animals are at greater risk (Smith et al. 1994). In most reports carcinogenicity seems to be a consequence of the oestrogenic effects of ZEA (Creppy 2005), though genotoxicity due to the formation of DNA adducts has been reported too (Pfohl-Leszkowicz et al. 1995). Table 9.1 lists the producers of ZEA according to Frisvad and Thrane (2000).

9.2.3 *Fumonisins*

Fumonisins are structurally similar to the sphingoid base backbone of sphingolipids and the structure and biosynthesis of fumonisin B_1 (FB_1; Fig. 9.2) and related compounds have been elucidated by ApSimon (2001). They were discovered fairly recently (Gelderblom et al. 1988), and about 15 different fumonisins have been described, the most important food contaminants being FB_1 and fumonisin B_2. The fumonisins are the causative agents of two animal diseases, equine leucoencephalomalacia and porcine pulmonary oedema. The main concerns, however, are reports on carcinogenicity in rodents (Eriksen and Alexander 1998; Creppy 2005) and the possible role of FB_1 in development of oesophageal cancer in humans (WHO 2000; Yoshizawa et al. 2005). Table 9.1 lists the producers of fumonisin according to Frisvad and Thrane (2000), Marin et al. (2004) and Seifert and Lévesque (2004). Seifert and Lévesque (2004)list several other producers of fumonisins but state that *Fusarium verticillioides* (former *F. moniliforme*) and *F. proliferatum* are the only significant producers on agricultural products.

9.2.4 *Aflatoxins*

The aflatoxins are highly oxygenated, heterocyclic compounds with closely similar chemical structure. They were first described in 1960, when more than 100,000 young turkeys, ducklings and pheasants died from the hitherto unknown 'Turkey X disease'. Careful investigations revealed that the disease was linked to intake of Brazilian peanut meal contaminated by *Aspergillus flavus* and a toxin to be named after its producer—aflatoxin (Eaton and Groopman 1994). The four major aflatoxins ranked after acute toxicity, carcinogenicity and mutagenicity, with aflatoxin B_1 (AFB_1) being the most critical, are AFB_1 > aflatoxin G_1 (AFG_1) > aflatoxin B_2 (AFB_2) > aflatoxin G_2 (AFG_2). AFB_2 and AFG_2 are dihydroxy derivatives, and aflatoxin M_1 (AFM_1) and aflatoxin M_2 are 4-hydroxy derivates of AFB_1 and AFG_1, respectively (Fig. 9.2). AFM_1 is produced by metabolic hydroxylation of AFB_1 in the liver and is excreted in the milk of lactating animals, including dairy cattle and humans. Although AFM_1 is less carcinogenic and mutagenic than AFB_1, its occurrence in milk makes it more threatening to human health than AFB_1 (Creppy 2005). Aflatoxins are toxic to most animal species (Eaton and Groopman 1994), AFB_1 being the most potent hepatocarcinogen known in mammals. Two species of *Aspergillus* are regarded the main producers of aflatoxins, i.e. *A. flavus* and *A. parasiticus* (Geiser et al. 2000) (AB and AG). Another producer is the agronomically less important *A. nomius* (Frisvad and Thrane 2000; Seifert and Lévesque 2004). Recently several new aflatoxin producers outside the section *Flavi* were described (Frisvad et al. 2005) but their agricultural significance remains to be elucidated.

9.2.5 Ochratoxins

The ochratoxins are closely related derivatives of isocoumarin linked to L-*ß*-phenylalanine, of which ochratoxin A (OTA; Fig. 9.2) and ochratoxin B can be found on agricultural commodities, the former by far the more important. The history of ochratoxin research was outlined by Krogh (1987), who revealed OTA as the causal agent of the swine disorder mycotoxic porcine nephropathy (Krogh 1978). OTA is nephrotoxic to all animal species tested and most likely to humans, who show the longest half-life time among tested species (Creppy 1999). OTA acts through several molecular pathways and is reported to be teratogenic, immunotoxic, mutagenic, carcinogenic and possibly genotoxic (Boorman 1989; WHO 1990; Smith et al. 1994; Creppy 1999, 2005). OTA was associated early with the human renal disorder, Balkan endemic eephropathy (BEN) and tumours of the urinary tract as reviewed by Pfohl-Leszkowicz et al. (2002). Recently, another endemic kidney disease has been linked to OTA-contaminated food (Creppy 1999; Wafa et al. 2004). Owing to major taxonomical achievement in recent years, scientists today agree that OTA is produced by two species of *Penicillium*, i.e. *P. nordicum* on meat and cheese and *P. verrucosum* on grain (Larsen et al. 2001; Seifert and Lévesque 2004). The list of *Aspergillus* species producing OTA has also changed (Krogh 1978; Frisvad and Thrane 2000; Seifert and Lévesque 2004), including today some strains of *A. niger*, which are currently believed to be the main OTA producer in wine, and *A. ochraceus*, which is regarded as the main OTA producer in coffee and cocoa (Frank 1999; Frisvad and Thrane 2000). Other producers are *A. alliaceus* and *A. muricatus*, now allocated to *Petromyces alliaceus* and *Neopetromyces muricatus*, respectively (Frisvad and Thrane 2000). Recently, new OTA-producing species have been described (Frisvad et al. 2004) but their importance is unknown.

9.2.6 Patulin

Patulin is a water-soluble lactone (Fig. 9.2). It is known under several names and was first discovered as an antibiotic during the 1940s (Moake et al. 2005). It has broad-spectrum effects on fungi and bacteria and many studies have demonstrated patulin to possess mycotoxic properties. Results point to both acute and chronic effects, although evidence of carcinogenic effects is inconclusive (Smith et al. 1994; Moake et al. 2005; Schumacher et al. 2005). Yet Moake et al. (2005) state that there is little doubt as to the potential hazard inherent in the contamination of food products by patulin. Patulin is produced by fungi belonging to *Penicillium, Paecilomyces* and *Aspergillus*, including *Penicillium carneum, Penicillium glandicola* and *Paecilomyces variotii* (growing in silage) and *Penicillium expansum*, which is the major producer of patulin in apples (Samson et al. 2000).

9.3 Routes by Which Mycotoxins Enter the Soil Environment

Overwhelmingly many experiments and surveys have been performed on mycotoxins regarding their toxicologicy, methods of analysis and occurrence in feed and food commodities. In comparison, surprisingly few studies deal with the environmental fate and implications of these compounds. Sections 9.3, 9.4 and 9.5.3 outline some of the mechanisms by which mycotoxins may enter or leave the soil ecosystem (summarised in Fig. 9.3).

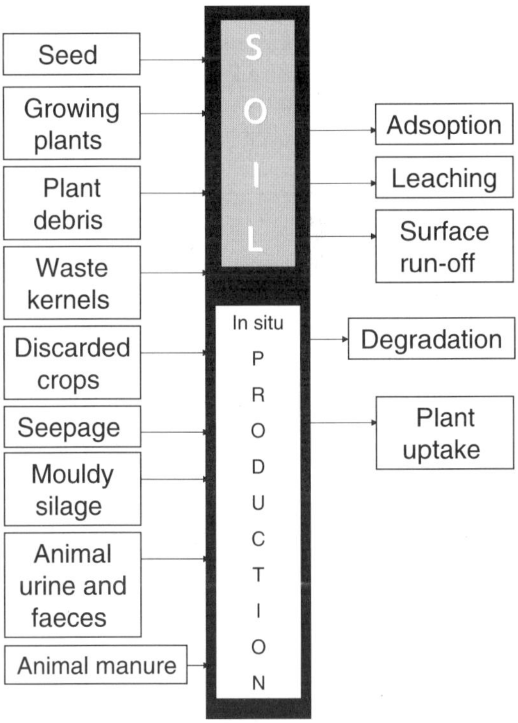

Fig. 9.3 Environmental fate of mycotoxins. The main routes by which the compounds are added (Sects. 9.3, 9.5.3) and removed (Sect. 9.4) from the soil envrironment are shown

9.3.1 Seed, Growing Plants and Plant Debris

Seed ought to be free of toxins and toxigenic fungi but there are exceptions, e.g. the FB_1-producing *F. verticillioides*. This seed-transmitted pathogen in maize can colonise the root system (Cavaglieri et al. 2005) as well as infect systemically and colonise the entire maize plant (Munkvold 2003). No FB_1 analyses of the plant material were made in these studies, and the amount of FB_1 entering the soil via contaminated maize plants is unknown, as also pointed out by Williams et al. (2003). Seed may also be contaminated by *P. verrucosum* and OTA, especially in organic farming, where fungicidal seed treatment is banned. Elmholt (2003) reported that home-grown seed batches from farms with insufficient drying facilities contained up to 100% kernels infested by *P. verrucosum* and that the fungus was present in soil at these farms. OTA in the standing crop was not found (Elmholt 2003), and preharvest OTA production in small-grain cereals is generally not considered important (Battaglia et al. 1996; Scudamore and Wilkin 1999). Yet there are reports that OTA can be produced in the field (Hokby et al. 1979), especially in wet harvest years (Elmholt and Rasmussen 2005). In these cases, OTA might enter the soil environment but this awaits further study.

During the growing season and following harvest, mycotoxins may enter the soil via contaminated crop plants as mentioned above for FB_1 (Madden and Stahr 1993; Williams et al. 2003). Hestbjerg et al. (2002a) found that barley seedlings infested with different strains of *F. culmorum* contained 0.2–50 ppm DON (median 17 ppm). Lopez et al. (1997) found that stems and leaves of maize retrieved from the field several months after harvest contained ZEA (3 mg kg^{-1}). Logrieco et al. (1998) presented the first report on the natural occurrence of FB_1 in asparagus, the average levels for crowns and stem samples being 7.4 and 0.83 ppm, respectively. Similar levels were found by Seefelder et al. (2002) in naturally infected spear samples and some of this fumonisin will most likely enter the soil with plant debris. Sinha and Savard (1997) inoculated heads of wheat with *F. graminearum* and found the median DON concentration to be 93 ppm in the rachis, 50 ppm in the chaff, 25 ppm in the kernels and 16 ppm in the peduncle. Once it reaches the rachis, they argue that DON can be translocated throughout the wheat head and into the peduncle through the phloem, along with the photosynthates and nutrients as a bulk solution. They conclude that food and feed safety would be improved if these parts of the heads are removed from the commodity as they contain much DON. The consequence of leaving them in the field is, however, that substantial amounts of DON are instead being added to the soil environment.

In any field or orchard some part of the crop will end up in the soil environment. Although small by weight, this fraction will often contain more fungal biomass and mycotoxins than the bulk. Ergosterol concentrations—an indicator of fungal biomass—are generally higher in small kernels and impurities than in bulk grain (Regner et al. 1994). Perkowski (1998) found that the smallest fraction of barley kernels contained more than 75% of the total DON. Using a monoclonal antibody technique designed for small samples, Sinha and Savard (1997) found average

DON concentrations below 1 ppm in healthy looking kernels, 1.8 ppm in shrivelled kernels, 174 ppm in white tombstone and 275 ppm in pink tombstone kernels. 'Tombstone kernels' are small, shrivelled and light. They are easily blown out the back of a combine, improving the quality of the harvested grain but serving at the same time as a source of contaminating soil with mycotoxins and mycotoxingenic fungi. Abramson et al. (1987) reported that 53 suspect samples contained between 0.2 and 5.4% tombstone kernels in the harvested grain, indicating that quite large amounts of these kernels had been produced, of which many no doubt had been left in the field. Sinha and Savard (1997) studied 30 infected plants and found that tombstone kernels constituted from 0 to 81%, with a median value of 14%. Lopez et al. (1997) found that waste grain of maize retrieved from the field several months after harvest contained DON (0.7 mg kg^{-1}) and T-2 (4.1 mg kg^{-1}).

Likewise, *A. flavus* invasion and preharvest aflatoxin formation are higher in small and immature peanut kernels than in mature ones (Dorner et al. 1989). Wotton and Strange (1987) hypothesised that immature kernels lose their capacity for producing phytoalexins. This should lead to earlier contamination as some phytoalexins inhibit germination and growth of *A. flavus*. Olanya et al. (1997) found that waste maize around cribs and unloading ports of storage bins was often heavily infested with *A. flavus* as was the soil under these deposits. The authors demonstrated a linear dispersal gradient of airborne conidia from these deposits and argued that they serve as point sources of *A. flavus* inoculum in the agroecosystem. There is ample evidence that soil population densities of aflatoxigenic fungi may increase manyfold following harvest especially in drought years, and infected drought-stressed plants of maize and peanut will replenish the soil with fungal inoculum and aflatoxin (Horn 2003). The classic study by Anderson et al. (1975)—the first to demonstrate preharvest production of aflatoxin—found that some kernels contained up to 600 ppm aflatoxin. In this respect, discarded crops present a special problem. US farmers have commonly incorporated maize exceeding the 20 ppb aflatoxin limit into the soil and this will add significant amounts of both toxin and fungal inoculum to the soil (Angle and Wagner 1980, 1981).

Decayed apples with rotted spots may contain high amounts of patulin (Xu and Berrie 2005). Kadakal et al. (2005) compared raw juice from sound apples and from apples with surface decays of 10, 60 and 100%, finding average patulin concentrations of 1.9, 179, 599 and 861 µg L^{-1}, respectively. Apples with rotted spots will often be left on the ground, discarded by animals and humans, and the patulin will end up in the soil environment.

9.3.2 *Seepage from Ensiled Forage Crops*

Ensiling is commonly used to preserve green forage crops by lactic fermentation under anaerobic conditions. The anaerobic phase is preceded by an aerobic phase, the duration of which depends on the ensiling conditions. Mycotoxins can be produced during this phase if toxigenic fungi are present and the temperature and moisture conditions are appropriate.

Mycotoxins produced prior to harvest or during the aerobic phase of ensiling may enter the soil environment via seepage (silage effluent). This risk is particularly high if the crop is ensiled above the recommended moisture content and the silage facilities are prone to leakage. Seepage might be a source of particularly water-soluble mycotoxins like FB_1, and maize silage samples can contain FB_1 in low amounts (Yu et al. 1999; Kim et al. 2004). There are examples of mycotoxins being degraded during ensiling, as exemplified for patulin and other mycotoxins in the review by Karlovsky (1999). Rotter et al. (1990) reported that the decreasing pH led to breakdown of OTA and AFB_1, while ZEA and some of the trichothecenes were stable at low pH.

I have not found any results indicating the load of mycotoxins entering soil via seepage—which may in part be due to the analytically very complex matrix of silage (Yu et al. 1999)—but even where seepage is properly collected and stored, it may end up in the soil environment if mixed with animal waste products and used as animal manure. Another concern is that once the finished silage is laid open, oxygen will reenter in amounts depending upon mechanical handling and—if the finished silage is stored in the field—rain may increase the moisture in the open parts. These are factors that can lead to renewed fungal growth and mycotoxin production. Silage parts, which are visibly mouldy, will not be used as feed and instead will be discarded and likely end up in the soil environment.

9.3.3 Animal Excretion

Farm animal waste is a source of mycotoxins entering the soil, both via urine and faeces. That urine and faeces may contain mycotoxins is known from many toxicological studies, addressing absorption, metabolism and excretion of mycotoxins by farm animals. I shall present a few examples to illustrate that mycotoxin excretion may have environmental consequences in the soil environment.

Pigs are sensitive to DON concentrations in the diet exceeding 1 ppm. To study their response to lower and more commonly occurring concentrations, Dänicke et al. (2004) undertook a dose–response study with *F. culmorum* contaminated wheat. In accordance with several other studies, more DON was excreted in urine than in faeces, 44–60% of the ingested DON in urine compared with less than 1% in faeces. A study on NIV metabolism was performed with pigs that were fed 0.05 mg NIV kg^{-1} body mass (BM) added in crystalline form to the feed (Hedman et al. 1997). The results showed that up to 40% NIV was excreted, more than half in faeces, indicating poorer absorption than for DON. No metabolites were found. Recognising that monogastrics like humans and pigs are more sensitive to DON and OTA than ruminants, farmers are inclined to use mycotoxin-contaminated grain as feed for cattle and sheep (Cote et al. 1986; Blank et al. 2005), because the rumen microflora can reduce the epoxide group of DON to the less toxic metabolite DOM-1 and OTA to the less toxic ochratoxin α. But only part of the DON is metabolised, as shown for dairy cows that were fed maize at a dietary concentration of 66 mg kg^{-1}. Twenty percent of the mycotoxin was recovered in urine and faeces, 96% thereof as DOM-1 and 4% as DON (Cote et al. 1986). Blank et al. (2005) conducted a feeding trial with wheat that had been inoculated with *A. ochraceus*. The sheep were fed 0–28.5 µg OTA kg^{-1} BM day^{-1} and none of the animals showed health disturbances that could be related to OTA. The total excretion of OTA plus ochratoxin α was 74–80% of the OTA intake. Proportions of about 2–3% of the applied amount were excreted in faeces and 6–8% in urine (28.4-113 ng OTA mL^{-1} urine).

In several European countries, an increased number of sows are reared outdoors to improve animal welfare and reduce building costs. There are, however, side effects in the form of N and P losses to aquifers (Eriksen and Kristensen 2001), and as mycotoxins are excreted in urine and faeces, outdoor pig production may also be a source of mycotoxins entering the soil environment. Valenta et al. (1993) reported that diets of 100 µg OTA kg^{-1} resulted in concentrations of 14–26 µg OTA L^{-1} urine. A stocking density of 15 sows per hectare and a yearly production per sow of 2.19 t urine (J. Fernandez, Danish Institute of Agricultural Sciences, personal communication) would result in excretion of 0.5–0.9 g OTA ha^{-1} $year^{-1}$. For pigs offered DON-contaminated feed at a level of 0.23 mg kg^{-1} BM, Razzazi-Fazeli et al. (2003) reported the mean urine concentration to be 580 µg DON L^{-1} and 32 µg DOM-1 L^{-1}. Using the assumptions above, this would lead to a yearly excretion of 19 g DON ha^{-1}. Here, it should be noted that outdoor living animals distribute their wastes very heterogeneously with high concentrations around feeders (Eriksen and Kristensen 2001). This may lead to some areas in the field having significantly higher mycotoxin loads than others.

Estimates like these are very generalised. They will vary according to chemical and physical characteristics of both soil and mycotoxin, the composition and level of feed contamination, and the amount of waste added to the soil. I have only found one—also very generalised indeed—example in the literature, i.e. for FB_1. This mycotoxin is poorly absorbed, not metabolised and the majority rapidly excreted by animals (WHO 2000). For the US 1998 crop of 2.5 $\times 10^8$ t maize, Williams et al. (2003) estimated the amount of FB_1 entering the soil via animal waste to be 130–270 t, assuming that 60% of the harvested maize was used for feed, that it contained 1–2 ppm FB_1 and that 90% of the harvested crop ended up in litter, sewage or on the ground.

To my knowledge there are no studies available on the occurrence or fate of mycotoxins that are deposited in the field with slurry and farmyard and liquid manure.

9.4 Routes by Which Mycotoxins Are Immobilised or Removed from the Soil Environment

The fate of mycotoxins formed in situ or added to the soil with contaminated plants or animal waste will depend on interactions with climatic conditions, soil textural and structural characteristics, water fluxes, plant growth and the activity and composition of the soil biota. Experimental evidence is extremely sparse and mostly originates from laboratory experiments.

9.4.1 Adsorption

Mycotoxins may be immobilised in soil by adsorption to soil particles. Some aluminosilicate-containing clays adsorb several mycotoxins, including aflatoxins, ZEA and OTA. This is used as a means of detoxifying animal feed (Huwig et al. 2001). The ability of clay silicate minerals to adsorb, e.g., ZEA increases with increasing hydrophobicity of the clay (Lemke et al. 1998). Mortensen et al. (2006) found that adsorption of OTA and ZEA to clay delayed their degradation and more so in a clayey than in a sandy soil. A third gyttja soil with a high content of silt was expected to elicit stronger adsorption, resulting in slower degradation than the other two soils. Degradation was, however, fast in the gyttja soil and the authors speculated that its low C/N ratio had led to high microbial activity and rapid degradation of the toxins prior to adsorption.

Using leachate columns, Madden and Stahr (1993) found no FB_1 in eluates or soil extracts and concluded that FB_1 was either irreversibly bound to the silty clay loam investigated or chemically altered. Aflatoxin is also strongly adsorbed to soil, especially to loamy soils with high contents of montmorillonite (Angle and Wagner 1980; Mertz et al. 1981). Goldberg and Angle (1985) established the adsorption coefficients of AFB_1 in four different soils and found it to be highest in silty clay loam (238 mg kg^{-1}) and lowest in sandy loam (17 mg kg^{-1}).

The abovementioned experiments were performed by addition of the mycotoxin in an appropriate solvent to the soil. In nature, however, mycotoxins will most often be found in an organic matrix, e.g. plants or animal waste. This may alter their adsorption characteristics.

9.4.2 Leaching and Surface Runoff

Mycotoxin-contaminated plant material and animal waste, left in the field, are potential contaminants of groundwater, especially if they contain toxins that are highly soluble in water like fumonisins (WHO 2000) and patulin. As mentioned already, Madden and Stahr (1993) could not recover FB_1 from eluates or soil extracts. Williams et al. (2003) argued that other soil types might adsorb differently and set up leachate columns to determine the movement of FB_1 through different soil types. Naturally FB_1 contaminated maize or water extracts thereof were placed on top of the columns. The results showed that FB_1 leaching was only slightly retarded in 100% sand columns, whereas approximately 60, 50 and 20% of the FB_1 was recovered in the column leachate when sand was mixed with 50, 75 and 100% sandy loam, respectively. The results indicated that the nature of binding in the sandy loam was ionic and that FB_1 could be released and become biologically available, e.g. via ionic interactions with soil constituents in acid soils. Williams et al. (2003) concluded that FB_1 from maize debris in field conditions might enter the ground water under certain environmental conditions.

High adsorption to the soil matrix is related to a reduced mobility of a substance (Vereecken 2005). Using columns with four different soil types, Goldberg and Angle (1985) showed that added AFB_1 and its derivatives AFB_2 and AFG_2 were retained in the upper 20 cm of saturated soil, 80% thereof in the upper 2.5 cm. No differences were found in leaching between the soils despite a large difference in adsorption coefficients, and no aflatoxin was found in the leachate. The lack of leaching from the sandy soil was ascribed to the soil being more tightly packed in this column, thereby restricting water and aflatoxin movement. Goldberg and Angle (1985) concluded that the risk of aflatoxin leaching is low because of (1) lack of extensive movement under saturated conditions, (2) relatively high adsorption coefficients in soil and (3) general low solubility in water. Madden and Stahr (1993) concluded that burial of discarded, contaminated crops in soil with 20% silty clay loam will reduce AFB_1 and prevent AFB_2 from leaching, while 50% silty clay loam will prevent both aflatoxins from leaching.

Although compounds like ZEA and OTA consist of aromatic structures, the hydroxyl or carboxylic groups make them somewhat soluble in water (31.8 mg L^{-1} for ZEA and 0.987 mg L^{-1} for OTA), increasing their leaching potential (Mortensen et al. 2006). No literature was found on the risk of trichothecenes leaching to the ground water, probably because of their very low solubility in water.

In general, high adsorption is related to reduced mobility (Goldberg and Angle 1985). There is, however, increasing evidence that even strongly adsorbing chemicals may be rapidly transported to drainage and ground water under appropriate environmental conditions. This was exemplified in a recent review on the herbicide glyphosate, for which sorption properties as well as results with packed columns indicate minimal risk of leaching (Vereecken 2005)–just as for AFB_1. Yet there are several findings that glyphosate can be leached from both lysimeter and field-scale studies and from drainage and ground-water surveys (Jonge et al. 2000; Vereecken 2005). Jonge et al. (2000) found 50–150 times more ^{14}C-glyphosate being leached from structured sandy loam soil than from structureless coarse sandy soil especially when glyphosate was added to the irrigation water and not allowed to adsorb prior to the irrigation event. They concluded that glyphosate may leach from the topsoils, where pronounced macropore (preferential) flow occurs shortly after application of glyphosate. Recent experiments with intact soil monoliths (diameter 60 cm, length 100 cm) demonstrated that oestrogens in slurry, deposited on the top of the monoliths, were leached with irrigation and natural precipitation to 1-m depth in both a sandy and a loamy soil. Their concentration in the leachate was sufficiently high to affect the endocrine system of aquatic wildlife (Laegdsmand and Andersen 2005). If the slurry contains excreted mycotoxins and the compounds escape adsorption, they might well be present in the soil water under conditions facilitating preferential flow. Field investigations have also shown that chemicals may reach subsurface drains by particle-facilitated transport, i.e. the compound is so to say transported on the back of soil colloids, as exemplified for the fungicide prochloraz (Villholth et al. 2000). This mechanism will depend on the sorption and degradation properties of the compound.

Glyphosate has been found more often in surface water than in ground water, and Vereecken (2005) stressed that glyphosate in surface waters may also be a result of surface runoff. I have found no results relating to the risk of macropore flow and surface runoff of mycotoxins. Both mechanisms might, however, be highly relevant, regarding contributions from discarded crops, animal waste and seepage from improperly handled silage.

9.4.3 Degradation

Pure or mixed cultures of mycotoxin-degrading organisms are sought after for biological detoxification of agricultural commodities (Karlovsky 1999). Some of these originate from soil. For example, T-2 toxin can be assimilated by a pure culture of *Curtobacterium* (Ueno et al. 1983) and by mixed communities (Beeton and Bull 1989). The mixed populations were more efficient than the pure cultures by themselves. It has taken more effort to find DON degraders (Karlovsky 1999) but there are reports on a strain from field soil belonging to the *Agrobacterium-Rhizobium* group (Shima et al. 1997) and a strain belonging to a new genus of *Alphaproteobacteria* from a spontaneously infected medium (Karlovsky 1999; Völkl et al. 2004). Völkl et al. (2004) found no DON degraders among 1,000 mixed and pure cultures from soils and cereal ears. Yet DON does not appear to accumulate in soil even though it is chemically stable (Völkl et al. 2004). This indicates some biological degradation, and the authors hypothesised that DON may be degraded by non-cultivatable organisms. Fumonisins are heat- and light-stable (WHO 2000) and only limited evidence of microbial degradation has been presented (Karlovsky 1999; WHO 2000). There are several reports on microbial degradation of aflatoxin in vitro, as reviewed by Karlovsky (1999). These relate primarily to *Flavobacterium aurantiacum* but also to fungi, e.g. *Trichoderma viride* (Bean et al. 1986). Recently *Pleurotus ostreatus*, reknown for its ability to degrade polycyclic aromatic hydrocarbons, was reported to produce an extracellular enzyme that cleaves the lactone ring of AFB_1 (Motomura et al. 2003). Several fungal species are reported to degrade OTA and ZEA in vitro, including isolates from soil (Varga et al. 2005).

It is likely that mycotoxins will be subjected to microbial degradation in the soil environment but available results are very sparse. Angle and Wagner (1980) showed that AFB_1 added in methanol to a silty loam was reduced to the much less toxic AFB_2 and AFG_2 within a few days, indicating the process to be chemical. AFB_2 and AFG_2 were degraded more slowly and probably microbially with levels below the limit of detection after 77 and 49 days, respectively. Microbial decomposition of ^{14}C-AFB_1 was measured as $^{14}CO_2$, and 14% was degraded after 112 days. Wheat straw decreased the degradation rate probably owing to some binding mechanism (Angle and Wagner 1980). It was later shown that ^{14}C-AFB_1 degraded much more slowly in a silty clay loam than in a silty and a sandy loam, indicating the formation of a conjugate between AFB_1 and soil organic matter or clay (Angle 1986). Mortensen et al. (2006) found that both OTA and ZEA were degraded very fast. The degradation data were well fitted by the sum of two first-order reactions, representing an initial very fast degradation and a second, slower transformation. This was interpreted as two pools with different bioavailabilities. A high microbial activity in the rhizosphere was probably the reason why they found less ZEA and OTA in soil planted with barley than in soil without plants (Mortensen et al. 2006). Patulin added to different soil types in a concentration of 400 ppm was degraded within 8 days, somewhat faster in autoclaved than in non-autoclaved soil, and Ellis et al. (1980) deduced from their results that both non-biological and biological degradation took place.

9.4.4 Uptake in Plants

There are a few reports showing that intact plant roots can take up radiolabelled toxins. Day and Mantle (1980) showed that ^{14}C-labelled verrucologen produced by *Penicillium estinogenum* was taken up by two-leaf-stage bean plants (*Phaseolus vulgaris* L.). The toxin was translocated to the shoot in amounts that—if extrapolated into a pasture context—exceeded a level that can make forage tremorgenic. Using a similar method, Mantle (2000) demonstrated OTA uptake by intact roots of small coffee plants but stressed the need to demonstrate whether this occurs in nature. He argued that OTA in coffee is produced by *A. ochraceus* and this fungus is soil-borne according to Frank (1999). However, Frank (1999) stressed that although early *A. ochraceus* contamination most likely occurs in the coffee rhizosphere, the actual route of infection is unknown, and he did not mention the possibility of the toxin itself being taken up by the roots. Mertz et al. (1981) found that lettuce seedlings that had been planted in a loam soil with added AFB_1 (33–267 ppm) could take up the compound. The amounts recovered were less than 1%, probably because the toxin had been strongly adsorbed to the loam.

9.5 Mycotoxigenic Fungi and Mycotoxins in the Soil Environment

Soil is considered the ultimate repository of most mycotoxin-producing fungi (Lillehoj and Elling 1983). Depending on the climatic and edaphic conditions of the given area, arable soil is a more or less hostile environment, as reviewed by Domsch et al. (1983), subjected to fluctuations in temperature, water potential and H^+ ion concentrations, physical disturbance and reduced gas exchange. Furthermore, nutrient limitation may lead to a high degree of stress and competition in capture and combat for resources (Pugh 1980). This section will exemplify how some of these aspects affect survival and growth of mycotoxigenic fungi in the soil environment, possible in situ production of mycotoxins, and interactions between mycotoxigenic fungi and inherent soil organisms.

9.5.1 Survival of Mycotoxigenic Fungi in Soil

The majority of *Fusarium* species are soil-borne or both soil-borne and airborne e.g. *F. culmorum* and *F. verticillioides* (= *F. moniliforme*). The soil-borne mode of existence seems to some extent dependent on the ability to produce chlamydospores for survival (Burgess 1981). Overwintered stubble of maize and small-grain cereals provides inoculum for *F. graminearum* (teleomorph *Gibberella zeae*), *F. culmorum* and *F. avenaceum*, the three main species responsible for foot rot (seedling blight) and head blight of small-grain cereals in Europe (Bottalico and Perrone 2002) and North America (Gilbert and Fernando 2004). During the last few decades, conservation tillage leaving a substantial cover of crop residues on the soil surface has become still more important in many countries owing to reduced costs and reduced risk of soil and water loss (Carter et al. 2004). There is a simultaneous risk, however, that survival, sporulation and dispersal of toxigenic fusaria on surface residues will increase in comparison with the case for ploughed-under residues (Krebs et al. 2000; Gilbert and Fernando 2004). For example, Krebs et al. (2000) showed that no-tilled maize as a preceding crop led to much higher DON contents in the following wheat crop than ploughed maize. Although Gilbert and Fernando (2004) recognised the importance of tillage practice in the survival of inoculum, they pointed to weather conditions around flowering as being more important in head blight development.

FB_1-producing strains of *Fusarium* can survive for many months in buried maize stalk residues and especially in surface residues. Cotten and Munkvold (1998) found that residue size and residue depth had significant effects on survival and that the fungi survived longer in surface residues than in buried residues. Survival decreased linearly over time in the buried residues. A linear model could not be used for surface residues, probably owing to fluctuations in temperature and moisture and recolonisation by inherent fungi.

The ecology and the pathology of the aflatoxigenic species *A. flavus* and *A. parasiticus* have been extensively reviewed (Wicklow 1995; Horn 2003). These fungi are widely distributed in soil in the maize-producing regions of the USA. Sclerotia of *A. flavus* dispersed by the combine during harvest or in diseased plant material serve as resistant overwintering inoculum (Wicklow et al. 1984). Jaime-Garcia and Cotty (2004) showed that even 2-year-old cobs contained 45 times more propagules than the surrounding soil. Zummo and Scott (1990) showed that sclerotia of *A. flavus* survived better in the soil environment than those of *A. parasiticus*. Horn (2003) speculated that sclerotia may be of greatest importance in natural habitats or fallow fields where soil populations of aflatoxigenic fungi are low and where preferred substrates such as seeds are rare or not immediately available. Not only sclerotia but also conidia of *A. flavus* can survive the winter and serve as inoculum for grain infection (Zummo and Scott 1990). The population fluxes in soil depend on the combination of conidium mortality and influx of conidia from infected crops and/or colonisation of soil organic matter (Horn 2003). In 1988 there were very high temperatures and drought in eastern Iowa and many fields produced maize with aflatoxin contents above 20 ppb (Shearer et al. 1992). Forty of these fields were investigated over a 3-year period. The soil population of *A. flavus/A. parasiticus* isolates declined from about 1,200 colonies per gram of soil in 1988 to 400 colonies per gram of soil in 1990. In opposition to the southeastern regions of the USA, where sclerotia are considered the most important survival structure, the results of Shearer et al. (1992) indicated that other survival mechanisms (hyphae or condidia) may be more important in the Midwest.

The low abundances of naturally occurring *P. verrucosum* propagules indicate that this species is an ephemeral invader brought into the soil during soil management (Elmholt 2003). The fungus can, however, survive in soil for many months and proliferate even without addition of nutrient resources in the form of waste grain (Elmholt and Hestbjerg 1999). Increases in conidial abundance were primarily observed in autumn and spring months, while the dry summer months showed a major decrease in abundance.

Zardari and Clarke (1988) found that propagule numbers of *P. expansum* decreased slowly in unsterilised soil. *P. expansum* is a ubiquitous fungus whose conidia are commonly found in orchard soil and orchard litter, and it is therefore generally believed that inoculum is not a limiting factor for development of this pathogen (Xu and Berrie 2005).

9.5.2 Saprotrophic Growth of Mycotoxigenic Fungi in Soil

Many soil-borne species of *Fusarium* are facultative saprotrophs. They are capable of a parasitic life style and mostly with a wide host spectrum, but their prime mode of existence is saprotrophic (Burgess 1981). The physiological functioning of these fusaria has been intensively studied mostly in the attempt to elucidate the factors contributing to disease, including growth at different temperatures, moisture and gaseous composition. Most studies have been performed under laboratory conditions, yet many results are relevant for soil conditions. It is beyond the scope of this chapter to go into detail and the reader is referred to reviews on these topics (Frisvad and Samson 1991; Domsch et al. 1993; Parry et al. 1995; Bottalico and Perrone 2002; Munkvold 2003; Gilbert and Fernando 2004; Marin et al. 2004).

Hestbjerg (1999) hypothesised that *Fusarium* intraspecies variation might relate to the existence of ecotypes that had developed in response to climatic and edaphic selection regimes. She studied the growth and mycotoxin-producing ability of soil-borne isolates of *F. culmorum* and *F. equiseti* from eight sites in five countries. *F. culmorum* isolates revealed an irregular growth pattern on the nutrient poor special nutrient-poor agar (SNA) but a very regular growth on the more nutrient rich potato sucrose agar (PSA). This was interpreted as an adaptation to more nutrient rich environments like living plants or fresh plant residues. The 45 *F. culmorum* isolates from the Norwegian N9 site, however, differed by having a regular growth pattern on both PSA and SNA and showing remarkably little variation in growth rate on PSA at 15 °C and large variations at 25 °C. The ecological explanation could be a stronger selection pressure towards growth at low temperatures, and the isolates from N9 may represent an ecotype within this species. The different results obtained at the two temperatures underline that ecological conclusions drawn from laboratory studies should be treated with care, as they will depend on given cultural conditions. *F. equiseti* on the other hand revealed an irregular growth pattern on PSA but a regular growth pattern on SNA, indicating adaptation to a saprotrophic mode of existence and probably a preference for nutrient-poor resources.

In 1975, aflatoxin contamination was reported for the first time in preharvest maize at all stages of development and maturity (Anderson et al. 1975). Until then aflatoxigenic fungi had been regarded as 'storage fungi'. The discovery that these fungi had part of their life cycle in the field initiated considerable research on their ecological and pathological behaviour in the field ecosystem (Wicklow 1995; Horn 2003). *A. flavus* may actively colonise residues of maize (Zummo and Scott 1990) and colonise and proliferate on ploughed-down residues of green manure rye and peanut fruits (Griffin and Garren 1976). Griffin and Garren (1976) found the highest abundance of *A. flavus* in soils with high amounts of clay and organic matter and a resulting high water content at field capacity. Soil type seemed more important to growth and sporulation of *A. flavus* than residue type. Furthermore the highest activity of the fungus was seen during the summer months probably owing to the fairly high optimum temperature of conidial germination (35 °C). Wicklow (1995) reported that sclerotial germination occurred in maize fields just prior to silking. This was probably promoted by high surface moisture, provided by the shading canopy, since sclerotia on an adjacent fallow ground did not germinate. *A. parasiticus*, which primarily attacks subterranean peanuts, seems better adapted to soil growth than *A. flavus*, and Horn (2003) presented several examples that soil populations increased beyond the level that can be expected from the sole presence of peanuts.

Though *P. verrucosum* is regarded a storage fungus (Abramson 1998), it has been argued that more focus should be put on its possible role in the field ecosystem (Lillehoj and Elling 1983; Miller 1995). Elmholt (2003) found *P. verrucosum* in 14 of 76 Danish fields and these findings are the first reports on the fungus being isolated from arable soil. Most soils contained only few propagules (100 to 300 cfu g^{-1} soil). When dilution plating is used, species with such low abundance will not be detected on general isolation media, and this may be why this species is normally not reported in arable soil (Elmholt and Kjøller 1989; Elmholt et al. 1993). The findings of Elmholt (2003) can probably be ascribed to using the selective and *P. verrucosum* diagnostic Dichloran yeast extract sucrose 18% glycerol (DYSG) agar, which was recommended for isolation of this species from foods, feeds (Frisvad et al. 1992) and soil (Elmholt and Hestbjerg 1996; Elmholt et al. 1999). Elmholt et al. (1999) showed that *P. verrucosum* could be detected in conidial concentrations below 200 cfu g^{-1} soil even when constituting no more than 0.3% of the total fungal colony-forming units of the soil tested. The experiments were performed with infested soils and the precision of propagule recovery was very good when using a proper mathematical model to calculate fungal abundance

(Elmholt et al. 1999). Later studies have confirmed that DYSG agar is extremely useful in ecological studies of *P. verrucosum* (Elmholt and Hestbjerg 1999; Elmholt 2003; Elmholt and Rasmussen 2005). Selective and diagnostic media have also been developed for the aflatoxigenic species (Beuchat 1995). The low abundances indicate that *P. verrucosum* is normally an ephemeral soil invader brought into the soil during soil management, but two soils had high frequencies (Elmholt 2003). In one, distinct spatial variations indicated that the fungus had established and proliferated in the soil environment. This is in accordance with Elmholt and Hestbjerg (1999), who conducted a field experiment that showed *P. verrucosum* is able to proliferate in soil both with and without addition of nutrient resources in the form of wheat grains.

Paster et al. (1995) studied growth and patulin production in three different strains of *P. expansum*. At temperatures from 0 to 30 °C, all strains were able to grow and produce lesions of 45 mm on apples and pears in 12–52 days.

9.5.3 Are Mycotoxins Produced In Situ in the Soil?

Are mycotoxins produced in situ in the soil by actively growing fungi? If yes, what is the purpose? These questions have intrigued many scientists but firm answers are still lacking, probably in part owing to methodological problems. If produced in situ—for example in resource competition between two organisms—the toxin will most likely be produced in amounts below the detection limit in the bulk sample of soil that is analysed. Elmholt and Mortensen (2003) performed a preliminary experiment with four oat samples and four spring wheat samples that were sown in the field. They were all heavily contaminated by *P. verrucosum* and some contained OTA at high levels. Yet neither bulk nor rhizosphere soil contained OTA in detectable amounts. Mantle (1998) speculated that trace amounts of OTA are generated in soil through saprophytic *Aspergillus* activity but it still remains to be demonstrated that OTA can be produced under soil conditions.

To my knowledge, there is only one example where the mycotoxins addressed in this chapter have been detected in soil, i.e. Norstadt and McCalla (1969), who studied microbial populations in ploughed and stubble-mulched soil. They found patulin in one soil sample (1.5 ppm) and in two wheat straw samples (40 and 75 ppm), all from subtilled plots. The authors assumed that the patulin was used in a microbial battle for the wheat straw resource, as plate counts of *Penicillium urticae*—the supposed producer—increased in number, while those of its antagonist, *Trichoderma* sp., decreased. Much in the same line, Horn (2003) proposed that aflatoxin production might be important in soil habitats rather than being a selective advantage in crops. He argued that the ecological niches in soil are more diverse and strains may be partitioned in some manner according to their aflatoxin-producing ability. Section 9.3.1 addressed the large variations in mycotoxin contents in seeds. Large variations might well be found in soil resources as well. Results with inappropriately stored maize showed how organic resources with high contents of mycotoxins—in this case kernels—were found adjacent to kernels with no detectable mycotoxin although the kernels were interconnected by mould growth (Shotwell et al. 1975). This indicates that very locally occurring conditions control mycotoxin production. For example, 20 kernels that were placed in physical sequence as posted below contained ZEA in the following amounts:

ND–ND–ND–ND–ND–94 ppm–ND–ND–ND–ND–ND–ND–ND–35 ppm–37 ppm–1,700 ppm–ND–530 ppm–ND–ND,where ND means not detected, i.e. a kernel without detectable amounts of ZEA. No kernels contained both ZEA and aflatoxin, indicating that either *Fusarium* or *Aspergillus* had ousted the other (Shotwell et al. 1975).

In vitro studies have shown that mycotoxin production is highly dependent on moisture level and temperature. In general, the critical moisture level for toxin production is often higher than for growth (Frisvad and Samson 1991; Marin et al. 2004) and the temperature range is narrower (Esteban et al. 2004). Sauer and Burroughs (1980) illustrated the complexity of mycotoxin formation by showing how even small variations in interseed humidity were critical to growth of *A. flavus* and aflatoxin production. Marin et al. (2004) very elegantly demonstrated how germination, growth and fumonisin production in *F. verticillioides* and *F. proliferatum* depended on temperature and water activity (a_w). FB_1 was produced over a much narrower moisture and temperature range than required for germination and growth of *F. proliferatum* and especially *F. verticillioides*. Ryu et al. (1999) furthermore demonstrated that FB_1 production in rice was stimulated by cycling temperatures, though differently for the two species *F. verticillioides* and *F. proliferatum*.

OTA production is highly dependent on the interaction between water availability, temperature and time (Northolt et al. 1979; Lillehoj and Elling 1983; Müller and Boley 1992), with water availability being more limiting to OTA production than temperature (Christensen et al. 1992; Lindblad et al. 2004). Lindblad et al. (2004)pointed to 17 18% moisture as a critical lower limit for OTA production. We found that even at 2 °C, *P. verrucosum* grew well and formed OTA if provided with sufficient grain moisture and time (Fig. 9.4), while others set 4 °C as the lower limit for OTA production (Lillehoj and Elling 1983; Müller and Boley 1992). Lillehoj and Elling (1983) reported 4–31 °C as the temperature interval for OTA production by *P. verrucosum* and 12–37 °C for *A. ochraceus*, explaining why the former dominates in temperate areas of the world and the latter in subtropical and tropical areas.

Nutrient availability plays a significant role in fungal physiological functioning and production of secondary metabolites. In the attempt to elucidate the possible role of mycotoxins in the soil environment, it is important to think of fungal metabolism in an ecological context—from a fungal point of view so to say (Hestbjerg 1999). To do so it is necessary to look critically at the cultural conditions we use in our laboratories. Many ecology studies of toxigenic fungi have been based on synthetic media, often chosen out of convenience and tradition in a given laboratory. There are, however, many examples showing that there choice of medium has a substantial effect on fungal growth and production of mycotoxins (Filtenborg et al. 1990; Frisvad et al. 1992; Beuchat 1995). For example, Esteban et al. (2004) found that *A. niger* formed more OTA on yeast extract sucrose (YES)

Fig. 9.4 *Penicillium verrucosum* growing on kernels at low temperature (2 °C)

agar than on Czapek yeast autolysate agar (CYA), while *A. carbonarius* reacted oppositely. No explanation for this was offered. Most laboratory media are rich in nutrients and energy. Soil, however, is an environment which is often characterised by some degree of substrate limitation (Pugh 1980).

Hestbjerg et al. (2002b) showed that 20 of 24 *F. culmorum* isolates produced ZEA on YES agar (2.2 g N L^{-1}; C/N ratio 33), while only one did so on PSA (0.1 g N L^{-1}; C/N ratio 84). DON was only produced on the nitrogen-rich YES agar. Some *F. equiseti* on the other hand produced ZEA on PSA but no DON or ZEA on YES agar. For comparison, a soil organic matter (SOM) agar (0.37 g N L^{-1}; C/N ratio 18) was tested. It consisted of 2% SOM and 2% agar in a soil extract. SOM was obtained by washing and sieving soil from a wheat field. *F. equiseti* did not support production of known metabolites on SOM agar, while *F. culmorum* produced chrysogine on SOM agar. These results show that the potential of secondary metabolism must be related to the medium provided. Some of the significant ecological aspects discussed by Hestbjerg (1999) were related to differences in fungal utilisation of the laboratory media. These aspects would have been missed if only the overall profile of metabolites had been recorded. The SOM agar medium was introduced to obtain a more natural medium for laboratory studies of fungal ecological relationships, and future research ought to be more aware of this aspect.

The influence of organic resource was also shown by Llorens et al. (2004), who studied the influence of temperature, moisture and strain on ZEA production in different small-grain cereals. They found that strains of *F. culmorum* produced more ZEA on wheat than on maize and rice, while the opposite was found for strains of *F. graminearum*. Micronutrients play a significant role in formation of mycotoxins too, e.g. OTA, as reviewed by Lillehoj and Elling (1983). Filtenborg et al. (1990) showed how YES agar, containing different brands of yeast extract, supported mycotoxin production very differently. The exact reason was not found but amendments of $MgSO_4$, $ZnSO_4$ and $CuSO_4$ to the medium increased the production of, e.g., patulin by *P. expansum* and OTA by *P. verrucosum*.

Distinct chemotypes are found in several species of *Fusarium*, e.g. *F. culmorum* and *F. graminearum* (Gang et al. 1998; Bakan et al. 2001, 2002; Bottalico and Perrone 2002). For example, some isolates of *F. culmorum* are DON and acetyl-DON producers, while others produce NIV and fusarenon-X. To some degree, the chemotypes seem to differ in geographical origin; the NIV chemotype has, for example, not been found in the USA (Lee et al. 2001). However, Bakan et al. (2001) could not relate the chemotypes to different regions in France and in some soils the two chemotypes coexist (Hestbjerg et al. 2002b). Different chemotypes also exist for several species of *Penicillium* (Frisvad and Filtenborg 1989). For example, *P. verrucosum* was divided into chemotype I found on processed meat products and chemotype II found on cereals. The former was later ascribed to a separate species, *P. nordicum* (Larsen et al. 2001).

Toxin production within isolates of the same species grown in similar cultural conditions may also vary much. Gilbert et al. (2002) showed how 15 isolates of *F. graminearum* differed in their ability to produce trichothecenes. Production of DON varied from 0.2 to 249 ppm, while the production of 15-acetyl-DON varied from 0.5 to 44.6 ppm. Disease severity in growing wheat and ergosterol and trichothecene production in vitro were not correlated. Gang et al. (1998) found that production of DON by 42 isolates of *F. culmorum* varied from 0.7 to 60.3 mg kg^{-1} rye in the field and from 0.4 to 376.3 mg kg^{-1} rye in vitro. In the field, these differences were correlated with disease development but the authors pointed out that isolates that produced high DON contents also produced high contents of ergosterol in the plants. Therefore, differences in DON content may reflect different amounts of fungal biomass in the host. Other examples of strain differences are reported for DON (Snijders and Perkowski 1990; Bakan et al. 2001, 2002) and OTA (Esteban et al. 2004).

9.5.4 Interaction of Mycotoxigenic Fungi with Other Organisms in the Soil Environment

If mycotoxigenic fungi are to establish themselves in the soil environment, they must compete successfully with all living organisms that claim the same physical habitat and nutrient resources. Species with anamorphs in *Fusarium, Penicillium* and *Aspergillus* are generally characterised by high sporulation and fast growth.

Citing Pugh (1980), 'they are ephemeral colonists of ephemeral substrates', often residing for a long time as dormant propagules in the soil, whereupon they germinate and grow rapidly to exploit added resources like root exudates or ploughed-in organic matter. This characterisation is very appropriate when comparing them withsoil fungi with very different life strategies (e.g. the *Basidiomycetes*), but from the literature it is evident that also closely related species within these genera can have different strategies. This was recently exemplified by Marin et al. (2004) in their review on the ecophysiology of fumonisin-producing species of *Fusarium, F. verticillioides* and *F. proliferatum*. They outline how the two species react differently to interactions between abiotic and biotic environmental factors, a_w and temperature, and which role fumonisins might play in colonisation and defence of captured nutrient resources. Apart from effects on soil-dwelling vertebrates, some mycotoxins have antibiotic and phytotoxic effects. A few examples of interactions of mycotoxigenic fungi with microorganisms, plants, and animals are given next.

9.5.4.1 Microorganisms

Marin et al. (2004) distinguished between the two competition situations, primary resource capture and resource combat, which call for different life strategies. Primary colonisers grow quickly and have rapidly germinating spores. Regarding growth rate, both *F. verticillioides* and *F. proliferatum* grow faster than competing species, e.g. *F. graminearum*, over a wide range of a_w and temperatures. However, they do not germinate under 0.88a_w, which narrows their niche in comparison with, e.g., *Eurotium*. However, primary resource capture is not always necessary for mycotoxigenic fungi in soil. As discussed in Sect. 9.3.1, many toxigenic fungi will act as soil invaders, being present and sometimes actively growing on plant material that is brought into the soil. These fungi have the competitive advantage of being prior colonisers in the sense of Bruehl and Lai (1966). When they enter the soil they will have to combat for—rather than capture—their resource. Essential in this combat are diffusive or volatile substances with antibiotic effects as well as the ability of the fungus to compete for nutrient resources upon hyphal contact. Marin et al. (2004) argued that production of fumonisins might be more advantageous in resource combat than in resource capture. They cited a number of studies on competition between *F. verticillioides* and *F. graminearum* or *F. verticillioides* and *A. parasiticus*, but concluded that at present FB_1 production cannot be directly linked with competition strategy.

A recent study by Utermark and Karlovsky (2007) provides a good example of mycotoxins acting in resource combat. The authors demonstrated how ZEA exerted a toxic effect on several filamentous fungi and that the effect declined in the order ZEA > α-zearalenol > β-zearalenol. Strains of *Gliocladium roseum* normally produce a lactonase, which catalyses the hydrolysis of ZEA, and these strains are not sensitive. As part of the experiment, mutants without the *zes2* lactanase gene were produced, and these strains were more sensitive to ZEA, suggesting that the biological function of this mycotoxin is most likely associated with fungal defence agaist mycoparasites (Utermark and Karlovsky 2007).

Most examples of interactions between mycotoxigenic fungi and inherent microbial soil populations relate to biocontrol, i.e. the ability of non-toxigenic species/strains to oust toxigenic species by inhibiting their growth and/or mycotoxin production. Most of these studies have been performed in the laboratory. For example, fusarium head blight causing fungi that survive on straw residues are inhibited by, e.g., *Trichoderma harzianum* (Gilbert and Fernando 2004). Likewise sclerotia of *A. flavus* are colonised by fungi in the soil and there have been attempts to use this mechanism in biocontrol (Will et al. 1994). Bean et al. (1986) found that *T. viride* significantly reduced the amount of aflatoxin produced by *A. parasiticus* in liquid media. As both fungi grew well, the authors hypothesised that they competed for some nutrient that is required for toxin production.

Trichothecenes have some antibiotic properties (Vesonder et al. 1981; Madhyastha et al. 1994) but it is not known whether these compounds play a role in resource competition. Angle and Wagner (1980) hypothesised that the high toxicity and mutagenic properties of AFB_1 contributed to its relatively slow degradation in soil. In follow-up experiments they found that soil populations of bacteria were more sensitive to AFB_1 than fungi although the inhibitory effects were reversible within the 6-week experimental period (Angle and Wagner 1981). The mutagenic properties were revealed as higher frequencies of revertant colonies per plate of *Rhizobium japonicum*, when concentrations of AFB_1 in the Petri dishes exceeded 100 ppb. The overall microbial activity (CO_2 evolution), however, was not affected significantly by AFB_1, nor were the nitrifying bacteria. Norstadt and McCalla (1969) proposed that patulin is used in the microbial battle for wheat straw resources as plate counts of the patulin-producing *P. urticae* and its antagonist *Trichoderma* sp. showed a reverse trend. Nicoletti et al. (2004) studied three isolates of *P. expansum* and found a correlation between their production of patulin and their ability to inhibit the growth of *Rhizoctonia solani*, the etiological agent of damping-off of tobacco. Stressing that results were based on only three strains of *P. expansum*, the authors suggested that antagonism of *P. expansum* towards *R. solani* might be related to their patulin production.

9.5.4.2 Animals

In 1977, Janzen (1977) presented a thought-provoking paper on ‘Why fruits rot, seeds mold and meat spoils’. He argued that rather than being ‘a metabolic accident’, the mycotoxin-producing ability is an evolutionary adaptation enabling its producer to protect itself and claim its resource in combat with larger arthropods or rodents. This assumption was supported by Wicklow (1995) although it is difficult to confirm experimentally (Marin et al. 2004).

There are numerous examples of interactions between mycotoxigenic fungi and insects in aboveground plants and stored plant products, e.g. for OTA (Blank et al. 1995) and aflatoxin, the latter being a potent insecticide (Sinha and Sinha 1992;

Wicklow 1995; Schatzki and Ong 2000). *A. flavus* is known as an insect pathogen, able to attack several lepidopterous insects and causing, e.g., the koji-kabi disease of silkworms (Wicklow 1990; Vineet et al. 2004). In the soil ecosystem, peanuts are infected by fungi through direct contact with soil and it is well established that damage of pods and seeds by various insects makes the peanut crop highly susceptible to invasion by aflatoxigenic fungi, especially if the plants are drought-stressed (Horn 2003). Also nitidulid beetles serve as vectors for *A. flavus*, dispersing fungal propagules by several metres from waste grain deposits of corn (Olanya et al. 1997).

Although a toxin-producing ability would seem a competitive advantage, there are exceptions. This is beautifully exemplified by the ancient symbiosis between the banner-tailed kangaroo rat (*Dipodomys spectabilis*) and the fungal species colonising its seed caches—a small-scale subterranean grain store. While humans have stored seed for some 7,000 years, this desert rodent species has done so for at least 10,000,000 years, leaving more time for coevolution (Frisvad et al. 1987). Wicklow (1995) described how the foraging behaviour and active management of the seed caches by the kangaroo rat have resulted in a 'domesticated' population of specific mould varieties. These specific varieties reside in the burrows and seem to colonise the seed when manipulated by the rodent. They are varieties of well-known ubiquitous species, and the interesting fact is that they produce antibiotics but apparently not mycotoxins (Frisvad et al. 1987). For example, *Penicillium chrysogenum* var. *dipodomyis* does not produce PR toxin and roquefortine C known from *P. chrysogenum*, and *P. aurantiogriseum* var. *neoechinulatum* does not produce the potent nephrotoxins xanthomegnin and viomellein and the tremorgenic penitrem A. Frisvad et al. (1987) speculated that the rough-walled stipes of these two varieties are a mechanical defence against arthropod predation and that the antibiotics may play a role in protecting the kangaroo rat from parasites or infectious microorganisms.

Several mycotoxins from the same or from different fungal species may occur simultaneously in plant products (Speijers and Speijers 2004) or in the soil environment. If these compounds elicit a synergistic toxic response the hazard to affected organisms will be increased. Sansing et al. (1976) showed that combinations of OTA and citrinin and of OTA and penicillic acid, respectively, decreased the LD_{50} in mice when compared with the individual toxins. The synergism between OTA and citrinin was later verified, as reviewed by Speijers and Speijers (2004), and was recently demonstrated by Bernhoft et al. (2004). Synergism between OTA and FB_1 has also been reported (Creppy et al. 2004).

Tajima et al. (2002) proposed a tiered statistical approach with a sequence of test stages as suitable strategy for investigating interactions between mycotoxins. They tested the effect of DON, NIV, T-2, ZEA and FB_1 on the inhibition of DNA synthesis in mammalian L929 cells and found the interaction between mycotoxins more complex than expected. Most effects were additive or less than additive but they also found synergistic effects, e.g. between T-2 and NIV. Creppy et al. (2004) pointed to the crucial importance of doses and concentrations when testing combined effects of mycotoxins and stressed the importance of considering whether the chosen combinations and doses are realistic for the commodity or environment in question. This was supported by Tajima et al. (2002), demonstrating that experiments need to be carried out covering the whole dose–response curve owing to differences in response at different dose levels.

9.5.4.3 Plants

It is well documented that some trichothecenes play a role in pathogenesis of above-ground plant parts and that *Fusarium* isolates producing these compounds have a selective advantage over those that do not (Desjardins et al. 1993; Proctor et al. 1995; Desjardins and Hohn 1997; Mesterhazy et al. 1999; Xu and Berrie 2005). Hestbjerg et al. (2002a) found that all tested isolates of *F. culmorum* caused seedling blight, and that there was significant correlation between DON concentration in the seedlings and disease severity. This correlation was also found between fusarium head blight and DON (Gang et al. 1998). The DON chemotype of *F. culmorum* appears more phytotoxic than the NIV chemotype (Gang et al. 1998; Miedaner and Reinbrecht 2001).

Fumonisins also possess phytotoxic properties, as reviewed by McLean (1996), but the possible role of fumonisins in plant pathogenesis is dubious (Munkvold 2003). Patulin has phytotoxic effects, reducing seed germination and growth of wheat seedlings (Ellis et al. 1980).

9.6 Conclusions

In conclusion, mycotoxins and their producers affect all three aspects of soil quality: productivity, environment and health. As an example of productivity effects, the trichothecene-producing *F. culmorum* and *F. graminearum* infect small-grain cereals and experimental evidence confirms that the trichothecenes play a role in development of, e.g., fusarium head blight—an economically very important plant disease. Other examples are maize infections by *A. flavus* and *F. verticillioides* and peanut infections by *A. parasiticus*. Most toxinogenic fungi can survive in soil for several months and many of them can grow saprothophically if provided with proper nutrient resources and environmental conditions of temperature, soil moisture, etc. These demands will of course vary from species to species as also demonstrated earlier. Mycotoxins can be formed in plants prior to harvest and enter the soil via waste grain, stubbles and roots. Among these are trichothecenes, fumonisins and aflatoxins. Crops that are heavily infected by *F. graminearum/F. culmorum* or *A. flavus* can contain a high percentage of kernels with substantial contents of DON and aflatoxin, respectively—so high that the crop is sometimes rejected for consumption–the human and animal health aspect of soil quality.

Instead of being used for food and feed, such crops are left in the field and ploughed under, leading to a trade-off with the environmental aspect of soil quality, as mycotoxins may affect the soil biota. Studies on the fate of mycotoxins in the soil environment are very sparse–practically non-existent–perhaps because chemical properties and initial soil column experiments with aflatoxin and fumonisin indicated that mycotoxins would be adsorbed to the soil matrix and not liable to leaching. However, recent studies on pesticides with similar chemical properties show that these can be leached to the ground water as has also been shown for oestrogens in swine slurry. The reason is that water flow under field conditions will often be much more complex than in packed soil columns. Water flow in the field is subjected to preferential and/or colloid-facilitated flow, increasing the risk of mycotoxin contamination of the drinking water resources—another aspect of potential human health effects of mycotoxins entering the soil environment. Mycotoxins formed before and after harvest may enter the soil if they are fed to the animals and excreted undegraded in animal urine and faeces. Excretion products can be deposited in the field by outdoor-reared animals or via animal fertilisers.

Although mycotoxins may enter the soil and water environment by several routes I have not found any reports trying to quantify the contributions from different sources. This lack of knowledge is surprising when considering the worldwide interest in the toxicological properties of compounds like aflatoxins, trichothecenes, fumonisins and ochratoxin (WHO 1990, 2000).

I would like to draw attention to another field of research, where Kolpin et al. (2002) also wondered how surprisingly little was known of the environmental fate of another group of organic chemicals—the synthetic pharmaceuticals and hormones. Instead of making a risk assessment by studying the fate of single compounds in soil, they approached the problem by performing a nationwide survey of a broad suite of 95 organic wastewater components in streams across the USA. The aim was to elucidate which of these compounds—if any—actually constituted a toxicological problem and then to conduct detailed studies with these compounds. In a similar manner, mycotoxins could be monitored in drain water, ground water and surface water and could perhaps be included in some of the on-going surveys that are already being performed on pesticides. In addition, there is a need to elucidate the fate of mycotoxins in slurry and manure.

Acknowledgements This work was partly financed by grants from the Ministry of Food, Fisheries and Agri Business and was performed within the context of the Danish Research Centre for Organic Farming, DARCOF (project PREMYTOX).

References

Abramson D (1998) Mycotoxin formation and environmental factors. In: Sinha KK, Bhatnagar D (eds) Mycotoxins in agriculture and food safety. Dekker, New York, pp 255–277

Abramson D, Clear RM, Nowicki TW (1987) *Fusarium* species and trichothecene mycotoxins in suspect samples of 1985 Manitoba wheat. Can J Plant Sci 67:611–619

Anderson HW, Nehring EW, Wichser WR (1975) Aflatoxin contamination of corn in the field. J Agric Food Chem 23:775–782

Angle JS (1986) Aflatoxin decomposition in various soils. J Environ Sci Health B 21:277–288

Angle JS, Wagner GH (1980) Decomposition of aflatoxin in soil. Soil Sci Soc Am J 44:1237–1240

Angle JS, Wagner GH (1981) Aflatoxin B1 effects on soil microorganisms. Soil Biol Biochem 13:381–384

ApSimon JW (2001) Structure, synthesis, and biosynthesis of fumonisin B1 and related compounds. Environ Health Perspect 109(Suppl 2):245–249

Bakan B, Pinson L, Cahagnier B, Melcion D, Semon E, Richard MD (2001) Toxigenic potential of *Fusarium culmorum* strains isolated from French wheat. Food Addit Contam 18:998–1003

Bakan B, Giraud-Delville C, Pinson L, Richard-Molard D, Fournier E, Brygoo Y (2002) Identification by PCR of *Fusarium culmorum* strains producing large and small amounts of deoxynivalenol. Appl Environ Microbiol 68:5472–5479

Battaglia R, Hatzold T, Kroes R (1996) Occurrence and significance of ochratoxin A in food. Food Addit Contam 13(Suppl):1–3

Bean GA, Reise RW, Goldberg BS, Angle JS (1986) The fate of aflatoxins in soil. In: Barry S, Houghton DR (eds) Biodeterioration 6. Papers presented at the 6th international biodeterioration symposium. CAB International, Wallingford, pp 269–272

Beeton S, Bull AT (1989) Biotransformation and detoxification of T-2 toxin by soil and freshwater bacteria. Appl Environ Microbiol 55:190–197

Bennett JW (1987) Mycotoxins, mycotoxicoses, mycotoxicology and mycopathologia. Mycopathologia 100:3–5

Bennett JW, Bentley R (1989) What's in a name? Microbial secondary metabolism. Adv Appl Microbiol 34:1–28

Bernhoft A, Keblys M, Morrison E, Larsen HJS, Flåøyen A (2004) Combined effects of selected *Penicillium* mycotoxins on in vitro proliferation of porcine lymphocytes. Mycopathologia 158:441–450

Beuchat LR (1995) Media for detecting and enumerating yeasts and moulds. In: Corry JEL, Curtis GDW, Baird RM (eds) Culture media for food microbiology. Elsevier, Amsterdam, pp 229–242

Bhat RV, Beedu SR, Ramakrishna Y, Munshi KL (1989) Outbreak of trichothecene mycotoxicosis associated with consumption of mould-damaged wheat products in Kashmir Valley, India. Lancet 1:35–37

Blank G, Goswami N, Madrid F, Marquardt RR, Frohlich AA (1995) Evaluation of *Tribolium castaneum* (Herbst) (Coleoptera: Tenebrionidae) excreta on ochratoxin production in stored wheat. J Stored Prod Res 31:151–155

Blank R, Rolfs J-P, Südekum KH, Frohlich AA, Marquardt RR, Wolframm S (2005) Effects of chronic ingestion of ochratoxin A on blood levels and excretion of the mycotoxin in sheep. J Agric Food Chem 51:6899–6905

Boorman GA (1989) Toxicology and carcinogenesis studies of ochratoxin A (CAS no. 303-47-9) in F344/N rats (gavage studies). Natl Toxicol Program Tech Rep Ser 358: 1–142

Bottalico A, Perrone G (2002) Toxigenic *Fusarium* species and mycotoxins associated with head blight in small-grain cereals in Europe. Eur J Plant Pathol 108:611–624

Bruehl GW, Lai P (1966) Prior-colonization as a factor in the saprophytic survival of several fungi in wheat straw. Phytopathology 56:766–768

Burgess LW (1981) General ecology of the fusaria. In: Nelson PE, Tousson TA, Cook RJ (eds) *Fusarium*: diseases, biology and taxonomy. Pennsylvania State University, University Park, pp 225–235

Carter MR, Andrews SS, Drinkwater LE (2004) Systems approaches for improving soil quality. In: Schjønning P, Elmholt S, Christensen BT (eds) Managing soil quality: challenges in modern agriculture. CAB International, Wallingford, pp 261–282

Cavaglieri LR, Andres L, Ibanez M, Etcheverry MG (2005) Rhizobacteria and their potential to control *Fusarium verticillioides*: effect of maize bacterisation and inoculum density. Antonie Van Leeuwenhoek 87:179–187

Christensen CM, Miller BS, Johnston JA (1992) Moisture and its measurement. In: Sauer DB (ed) Storage of cereal grains and their products. American Association of Cereal Chemists, St Paul, pp 39–54

Cote LM, Dahlem AM, Yoshizawa T, Swanson SP, Buck WB (1986) Excretion of deoxynivalenol and its metabolite in milk, urine, and feces of lactating dairy cows. J Dairy Sci 69:2416–2423

Cotten TK, Munkvold GP (1998) Survival of *Fusarium moniliforme, F. proliferatum*, and *F. subglutinans* in maize stalk residue. Phytopathology 88:550–555

Creppy EE (1999) Human ochratoxicosis. J Toxicol Toxin Rev 18:277–293

Creppy EE (2005) Update of survey, regulation and toxic effects of mycotoxins in Europe. Toxicol Lett 127:19–28

Creppy EE Chiarappa P, Baudrimont I, Borracci P, Moukha S, Carratu MR (2004) Synergistic effects of fumonisin B1 and ochratoxin A: are in vitro cytotoxicity data predictive of in vivo acute toxicity? Toxicology 201:115–123

Dänicke S, Goyarts T, Valenta H, Razzazi E, Bohm J (2004) On the effects of deoxynivalenol (DON) in pig feed on growth performance, nutrients utilization and DON metabolism. J Anim Feed Sci 13:539–556

Day JB, Mantle PG (1980) Tremorgenic forage and ryegrass staggers. Vet Rec 106:463–464

Desjardins AE, Hohn TM (1997) Mycotoxins in plant pathogenesis. Mol Plant Microbe Interact 10:147–152

Desjardins AE, Hohn TM, McCormick SP (1993) Trichothecene biosynthesis in Fusarium species: chemistry, genetics and significance. Microbiol Rev 57:595–604

Domsch KH, Jagnow G, Anderson TH (1983) An ecological concept for the assessment of side-effects of agrochemicals on soil microorganisms. Residue Rev 86:65–105

Domsch KH, Gams W, Anderson TH (1993) Compendium of soil fungi. IHW, Eching

Doran JW, Coleman DC, Bezdicek DF, Stewart BA (1994) Defining soil quality for a sustainable environment. SSSA special publication 35. Soil Science Society of America, American Society of Agronomy, Madison

Dorner JW, Cole RJ, Sanders TH, Blankenship PD (1989) Interrelationship of kernel water activity, soil temperature, maturity, and phytoalexin production in preharvest aflatoxin contamination of drought-stressed peanuts. Mycopathologia 105:117–128

Eaton DL, Groopman JD (1994) The toxicology of aflatoxins. Human health, veterinary, and agricultural significance. Academic, San Diego

Ellis JR, McCalla TM, Norstadt FA (1980) Soil effects on patulin disappearance and the effect of ammonia on patulin phytotoxicity. Soil Sci 129:371–375

Elmholt S (2003) Ecology of the ochratoxin A producing *Penicillium verrucosum*: Occurrence in field soil and grain with special attention to farming system and on-farm drying practices. Biol Agric Hortic 20:311–337

Elmholt S, Hestbjerg H (1996) Recovery and detection of deuteromycete conidia from soil: Exemplified by isolation of *Penicillium verrucosum* Dierckx on a selective and diagnostic agar medium. In: Jensen DF, Jansson HB, Tronsmo A (eds) Monitoring antagonistic fungi deliberately released into the environment. Kluwer, Dordrecht, pp 49–55

Elmholt S, Hestbjerg H (1999) Field ecology of the ochratoxin A-producing *Penicillium verrucosum*: survival and resource colonisation in soil. Mycopathologia 147:67–81

Elmholt S, Kjøller A (1989) Comparison of the occurrence of the saprophytic soil fungi in two differently cultivated field soils. Biol Agric Hortic 6:229–239

Elmholt S, Mortensen GK (2003) Kan OTA-dannende lagersvampe inficere kornet i marken? Forsk Økol Landbruk Nord 18–20

Elmholt S, Rasmussen PH (2005) *Penicillium verrucosum* occurrence and ochratoxin A contents in organically cultivated grain with special reference to ancient wheat types and drying practice. Mycopathologia 159:421–432

Elmholt S, Frisvad JC, Thrane U (1993) The influence of fungicides on soil mycoflora with special attention to tests of fungicide effects on soil-borne pathogens. In: Altman J (ed) Pesticide interactions in crop production: beneficial and deleterious effects. CRC, New York, pp 227–243

Elmholt S, Labouriau R, Hestbjerg H, Nielsen JM (1999) Detection and estimation of conidial abundance of *Penicillium verrucosum* in soil by dilution plating on a selective and diagnostic agar medium (DYSG). Mycol Res 103:887–895

Eriksen, GS, Alexander, J (1998) *Fusarium* toxins in cereals—a risk assessment. TemaNord 502. Nordic Council of Ministers, Copenhagen

Eriksen J, Kristensen K (2001) Nutrient excretion by outdoor pigs: a case study of distribution, utilization and potential for environmental impact. Soil Use Manag 17:21–29
Esteban A, Abarca ML, Cabanes FJ, Bragulat MR (2004) Effects of temperature and incubation time on production of ochratoxin A by black aspergilli. Res Microbiol 155:861–866
FAO (2004) Worldwide regulations for mycotoxins in food and feed. FAO, Rome
Filtenborg O, Frisvad JC, Thrane U (1990) The significance of yeast extract composition on metabolite production in *Penicillium*. In: Samson RA, Pitt JI (eds) Modern concepts in *Penicillium* and *Aspergillus* classification. Plenum, New York, pp 443–440
Frank M (1999) Mycotoxin prevention and decontamination. HACCP and its mycotoxin control potential: an evaluation of ochratoxin A in coffee production. 1999. Third joint FAO/WHOUNEP international conference on mycotoxins, Tunis, Tunisia. http://www.fao.org/es/ESN/food/quality_mycoconf_en.stm
Frisvad JC (1989) The connection between the penicillia and aspergilli and mycotoxins with special emphasis on misidentified isolates. Arch Environ Contam Toxicol 18:452–467
Frisvad JC, Filtenborg O (1989) Terverticillate penicillia: chemotaxonomy and mycotoxin production. Mycologia 81:837–861
Frisvad JC, Samson RA (1991) Filamentous fungi in foods and feeds; ecology, spoilage, and mycotoxin production. In: Arora DK, Mukerji KG, Marth EH (eds) Handbook of applied mycology, vol 3. Foods and feeds. Dekker, New York, pp 31–68
Frisvad JC, Thrane U (2000) Mycotoxin production by common filamentous fungi. In: Samson RA, Hoekstra ES, Frisvad JC, Filtenborg O (eds) Introduction to food- and airborne fungi. Centraalbureau Voor Schimmelcultures, Utrecht, pp 321–332
Frisvad JC, Filtenborg O, Wicklow DT (1987) Terverticillate penicillia isolated from underground seed caches and cheek pouches of banner-tailed kangaroo rats (*Dipodomys spectabilis*). Can J Bot 65:765–773
Frisvad JC, Filtenborg O, Lund F, Thrane U (1992) New selective media for the detection of toxigenic fungi in cereal products, meat and cheese. In: Samson RA, Hocking AD, Pitt JI, King AD (eds) Modern methods in food mycology. Elsevier, Amsterdam, pp 275–285
Frisvad JC, Thrane U, Filtenborg O (1998) Role and use of secondary metabolites in fungal taxonomy. In: Frisvad JC, Bridge PD, Arora DK (eds) Chemical fungal taxonomy. Dekker, New York, pp 289–319
Frisvad JC, Frank JM, Houbraken JAMP, Kuijpers AFA, Samson RA (2004) New ochratoxin A producing species of Aspergillus section Circumdati. Stud Mycol 50: 23–43
Frisvad JC, Skouboe P, Samson RA (2005) Taxonomic comparison of three different groups of aflatoxin producers and a new efficient producer of aflatoxin B-1, sterigmatocystin and 3-O-methylsterigmatocystin, *Aspergillus rambellii* sp nov. Syst Appl Microbiol 28:442–453
Gang G, Miedaner T, Schuhmacher U, Schollenberger M, Geiger HH (1998) Deoxynivalenol and nivalenol production by *Fusarium culmorum* isolates differing in aggressiveness toward winter rye. Phytopathology 88:879–884
Geiser DM, Dorner JW, Horn BW, Taylor JW (2000) The phylogenetics of mycotoxin and sclerotium production in *Aspergillus flavus* and *Aspergillus oryzae*. Fungal Genet Biol 31:169–179
Gelderblom WCA, Jaskiewicz K, Marasas WFO, Thiel PG, Horak RM, Vleggaar R, Kriek N-PJ (1988) Fumonisins—novel mycotoxins with cancer-promoting activity produced by *Fusarium moniliforme*. Appl Environ Microbiol 54:1806–1811
Gilbert J, Fernando WGD (2004) Epidemiology and biological control of Gibberella zeae/Fusarium graminearum. Can J Plant Pathol 26:464–472
Gilbert J, Abramson D, McCallum B, Clear R (2002) Comparison of Canadian *Fusarium graminearum* isolates for aggressiveness, vegetative compatibility, and production of ergosterol and mycotoxins. Mycopathologia 153:209–215
Goldberg BS, Angle JS (1985) Aflatoxin movement in soil. J Environ Qual 14:224–228
Griffin GJ, Garren KH (1976) Colonization of rye green manure and peanut fruit debris by *Aspergillus flavus* and *Aspergillus niger* group in field soils. Appl Environ Microbiol 32:28–32

Hawksworth DL (1991) Presidential address 1990. The fungal dimension of biodiversity: magnitude, significance, and conservation. Mycol Res 95:641–655

Hedman R, Pettersson H, Lindberg JE (1997) Absorption and metabolism of nivalenol in pigs. Arch Anim Nutr 50:13–24

Hestbjerg H (1999) Mycometabolites in the ecology of *Fusarium*—exemplified by characterisation of *F. culmorum* and *F. equiseti*. PhD thesis, University of Copenhagen

Hestbjerg H, Felding G, Elmholt S (2002a) *Fusarium culmorum* infection of barley seedlings: correlation between agressiveness and deoxynivalenol content. J Phytopathol 150:308–312

Hestbjerg H, Nielsen KF, Thrane U, Elmholt S (2002b) Production of trichothecenes and other secondary metabolites by *Fusarium culmorum* and *Fusarium equiseti* on common laboratory media and a soil organic matter agar: an ecological interpretation. J Agric Food Chem 50:7593–7599

Hokby E, Hult K, Gatenbeck S, Rutqvist L (1979) Ochratoxin A and citrinin in 1976 crop of barley stored on farms in Sweden. Acta Agric Scand 29:174–178

Horn BW (2003) Ecology and population biology of aflatoxigenic fungi in soil. J Toxicol Toxin Rev 22:351–379

Huwig A, Freimund S, Kappeli O, Dutler H (2001) Mycotoxin detoxication of animal feed by different adsorbents. Toxicol Lett 122:179–188

Jaime-Garcia R, Cotty PJ (2004) *Aspergillus flavus* in soils and corncobs in south Texas: implications for management of aflatoxins in corn-cotton rotations. Plant Dis 88:1366–1371

Janzen DH (1977) Why fruits rot, seeds mold, and meat spoils. Am Nat 111:691–713

Jonge H, Jonge LW, Jacobsen OH (2000) [14C]Glyphosate transport in undisturbed topsoil columns. Pest Manag Sci 56:909–915

Kadakal C, Nas S, Ekinci R (2005) Ergosterol as a new quality parameter together with patulin in raw apple juice produced from decayed apples. Food Chem 90:95–100

Karlovsky P (1999) Biological detoxification of fungal toxins and its use in plant breeding, feed and food production. Nat Toxins 7:1–23

Kim EK, Maragos CM, Kendra DF (2004) Liquid chromatographic determination of fumonisins B-1, B-2, and B-3 in corn silage. J Agric Food Chem 52:196–200

Kolpin D, Furlong E, Meyer M, Thurman E, Zaugg S, Barber L, Buxton H (2002) Pharmaceuticals, hormones, and other organic wastewater contaminants in U.S. Streams, 1999-2000: a national reconnaissance. Environ Sci Technol 36:1202–1211

Krebs H, Dubois D, Kulling C, Forrer HR (2000) Effects of preceding crop and tillage on the incidence of *Fusarium* spp. and mycotoxin deoxynivalenol content in winter wheat grain. Agrarforschung 7:264–268

Krogh P (1978) Causal associations of mycotoxic nephropathy. Acta Pathol Microbiol Scand Suppl 269:6–28

Krogh P (1987) Ochratoxins in food. In: Krogh P (ed) Mycotoxins in food. Academic, London, pp 97–121

Laegdsmand M, Andersen HR (2005) Can steroid estrogens from slurry leach to drains and groundwater? in: Abstracts SETAC Europe 15th annual meeting, Lille, France. Society of Environmental Toxicology and Chemistry, p 208

Larsen TO, Svendsen A, Smedsgaard J (2001) Biochemical characterization of ochratoxin A-producing strains of the genus *Penicillium*. Appl Environ Microbiol 67:3630–3635

Lee T, Oh DW, Kim HS, Lee J, Kim YH, Yun SH, Lee YW (2001) Identification of deoxynivalenol- and nivalenol-producing chemotypes of *Gibberella zeae* by using PCR. Appl Environ Microbiol 67:2966–2972

Lemke SL, Grant PG, Phillips TD (1998) Adsorption of zearalenone by organophilic montmorillonite clay. J Agric Food Chem 46:3789–3796

Lillehoj EB, Elling F (1983) Environmental conditions that facilitate ochratoxin contamination of agricultural commodities. Acta Agric Scand 33:113–128

Lindblad M, Johnsson P, Jonsson N, Lindqvist R, Olsen M (2004) Predicting noncompliant levels of ochratoxin A in cereal grain from *Penicillium verrucosum* counts. J Appl Microbiol 97:609–616

Llorens A, Mateo R, Hinojo MJ, Logrieco A, Jimenez M (2004) Influence of the interactions among ecological variables in the characterization of zearalenone producing isolates of *Fusarium* spp. Syst Appl Microbiol 27:253–260
Logrieco A, Doko B, Moretti A, Frisullo S, Visconti A (1998) Occurrence of fumonisin B1 and B2 in *Fusarium proliferatum* infected asparagus plants. J Agric Food Chem 46:5201–5204
Lopez TA, Escande A, Chayer R, Dosanto M, Gerpe O, Salomon ML (1997) *Fusarium crookwellense*-produced zearalenone in maize stubble in the field. N Z Vet J 45:251–253
Madden UA, Stahr HM (1993) Preliminary determination of mycotoxin binding to soil when leaching through soil with water. Int Biodeterior Biodegradation 31:265–275
Madhyastha MS, Marquardt RR, Masi A, Borsa J, Frohlich AA (1994) Comparison of toxicity of different mycotoxins to several species of bacteria and yeasts: use of *Bacillus brevis* in a disc diffusion assay. J Food Protect 57:48–53
Mantle PG (1998) Ochratoxin A in coffee. J Food Mycol 1:63–65
Mantle PG (2000) Uptake of radiolabelled ochratoxin A from soil by coffee plants. Phytochemistry 53:377–378
Marasas WFO, Nelson PE (1987) Mycotoxicology. Introduction to the mycology, plant pathology, chemistry, toxicology, and pathology of naturally occuring mycotoxicoses in animals and man. The Pennsylvania State University Press, University Park
Marasas WFO, Nelson PE, Toussoun TA (1984) Taxonomy of toxigenic fusaria. In: Lacey J (ed) Trichothecenes and other mycotoxins. Wiley, Chichester, pp 3–14
Marin S, Magan N, Ramos AJ, Sanchis V (2004) Fumonisin-producing strains of *Fusarium*: a review of their ecophysiology. J Food Protect 67:1792–1805
McLean M (1996) The phytotoxicity of *Fusarium* metabolites: an update since 1989. Mycopathologia 133:163–179
Mertz D, Edward T, Lee D, Zuber M (1981) Absorption of aflatoxin by lettuce seedlings grown in soil adulterated with aflatoxin B1. J Agric Food Chem 29:1168–1170
Mesterhazy A, Bartok T, Mirocha CG, Komoroczy R (1999) Nature of wheat resistance to *Fusarium* head blight and the role of deoxynivalenol for breeding. Plant Breed 118:97–110
Miedaner T, Reinbrecht C (2001) Trichothecene content of rye and wheat genotypes inoculated with a deoxynivalenol- and a nivalenol-producing isolate of *Fusarium culmorum*. J Phytopathol 149:245–251
Miller JD (1995) Fungi and mycotoxins in grain: implications for stored product research. J Stored Prod Res 31:1–16
Mirocha CJ, Christensen CM (1974) Oestrogenic mycotoxins synthesized by *Fusarium*. In: Purchase IFH (ed) Mycotoxins. Elsevier, Amsterdam, pp 129–148
Moake MM, Padilla-Zakour OI, Worobo RW (2005) Comprehensive review of patulin control methods in foods. Comp Rev Food Sci Food Saf 4:8–21
Mortensen GK, Strobel BW, Hansen HCB (2006) Degradation of zearalenone and ochratoxin A in three Danish agricultural soils. Chemosphere 62:1673–1680
Moss MO, Thrane U (2004) *Fusarium* taxonomy with relation to trichothecene formation. Toxicol Lett 153:23–28
Motomura M, Toyomasu T, Mizuno K, Shinozawa T (2003) Purification and characterization of an aflatoxin degradation enzyme from *Pleurotus ostreatus*. Microbiol Res 158:237–242
Müller HM, Boley A (1992) Investigations on refrigerated storage of wheat. 1. Ergosterol, ochratoxin A and citrinin after inoculation with *Penicillium verrucosum*. Arch Anim Nutr 42:351–363
Munkvold GP (2003) Epidemiology of *Fusarium* diseases and their mycotoxins in maize ears. Eur J Plant Pathol 109:705–713
Nicoletti R, Stefano Md, Stefano S, Trincone A, Marziano F (2004) Antagonism against *Rhizoctonia solani* and fungitoxic metabolite production by some *Penicillium* isolates. Mycopathologia 158:465–474
Norstadt FA, McCalla TM (1969) Microbial populations in stubble-mulched soil. Soil Sci 107:188–193

Northolt MD, Egmond HV, Paulsch WE (1979) Ochratoxin A production by some fungal species in relation to water activity and temperature. J Food Protect 42:485–490

Olanya OM, Hoyos GM, Tiffany LH, McGee DC (1997) Waste corn as a point source of inoculum for *Aspergillus flavus* in the corn agroecosystem. Plant Dis 81:576–581

Parry DW, Jenkinson P, McLeod L (1995) *Fusarium* ear blight (scab) in small grain cereals-a review. Plant Pathol 44:207–238

Paster N, Huppert D, Barkai-Golan R (1995) Production of patulin by different strains of *Penicillium expansum* in pear and apple cultivars stored at different temperatures and modified atmospheres. Food Addit Contam 12:51–58

Perkowski J (1998) Distribution of deoxynivalenol in barley kernels infected by *Fusarium*. Nahrung 42:81–83

Pfohl-Leszkowicz A, Chekir-Ghedira L, Bacha H (1995) Genotoxicity of zearalenone, an estrogenic mycotoxin: DNA adduct formation in female mouse tissues. Carcinogenesis 16:2315–2320

Pfohl-Leszkowicz A, Petkova-Bocharova T, Chernozemsky IN, Castegnaro M (2002) Balkan endemic nephropathy and associated urinary tract tumours: a review on aetiological causes and the potential role of mycotoxins. Food Addit Contam 19:282–302

Pitt JI, Samson RA (1990) Approaches to *Penicillium* and *Aspergillus* systematics. Stud Mycol 32:77–90

Proctor RH, Hohn TM, McCormick SP (1995) Reduced virulence of *Gibberella zeae* caused by disruption of a trichothecene toxin biosynthetic gene. Mol Plant Microbe Interact 8:593–601

Pugh GJF (1980) Strategies in fungal ecology. Trans Br Mycol Soc 75:1–14

Razzazi-Fazeli E, Bohm J, Jarukamjorn K, Zentek J (2003) Simultaneous determination of major B-trichothecenes and the de-epoxy-metabolite of deoxynivalenol in pig urine and maize using high-performance liquid chromatography-mass spectrometry. J Chromatogr B 796:21–33

Regner S, Schnürer J, Jonsson A (1994) Ergosterol content in relation to grain kernel weight. Cereal Chem 71:55–58

Rotter RG, Marquardt RR, Frohlich AA, Abramson D (1990) Ensiling as a means of reducing ochratoxin A concentrations in contaminated barley. J Sci Food Agric 50:155–166

Ryu D, Munimbazi C, Bullerman LB (1999) Fumonisin B-1 production by *Fusarium moniliforme* and *Fusarium proliferatum* as affected by cycling temperatures. J Food Protect 62:1456–1460

Samson RA, Hoekstra ES, Frisvad JC, Filtenborg O (eds) (2000) Introduction to food- and airborne fungi. Centraalbureau voor Schimmelcultures, Utrecht

Sansing GA, Lillehoj EB, Detroy RW, Miller MA (1976) Synergistic toxic effects of citrinin, ochratoxin A and penicillic acid in mice. Toxicon 14:213–220

Sauer DB, Burroughs R (1980) Fungal growth, aflatoxin production and moisture equilibration in mixtures of wet and dry corn. Phytopathology 70:516–521

Schatzki TF, Ong MS (2000) Distribution of aflatoxin in almonds. 2. Distribution in almonds with heavy insect damage. J Agric Food Chem 48:489–492

Schumacher DM, Metzler M, Lehmann L (2005) Mutagenicity of the mycotoxin patulin in cultured Chinese hamster V79 cells, and its modulation by intracellular glutathione. Arch Toxicol 79:110–121

Scudamore KA, Wilkin DR (1999) A study to determine whether on-floor ambient drying systems are conducive to the formation of ochratoxin A in grain. Project report 196. Home Grown Cereals Authority, London

Seefelder W, Gossmann M, Humpf HU (2002) Analysis of fumonisin B1 in *Fusarium proliferatum*-infected asparagus spears and garlic bulbs from Germany by liquid chromatography-electrospray ionization mass spectrometry. J Agric Food Chem 50:2778–2781

Seifert KA, Lévesque CA (2004) Phylogeny and molecular diagnosis of mycotoxigenic fungi. Eur J Plant Pathol 110:449–471

Shearer JF, Sweets LE, Baker NK, Tiffany LH (1992) A study of *Aspergillus flavus/parasiticus* in Iowa crop fields: 1988-1990. Plant Dis 76:19–22

Shima J, Takase S, Takahashi Y, Iwai Y, Fujimoto H, Yamazaki M, Ochi K (1997) Novel detoxification of the trichothecene mycotoxin deoxynivalenol by a soil bacterium isolated by enrichment culture. Appl Environ Microbiol 63:3825–3830

Shotwell OL, Goulden ML, Bothast RJ, Hesseltine CW (1975) Mycotoxins in hot spots in grains. I. Aflatoxin and zearalenone occurrence in stored corn. Cereal Chem 52:687–696

Sinha KK, Sinha AK (1992) Impact of stored grain pests on seed deterioration and aflatoxin contamination in maize. J Stored Prod Res 28:211–219

Sinha RC, Savard ME (1997) Concentration of deoxynivalenol in single kernels and various tissues of wheat heads. Can J Plant Pathol 19:8–12

Smith JE, Lewis CW, Anderson JG, Solomons GL (1994) Mycotoxins in human nutrition and health. EUR 16048 EN. European Commission Directorate-General XII, Agro-Industrial Division, E-2, Brussels

Snijders CHA, Perkowski J (1990) Effects of head blight caused by *Fusarium culmorum* on toxin content and weight of wheat kernels. Phytopathology 80:566–570

Speijers GJA, Speijers MHM (2004) Combined toxic effects of mycotoxins. Toxicol Lett 153:91–98

Tajima O, Schoen ED, Feron VJ, Groten JP (2002) Statistically designed experiments in a tiered approach to screen mixtures of *Fusarium* mycotoxins for possible interactions. Food Chem Toxicol 40:685–695

Tamm C, Breitenstein W (1980) The biosynthesis of trichothecene mycotoxins. In: Steyn PS (ed) The biosynthesis of mycotoxins. Academic, New York, pp 69–104

Thrane U (1989) *Fusarium* species and their specific profiles of secondary metabolites. In: Chelkowski J (ed) *Fusarium*—mycotoxins, taxonomy, and pathogenicity. Elsevier, Amsterdam, pp 199–225

Thrane U, Adler A, Clasen PE, Galvano F, Langseth W, Logrieco A, Nielsen KF, Ritieni A (2004) Diversity in metabolite production by *Fusarium langsethiae, Fusarium poae*, and *Fusarium sporotrichioides*. Int J Food Microbiol 95:257–266

Ueno Y (1983) Trichothecenes - Chemical, Biological and Toxicological Aspects. Kodansha, Tokyo

Ueno Y, Nakayama K, Ishii K, Tashiro F, Minoda Y, Omori T, Komagata K (1983) Metabolism of T-2 toxin in *Curtobacterium* sp. strain 114-2. Appl Environ Microbiol 46:120–127

Utermark J, Karlovsky P (2007) Role of Zearalenone Lactonase in Protection of Gliocladium roseum from Fungitoxic Effects of the Mycotoxin Zearalenone. Appl Environ Microb 73:637–642

Valenta H, Kuhn I, Rohr K (1993) Determination of ochratoxin A in urine and faeces of swine by high-performance liquid chromatography. J Chromatogr 613:295–302

Varga J, Peteri Z, Tabori K, Teren J, Vagvolgyi C (2005) Degradation of ochratoxin A and other mycotoxins by *Rhizopus* isolates. Int J Food Microbiol 99:321–328

Vereecken H (2005) Mobility and leaching of glyphosate: a review. Pest Manag Sci 61:1139–1151

Vesonder RF, Ellis JJ, Rohwedder WK (1981) Elaboration of vomitoxin and zearalenone by *Fusarium* isolates and the biological activity of *Fusarium*-produced toxins. Appl Environ Microbiol 42:1132–1134

Villholth KG, Jarvis NJ, Jacobsen OH, Jonge Hd (2000) Field investigations and modeling of particle-facilitated pesticide transport in macroporous soil. J Environ Qual 29:1298–1309

Vineet K, Singh GP, Babu AM (2004) Surface ultrastructural studies on the germination, penetration and conidial development of *Aspergillus flavus* Link:Fries infecting silkworm, *Bombyx mori* Linn. Mycopathologia 157:127–135

Völkl A, Vogler B, Schollenberger M, Karlovsky P (2004) Microbial detoxification of mycotoxin deoxynivalenol. J Basic Microbiol 44:147–156

Wafa H, Abid-Essafi S, Abdellatif A, Noureddine G, Abdelfettah Z, Farielle E, Creppy EE, Hassen B (2004) Karyomegaly of tubular kidney cells in human chronic interstitial nephropathy in Tunisia: respective role of ochratoxin A and possible genetic predisposition. Hum Exp Toxicol 23:339–346

WHO (1990) Environmental health criteria 105. Selected mycotoxins: ochratoxins, trichothecenes, ergot. WHO, Geneva

WHO (2000) Environmental health criteria 219. Fumonisin B_1. WHO, Geneva

Wicklow DT (1990) Adaptation in *Aspergillus flavus*. Ttans Mycol Soc Jpn 31:511–523

Wicklow DT (1995) The mycology of stored grain: an ecological perspective. In: Jayas DS, White NDG, Muir WE (eds) Stored-grain ecosystems. Dekker, New York, pp 107–249

Wicklow DT, Horn BW, Burg WR, Cole RJ (1984) Sclerotium dispersal of *Aspergillus flavus* and *Eupenicillium ochrosalmoneum* from maize during harvest. Trans Br Mycol Soc 83:299–303

Will ME, Wilson DM, Wicklow DT (1994) Evaluation of *Paecilomyces lilacinus*, chitin, and cellulose amendments in the biological control of *Aspergillus flavus* fungi. Biol Fert Soils 17:281–284

Williams LD, Bacon CW, Meredith FI, Franzluebbers AJ, Wyatt RD, Smith MA, Riley RT (2003) Leaching and binding of fumonisins in soil microcosms. J Agric Food Chem 51:685–690

Wotton HR, Strange RN (1987) Increased susceptibility and reduced phytoalexin accumulation in drought-stressed peanut kernels challenged with *Aspergillus flavus*. Appl Environ Microbiol 53:270–273

Xu XM, Berrie AM (2005) Epidemiology of mycotoxigenic fungi associated with *Fusarium* ear blight and apple blue mould: a review. Food Addit Contam 22:290–301

Yoshizawa T, Yamashita A, Luo Y (2005) Fumonisin occurrence in corn from high- and low-risk areas for human esophageal cancer in China. Appl Environ Microbiol 60:1626–1629

Yu W, Yu F, Undersander DJ, Chu FS (1999) Immunoassays of selected mycotoxins in hay, silage and mixed feed. Food Agric Immunol 11:307–319

Zardari M, Clarke DD (1988) Spore survival, germination and sporeling growth of *Penicillium expansum* in soil. Pak J Bot 20:251–257

Zummo N, Scott GE (1990) Relative aggressiveness of *Aspergillus flavus* and *A. parasiticus* on maize in Mississippi. Plant Dis 74:978–981

Part IV
Plant Metabolites

Chapter 10
Constitutive Secondary Plant Metabolites and Soil Fungi: Defense Against or Facilitation of Diversity

Franz Hadacek

10.1 Introduction

In general, we assign the term secondary metabolites to low molecular weight compounds produced by living organisms that apparently lack life-sustaining functions. Instead, they are assumed to contribute to the producing organism's survival in the ecosystem. By contrast, primary metabolites are practically indispensable (Hartmann 1996). The congruence of apparent accumulation of secondary metabolites in those organisms lacking an immune system, such as plants, fungi, or invertebrates, additionally supports the notion that the main function of these originally classified as waste products metabolites is defense against pathogens and predators. Besides, there exist also morphological defenses, such as cutin armor, thorns, or spines (Rubinstein 1992; Gershenzon 2002).

Plant–herbivore interactions represent highly apparent biotic interactions and their exploration has been a starting point to obtain insight into how secondary metabolites may act as agents in chemical defense (Dethier 1954; Fraenkel 1959; Ehrlich and Raven 1964; but see Kerner von Marilaun 1890). Secondary plant metabolites provide various cues for insects to locate their food plants and oviposit on plants that produce chemicals which can be tolerated, detoxified, or even sequestrated by the hatching larvae (Hilker and Meiners 2002; Nishida 2002). Moreover, predators of the herbivore's larvae may locate their host by plant-produced volatiles induced by components present in the oral secretions of the feeding larvae (Holopainen 2004). The seminal paper by Ehrlich and Raven (1964) on the coevolution between butterflies and their host plants has fundamentally stimulated research in exploring the role of secondary metabolites in biotic interactions (Stamp 2003). Some authors viewed the diversity of secondary metabolites found in plants as a result of a coevolutionary arms race between plants and their herbivore predators

Franz Hadacek
Department of Chemical Ecology and Ecosystem Research, Faculty of Life Sciences, University of Vienna, Althanstrasse 14, A-1090 Vienna, Austria
e-mail: franz.hadacek@univie.ac.at

P. Karlovsky (ed.), *Secondary Metabolites in Soil Ecology. Soil Biology 14.*

(Berenbaum 1983). At present, after 40 years of exploring this hypothesis, true cases of coevolution between organisms have only been demonstrated in specific cases and for limited time intervals, too rarely for unambiguous support (Farrell and Mitter 1993; Futuyma 2000).

Biotic interactions between plants and fungi, apart from visible disease symptoms caused by pathogenic strains, are less apparent. The discovery that antibiotics may be produced by molds, such as *Penicillium* isolates, originally stimulated the exploration of microbial secondary metabolite diversity (Gräfe 1992). Likewise, recognition that a good number of plant diseases may be caused by fungi has drawn the attention of phytopathologists to plant–microbe interactions, and to the inherent role of secondary metabolites. For example, hemibiotrophic pathogenic fungi may use phytotoxic secondary metabolites to overcome the hypersensitive defense response of their host plant (Scheffer 1991). Conversely, nonsusceptible host plants may suppress microbial attackers by producing phytoalexins, inducible secondary metabolites that inhibit the development of the invading microorganism (Barz 1997). Whereas fungi have developed specific mechanisms to actively invade plant organs and cells, such as appressorium formation and biosynthesis of cell wall degrading enzymes (Mendgen and Deising 1993), bacteria use a different strategy for pathogenicity: the development of certain population densities within the host tissue is required; using quorum-sensing agents, such as acyl homoserine lactones in gram-negative bacteria, allows bacteria to efficiently produce toxic metabolites in a coordinated fashion (Taga and Bassier 2003). Similarly as in plant–herbivore interactions, a coevolutionary scenario was proposed for plant–microbe interactions: Flor (1971) suggested that susceptibility and resistance of a host plant depended on the reciprocal development of avirulence genes in host plants and virulence genes in pathogens (gene-for-gene concept). Despite the similarity in the set-up of the hypotheses, a combined reassessment was only done a couple of years ago (Agrawal et al. 1999). Today, we have a glimpse of the complexity of the system: signal cascades involving salicylic acid (Shah 2003), jasmonic acid (Stratmann 2003), nitric oxide (Neill at al. 2003), reactive oxygen species, hydrogene peroxide and various oxygen radicals (Laloi et al. 2004), and ethylene (Guo and Ecker 2004) concertedly monitor and modulate the hypersensitive defense response reaction of a plant to the attack of different foliar predators as well as various other biotic and abiotic stresses (Felton et al. 1999; Maleck and Dietrich 1999; de Bruxelles and Roberts 2001). Similar signal cascades have also been detected in roots with one exception, salicylic acid mediated resistance (Okubara and Paulitz 2005).

Phytophagous insects are thought to have emerged in the Upper Carboniferous, definitely later than flower-visiting insects (Scott et al. 1992); thus, plant–herbivore interactions represent rather recent developments in the history of life compared with plant–fungus interactions, which are most probably more than 100 million years older. The 400-million-year-old fossils of the first vascular plants already show traces of fungal colonization (Remy et al. 1994). Fungi have participated in various symbiotic interactions that may have decisively facilitated the development of land plants. Fungi are also lichen symbionts—lichens were among the first colonizers that endured the harsh conditions on land during the earliest colonization phase (Sanders 2001); extant mycotoxin-producing molds as well as other major groups of fungi are suggested to have arisen from lichen symbiont ancestors (Lutzoni et al. 2001); soil decomposing and mycorrhizal fungi play an important role in carbon nutrient cycling (Eissenstat et al. 2000; Lindahl et al. 2002); ectomycorrhizal fungi are suggested to be able to mobilize plant nutrients directly from minerals (Landeweert et al. 2001); and all living plants are colonized by fungal endophytes that remain inconspicuous owing to the absence of disease symptoms but most certainly decompose their host in the end (Rodriguez and Redman 1997).

Constitutive secondary metabolites are formed during the ontogenetic differentiation of plant tissues and organs, and may often accumulate in comparatively large amounts, especially in perennial organs (McKey 1979; Herms and Mattson 1992). Once they were viewed as waste products but today the general notion prevails that this thesis cannot be maintained (Hadacek 2002). As plant traits subjected to selection pressure, secondary metabolites constitute a set of characters equally important for the fitness of their producer as morphological features, though (except for pigments) they are less accessible to direct observation. Higher plants constitute the primary producers in terrestrial ecosystems and, as a consequence, we may assume that their primary and constitutive secondary metabolites exert direct and indirect effects not only on their predators (Wittstock and Gershenzon 2002), but also on their decomposer communities (Wardle 2002). This extends the original approach of attempting to understand factors determining plant community structure (Tilman and Pacala 1993) that was predominantly focused on abiotic factors, such as nutrient availability or light, to biotic interactions. In this context, physiological mechanisms, such as plant defense and proportional contributions of plant primary and secondary metabolites, are still to be explored in much more detail (Wardle et al. 2004).

In this chapter I will concentrate on plant underground organs, such as roots and stolons and on fungi as potential predators and decomposers of plant roots. I will (1) explore current theories that support the notion that secondary metabolites have evolved as efficient defenses against predators and explore if this applies to antifungal defense and (2) I will attempt to present data and notions from the recent literature that suggest root secondary metabolites may have a function beyond pure chemical defense in maintaining plant biodiversity. In my opinion, such a chemodiversity hypothesis (Iason et al. 2005) is providing a more attractive framework to explore secondary metabolites in a biological context. Challenging frameworks constitute a stringent basis to stimulate further research.

10.2 Secondary Metabolites in Chemical Defense

10.2.1 Diversity and Biosynthesis of Secondary Metabolites

The number of existing seed plants is estimated to amount to about 420,000 (Govaerts 2001) and the number of existing secondary metabolites in plants may exceed 500,000 (Mendelsohn and Balick 1995). This suggests that most seed plant species should be capable of producing one unique secondary metabolite at least.

Our knowledge of enzymes involved in secondary metabolites is increasing rapidly. The development of molecular biological methods allows us to use gene sequence homologies to isolate genes which may be potentially involved in the pathways to specific secondary metabolites and characterize their function by heterologous expression (Pichersky and Gang 2000). The combined utilization of genomic, proteomic, and metabolomic methods is expected to further accelerate this process (Fridman and Pichersky 2005). The completely sequenced genome of *Arabidopsis thaliana* contains about 25,500 genes encoding proteins, 15–25% of which are may encode enzymes for secondary metabolism. We are far from knowing all genes leading to the major pathways, but hundreds of thousands of genes are most probably involved and a single plant species contains only a fraction of them (Somerville and Somerville 1999). Genes for secondary metabolism are often present only in a single copy in the genome—by contrast, genes for primary metabolism are usually present in multiple copies to safeguard against lethal consequences of mutations; and a single mutation of a gene involved in secondary metabolism can result in several novel secondary metabolites in the tissues where it is expressed (Jarvis and Miller 1996; Jarvis 2000). A further characteristic of these enzymes is that they may not be always highly substrate specific and that one enzyme may catalyze more than one reaction (Schwab 2003; Firn and Jones 2003).

10.2.2 Microbial Biodiversity and Soil Fungi

Estimates predict the presence of 10^8–10^9 bacteria, 10^5–10^8 actinomycetes, 10^5–10^6 fungi, 10^3–10^6 microalgae, 10^3–10^5 protozoa, 10^1–10^2 nematodes, and 10^3–10^5 other invertebrates in 1 g of soil. A square meter of soil may further contain 30–300 earthworms (Metting 1993). Actually there are more living organisms in 1 g of soil than there are human beings on our planet. Numerous studies have provided evidence that seeds harbor diverse microbial communities, not only on their surfaces but also within the embryo (Nelson 2004). Similarly, fungi may also be transmitted vertically (Saikkonen et al. 2004) and grass endophytes of the genus *Neotyphodium* (formerly classified as *Acremonium*) were shown to produce ergot alkaloids that may potentially contribute to host plant resistance against herbivore predators (Clay and Schardl 2002). Interactions with horizontally transmitted fungi start as early as the seeds are germinating. The spermosphere represents a soil layer of 5–10 mm surrounding the germinating seeds that is already affected by exudates containing sugars and sugar alcohols, amino acids, aliphatic and aromatic organic acids, fatty acids as well as various secondary metabolites, such as flavonoids, cinnamic acid derivatives, and terpenoids, which represent a vital carbon source for various heterotrophic soil biota. Furthermore, these compounds may represent signals for mutualists, such as nitrogen-fixing bacteria and mycorrhizal fungi. Concomitantly, endophytic fungi colonize the plant roots, some of which may turn into pathogens. These processes have undoubtedly fundamental effects on seedling survival and plant development. Surprisingly, research focusing on this developmental stage clearly lags behind the number of studies that are devoted to the rhizosphere of adult plants. The rhizosphere was defined by Hiltner (1904) as that volume of soil that is affected by the plant root. More recently, the rhizosphere is also understood to comprise the root (endorhizosphere) and the surrounding soil (ectorhizosphere), whereas the rhizoplane designates the surface of the root (Brimecombe et al. 2001). A lot of attention has been paid to arbuscular mycorrhizal fungi (AM fungi), a group of nearly ubiquitously occurring biotrophic zygomycetes that colonize young plant roots and increase plant access to rare or immobile soil minerals, particularly phosphorous. Originally, low host specificity was assigned to this group of cryptic fungi; however, recent findings demonstrate that each host plant attracts specific communities of AM fungi and thus suggest AM fungi exert more pronounced effects on plant succession and communities than hitherto assigned (Van der Heijden et al. 1998; Bever et al. 2001). In comparison with AM fungi, decomposer soil fungi, the majority of them being deuteromycetes propagating by asexual spores, are present in definitely larger species numbers in the rhizosphere. Attention is now rising owing to the increased interest in belowground food webs (Wardle et al. 2004). The quality of plant litter is also largely defined by secondary metabolites, for leaves, and probably much more so for underground organs, such as stolons and roots. Secondary metabolites affect the performance of decomposer fungi and thereby also the availability of nutrients (Souto and Pellisier 2002; Wardle 2002). Some secondary metabolites, such as polyphenols, may also directly affect nutrient availability (Hättenschwiler and Vitousek 2000).

10.2.3 Secondary Metabolites with Antifungal Biological Activities

When obtaining information about the occurrence of antifungal compounds in plants, we must be aware that most of them do not occur in the whole plant. Usually, there may be quite considerable differences in the patterns we find in different organs. Moreover, studies focusing on individuals indicate that the amounts and even the detectability of the various compounds may vary from individual to individual (Hadacek 2002). Figure 10.1 illustrates the extent of variability of accumulation patterns of secondary metabolites in various organs exemplified on the rutaceous tree *Glycomis trichanthera* occurring in tropical rainforests of Thailand and Malaysia (Vajrodaya et al. 1998); root bark, stem bark, and leaves accumulate pronouncedly different types of secondary metabolites (note the structures illustrated). It is evident that differential gene expression is responsible for the striking differences in the metabolite patterns of the single organs of this plant. In recent years, our understanding of epigenetic control mechanisms, such as DNA methylation, histone modification, micro RNA, and matrix/scaffold attachment regions of the DNA has grown substantially (Rapp and Wendel 2005). Comparing patterns of secondary metabolites is of paramount importance in chemotaxonomy. The puzzling absence of certain classes of secondary metabolites in phylogenetically

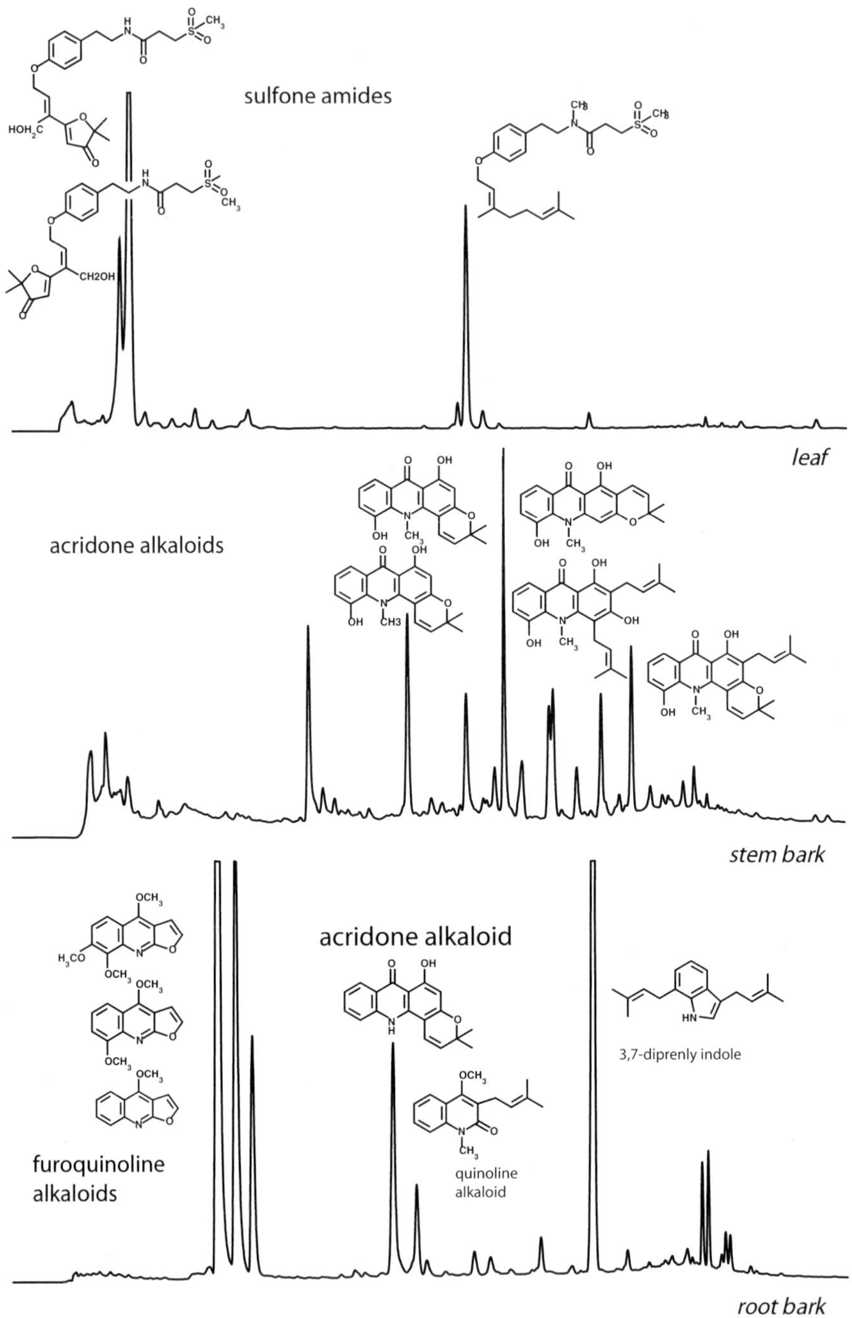

Fig. 10.1 Variation of secondary metabolites in the rutaceous tree *Glycosmis trichanthera*; high-performance liquid chromatography (HPLC)–diode array detection(DAD) profiles of leaf, stembark, and rootbark extracts with identified structures. (Data redrawn from Vajrodaya et al. 1998)

related groups compared with more ancestral groups may now be explained by assuming the involvement of the control mechanism outlined above (Wink 2003).

Grayer and Harborne (1994) reviewed the literature published between 1982 and 1993 reporting antifungal compounds from higher plants. This publication represents a basic survey of classes of secondary metabolites that may generate antifungal derivatives, either induced or constitutively expressed. The tabular overview presented in this paper indicates that antifungal compounds can be found in all plant organs and a wide range of plant families; the most intriguing occurrences are perhaps those found in leaf epicuticular waxes.

To my knowledge, no antifungal compounds are known from root surfaces (as opposed to exudates); this might constitute a hitherto unexplored source. However, Asiegbu (2000) reported that long-chained fatty acids from root surfaces stimulated the germination of fungal spores. Lipophilicity serves as an important signal for hyphal tips of fungi to produce hydrophobins, proteins that facilitate the adhesion of hyphae to lipophilic surfaces (Talbot et al. 1996). *Pyricularia grisea* (teleomorph *Magnaporthe grisea*) is one of the most damaging diseases on rice (blast) and also a paradigm system for exploring foliar pathogenicity, especially because of the formation of notable melanized infection structures, so-called appressoria. One recent finding was that this fungus may also colonize roots of rice (Sesma and Osbourn 2004). For successful penetration of root tissues, *Pyricularia* also produces infection structures, so-called hyphopodia, on root surfaces. Interestingly, fungal secondary metabolites have already been identified that inhibit the formation of appressoria (Thines et al. 2004). So far, no active plant secondary metabolites have been found with comparable activities. These results suggest that fungal competition for hosts may have led to the evolutionary selection of secondary metabolites inhibiting the formation of infection structures of competitors, whereas plants have not been faced by a comparable constraint. However, such interpretations have to be made with care. The assay techniques for assessing appressorium formation inhibition are successfully managed by far fewer research groups than antifungal bioautography, a technique that allows detection of antifungal compounds on developed thin-layer plates (Homans and Fuchs 1970) and which accounts for a considerable portion of discoveries of antifungal compounds reported in the literature.

On the other hand, fungal competition for carbon, especially that of decomposer fungi, seems to be a good environment for developing biologically active compounds; one of the most promising leads for a commercial fungicide, the strobilurins, was discovered in the culture filtrates of wood-decomposing basidiomycetes (Sauter et al. 1999). Fungal hyphae are characterized by tip growth and, as a consequence, antagonistic interactions between fungi are restricted to spatially extremely restricted compartments. Figure 10.2 illustrates this hyphal interference (Ikediugwu and Webster 1970; Ikediugwu 1976; Deacon 1997). However, the presence of an antifungal substance in the culture filtrate of fungus does not imply that it is produced in amounts needed for antifungal activity.

Plants and fungi, though closely interacting, are different living organisms, and despite the fact that they both produce secondary metabolites, they handle them in quite different fashions. Fungi produce a lot of compounds that are rapidly excreted,

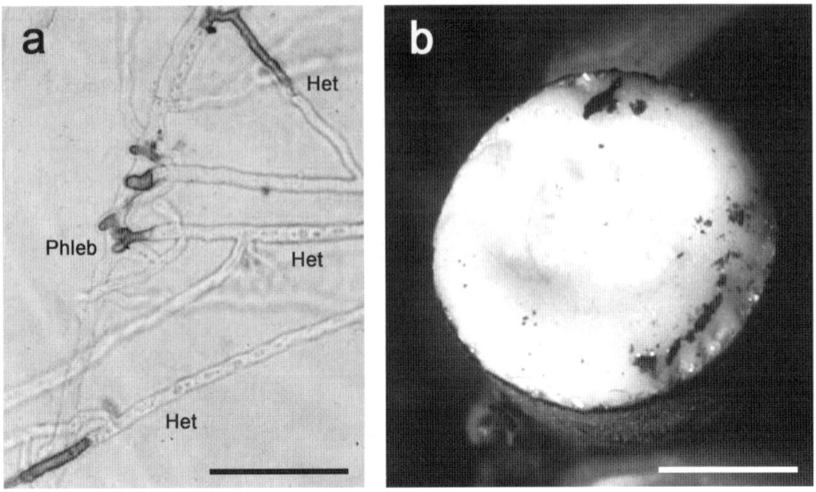

Fig. 10.2 a Hyphal interference of *Phlebiopsis gigantea* (*Phleb*, formerly called *Peniophora gigantea*), antagonizing *Heterobasidium annosum* (*Het*), one of the most damaging root pathogens of the Northern Hemisphere; treatment of agar cultures with neutral red contrasts those portions of the hyphae with affected membrane integrity—normal integrity excludes this dye. *bar* 1 mm. **b** Cross section of tap root of the umbellifer *Peucedanum cervaria*, the broad-leaved spignel, clearly shows resin oozing out of the tissue; a resin is defined by its chemical composition, usually a mixture of volatile and non-volatile secondary metabolites, in this case monoterpenes, triterpenes, and polyacetylenes. (**a** Modified from http://helios.bto.ed.ac.uk/bto/microbes/phleb2.jpg with permission from Jim Deacon)

partially with an effect on competing microbes, but efficient excretion mechanisms such as ABC transporters and other efflux pumps may also protect the producing organisms from harmful effects of its own metabolites, those of antagonists, or defense compounds from plants (Duffy et al. 2003). By contrast, plants themselves accumulate secondary metabolites in their tissues. According to their chemical properties, they are predominantly stored in the vacuole, in unspecialized and specialized cells (idioblasts), in specialized tissues such as oil ducts (laticifers) and secretory channels and resin canals, or in glandular hairs of aerial parts and root hairs belowground (Holloway 1982; Langenheim 2003; Nelson 2004). Volatile secondary metabolites may also be released when the ambient temperatures are high enough. Some plant species show also high proportions of exuded secondary metabolites on their leaf surfaces and in their rhizospheres. Table 10.1 lists terms that are commonly used in connection with accumulation phenomena of secondary metabolites in higher-plant tissues.

I will give some examples for antifungal secondary metabolites that are known to occur in roots:

Terpenoids constitute a major class of plant secondary metabolites that are catalyzed by a special group of enzymes, so-called terpene cyclases, which have been found to occur in a wide range of living organisms (Bohlmann et al. 2000; Wendt et al. 2000). Monoterpenes and sesquiterpenes are volatile and thus major constituents

Table 10.1 Characterization of resins, gums, mucilages, oils, waxes, and latex. (Modified from Langenheim 2003)

	Main components	Solubility	Secretory tissue
Resins	Mixture of volatile and nonvolatile secondary metabolites	Lipid-soluble	Canals, pockets, cavities, trichomes, epidermal cells
Gums	Polysaccharides	Water-soluble	Cavities
Mucilages	Polysaccharides	Water-soluble	Idioblasts, epidermal cells, trichomes, ducts, cavities
Oils (fats)	Fatty acids and glycerols	Lipid-soluble	None
Waxes	Esters of fatty acids with long-chain alcohols	Lipid-soluble	Unspecialized epidermal cells, epicuticular layers
Latex	Complex mixture of variable secondary metabolites, proteins, carbohydrates, etc.	Lipid-soluble	Laticifers

of plant odors. Besides, monoterpenes are also known for their antimicrobial activity, with increased oxygenation contributing to antifungal activity (Naigre et al. 1996; Langenheim 1994). Avenacin A-1, a triterpene glycoside, a saponin phytoalexin of some grass species, causes membrane depolarization and suppresses growth of a wide range of microorganisms (Osbourn et al. 2003). Saponins are widespread in dicotyl angiosperms as constitutive secondary metabolites. More oxygenated steroids may act as hormones, and β-sitosterol, one of the most commonly occurring triterpenes, confers stability to plant membranes (Fig. 10.3).

Chalcone synthases constitute another enzyme family that utilizes the shikimic acid pathway derived metabolite cinnamic acid to yield *flavonoids* and *stilbenes* , among others. Figure 10.4 illustrates some representative structures. Viewed in a wider context, chalcone synthases are classified as polyketide synthases, an enzyme family that also contributes to secondary metabolite diversity in bacteria and fungi (Schröder 2000). Roots do rather not represent organs with a reputation for antifungal flavonoids, quite contrary to leaf surfaces (Grayer and Harborne 1994). However, the dihydroflavonol (+)-catechin is known for its antifungal properties (Veluri et al., 2004; see also Chap. 11 by Bais et al.). Stilbene derivates, such as pinosylvin, are also known for their antifungal activity; already more than 10 years ago attempts were carried out to transform a stilbene synthase gene into crop plants to improve resistance to fungal pathogens (Fliegmann et al. 1992). Pterocarpan isoflavonoids are characteristic for legumes; pisatin is an often referred to phytoalexin in pea roots (Barz 1997); the cyclopenta[*b*]benzofuran rocaglaol belongs to a class of compounds that are characteristic for the genus *Aglaia*, Meliaceae (Brader et al. 1998). Not unexpected, there exist some congruencies between the structures presented in Fig. 10.4.

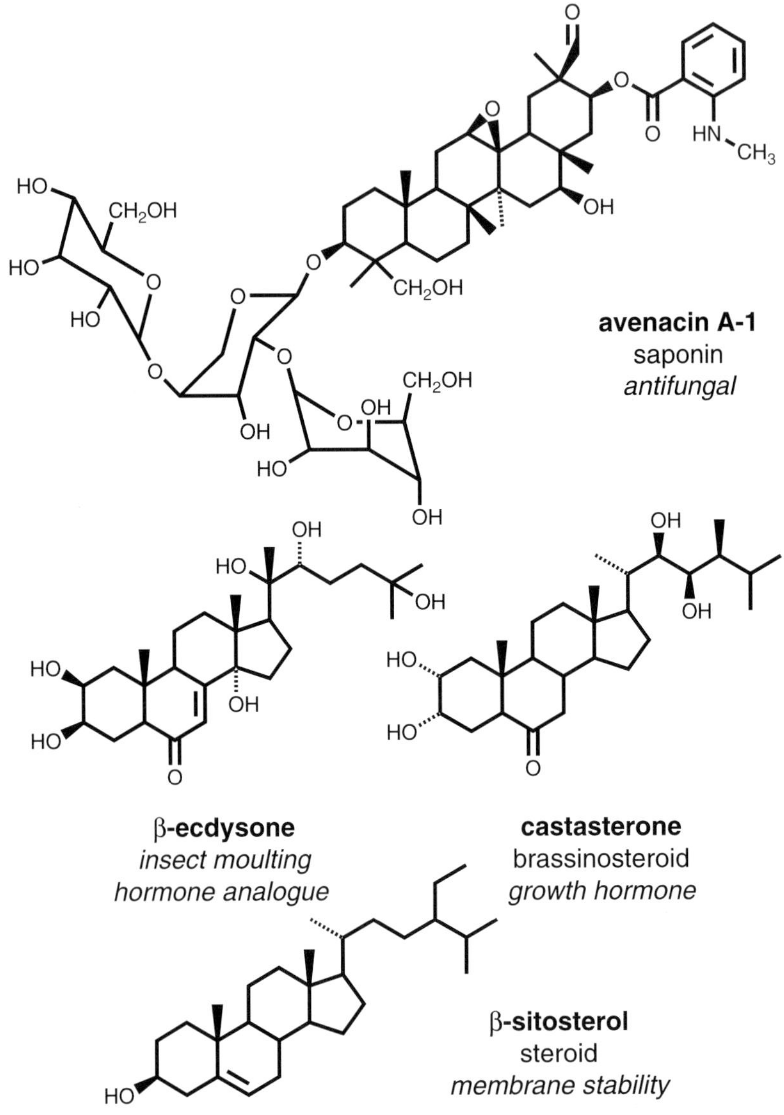

Fig. 10.3 Triterpenes and steroids from plants, with diverse biological activities

Polyacetylenes in higher plants represent derivatives of oleic acid and have attracted attention because of the high proportion of triple bonds in the molecule. They are characteristic for Apiaceae and Asteraceae; in the former, two conjugated triple bonds are possible, in the Asteraceae even up to five (Bohlmann et al. 1973). Falcarindiol is often referred to as antifungal compound (Kemp 1978).

The *naphtoquinone* juglone (Fig. 10.5) is synthesized via chorismate. This compound has received some attention as a phytotoxic allelochemical that has been made responsible for contributing to allelopathy of the walnut tree. Various

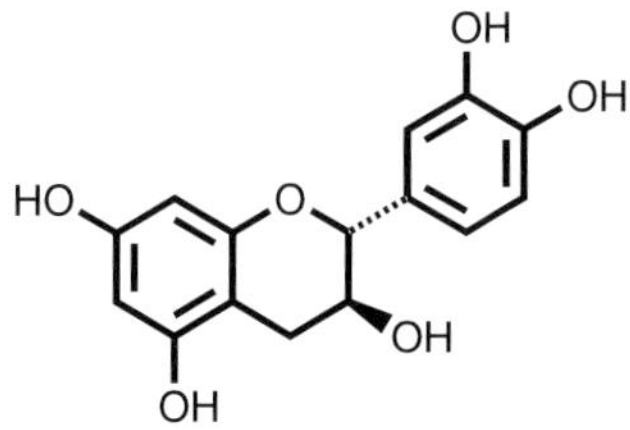

(+)-catechin
dihydroflavonol

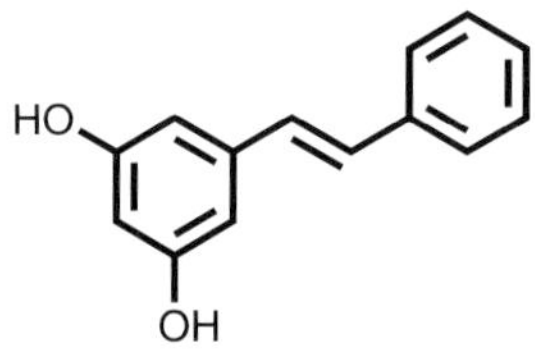

pinosylvin
stilbene

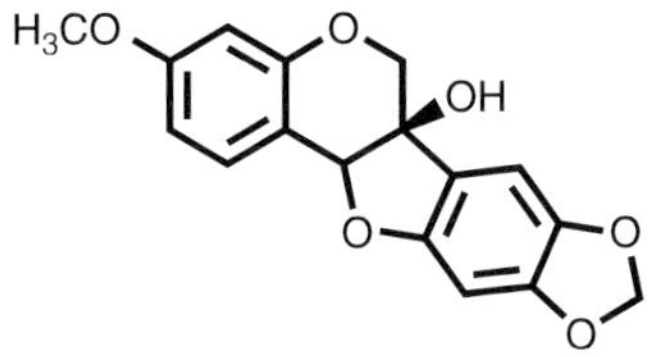

pisatin
isoflavone

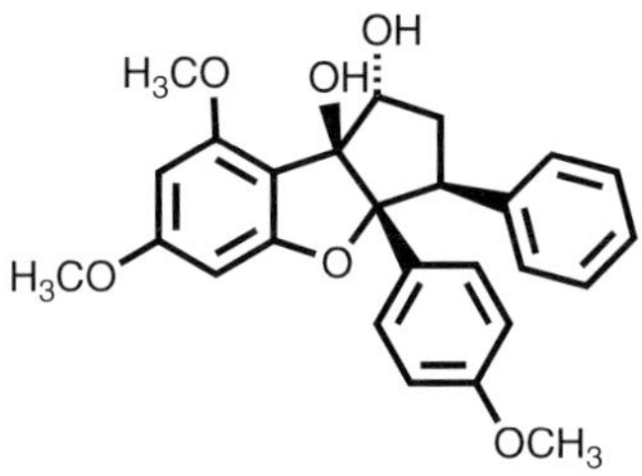

rocaglaol
cyclopenta[*b*]benzofuran

Fig. 10.4 Antifungal flavonoids, stilbenes, and cyclopenta[*b*]benzofurans from plants

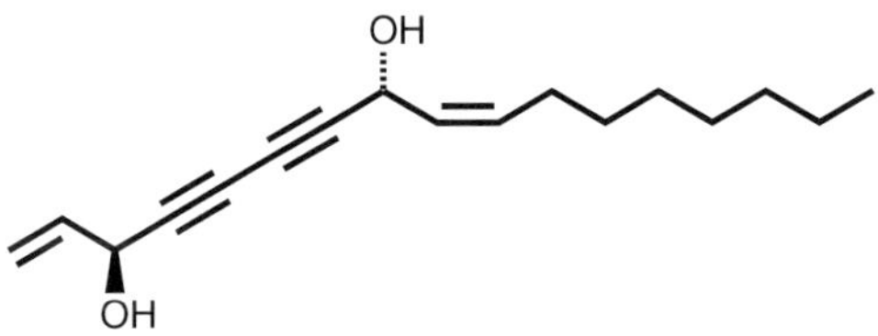

(+)-falcarindiol

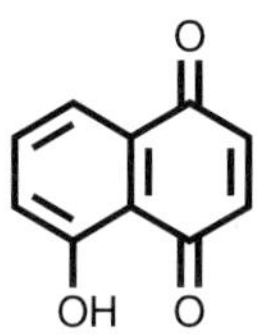

juglone

Fig. 10.5 Miscellaneous antifungal compounds from higher plants

prerequisite processes have already been proven for such an action actually to take place; however, the ultimate proof that juglone is actively taken up by the affected plant is still missing (Jose 2002).

The Brassicaceae have developed an especially efficient chemical defense system: *glucosinolates*. Derived from the amino acid precursors, glucosinolates are very polar secondary metabolites with good solubility to be compartmented in the vacuole. Onto damage of the tissue, myrosinase, an enzyme that was originally accumulated in specific cells in intact tissues, comes into contact with the glucosinolates and transforms them into aggressive volatile isothiocyanates, thiocyanates, and nitriles (Bones and Rossiter 1996). Various amino acids can serve as precursors for glucosinolates (Fig. 10.6 shows tryptophan). In the case of indole derivatives, the

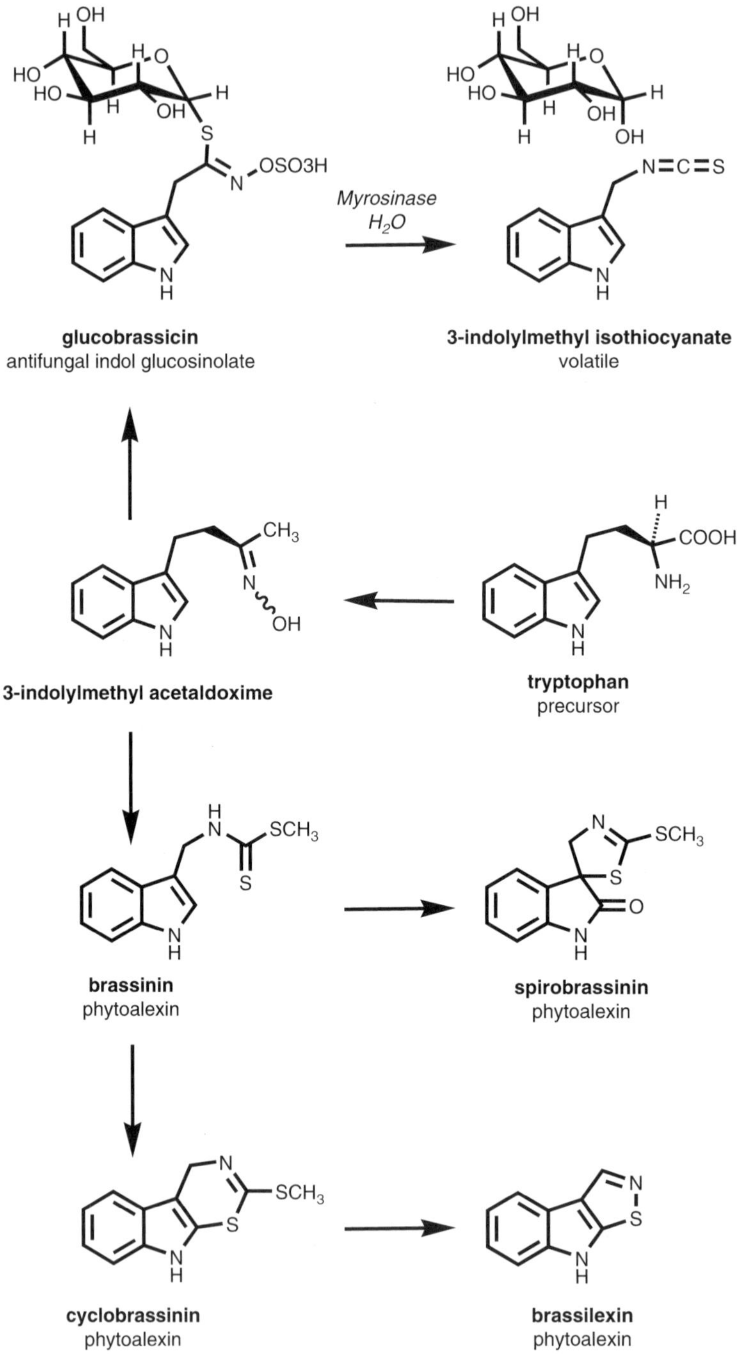

Fig. 10.6 Glucosinolates and indole phytoalexins from Brassicaceae

resulting 3-indolylmethyl acetaldoxime serves as a switch point to glucosinolates and indole phytoalexins, both of which confer resistance against fungi (Soledade et al. 2002).

10.2.4 Plant Defense Theories and Practical Efficiency

Plant defense theories aim to explain why most plants seem to be well protected from pathogen attack. The notion that plant secondary metabolites constitute traits that have evolved as efficient chemical defenses was established by the seminal studies of Dethier (1954), Fraenkel (1959), and Ehrlich and Raven (1964). Stamp (2003) reviews the four prevailing hypotheses.

McKey (1974, 1979) and Rhoades (1979) introduced the *optimal defense hypothesis*. Its basic predictions are (1) that organisms evolve and allocate defenses in a way that maximizes individual fitness and (2) that the development of these defenses is costly and diverts resources from other needs. Within the framework of this hypothesis, the following subhypotheses have been formulated:

1. The *plant's apparency subhypothesis* (Feeny 1975, 1976; Rhoades and Cates 1976) predicts that apparent plants, such as abundant tree species, are easily found by herbivores and thus invest in "quantitative defenses," such as tannins interfering with the nutrient uptake of the herbivores. By contrast, unapparent plants, such as casually occurring grasses and forbs, invest in qualitative defenses, e.g., glucosinolates in Brassicaceae (Fig. 10.6). However, support for this hypothesis was ambiguous, e.g., apparent plants were found to use qualitative defenses and unapparent plants to rely on quantitative defenses (Futuyma 1976).
2. The optimal defense within plant subhypothesis states that, within a plant, defenses are allocated in proportion to the risk of the particular organ to predator attack (McKey 1974, 1979; Rhoades 1979). Until recently, when experiments with genetically modified plants became feasible (Kessler et al. 2004), realistic tests of this hypothesis were limited. However, it seems reasonable to adopt the idea that defense allocation among plant parts reflects cost–benefit patterns in plant fitness (Stamp 2003). Consistently, we usually find comparatively high and thus costly accumulations of secondary metabolites in plant roots, which in many instances constitute the only true perennial organs of many grasses and forbs.
3. The *inducible defense subhypothesis* refers to secondary metabolite defenses produced in response to damage. As chemical defenses are costly, they should be reduced in the absence of the targets (Rhoades 1979). Although several phenomena, such as delayed induction compared with the expected immediate induction of defense compounds and the slow relaxation times compared with rapid relaxation in order to avoid costs (Karban and Baldwin 1997), do not comply with predictions of the optimal defense hypothesis, they fit the general statement of it.

4. The allocation cost of phenotypic defense subhypothesis predicts that allocation to other needs (e.g., growth and reproduction) is lowered when investments in chemical defenses are high (Rhoades 1979). Under stress, plants are expected to reduce costly defenses in favor of less costly but also less efficient defenses.

In summary, the results of numerous extant studies more or less support the various subhypotheses of the optimal defense view, though many studies indicated the existence of more complex patterns than originally envisioned. However, the consistencies detected stimulated the development of further hypotheses for plant defense (Stamp 2003).

The *carbon–nutrient balance hypothesis* represents a model explaining how the supply of carbon and nutrients in the environment influences the phenotypic expression of plant defenses (Bryant et al. 1983; Tuomi et al. 1988, 1991). The basic view is that the chemical defense of a plant accrues from a combination of baseline (proportional to growth) and flexible allocation (shunting of carbon surplus into defense), and that plant responses to changes in the carbon–nutrient ratio alter the phenotypic expression of a plant's defense genes. Assumptions include (1) that good supply of minerals results in a carbon allocation to growth, (2) nutrients limit growth more than photosynthesis, (3) herbivory selects for secondary metabolite defenses, and (4) defenses reduce herbivory. It predicts that (1) effects of changes in the carbon–nutrient levels do not translate into changes of defense levels in genotypes with low phenotypic plasticity in defense, e.g., trees, and (2) high phenotypic plasticity in defense correlates resource conditions to changes of the total defense level, i.e., excess carbon results in nonnitrogenous defense and nitrogen excess in accumulation of alkaloids and nonprotein amino acids (e.g., Fabaceae). However, testing this hypothesis requires an assessment of the baseline carbon allocation under optimal conditions and maximal growth, and, so far, nearly no study exists that actually fulfills this criterion (Stamp 2003). The carbon–nutrient balance hypothesis also provided a basis to develop the next hypotheses.

The *growth rate hypothesis* (Coley et al. 1985) makes the following predictions: (1) replacing resources lost to herbivory competition favors fast-growing plant species in high-resource environments, and slow-growing plant species in low resource environments; (2) fast-growing plants have more secondary metabolites (mobile, high turnover, reversible); in contrast, slow-growing plants invest in biopolymers, such as lignin and tannins. However, explorations of the growth rate hypothesis provided mixed results (Stamp 2003). Nevertheless, research related to the growth rate hypothesis supports the notion that resource availability may be more relevant than pressure from herbivores (Coley 1987).

The *growth–differentiation balance hypothesis* provides a framework for predicting how plants will balance allocation between differentiation-related processes and growth-related processes over a range of environmental conditions (Loomis 1932, 1953). Environmental factors, such as shortage of water and nutrients, cause slowing of growth much more than photosynthesis. Instead, the latter increases the resource pool (excess carbon) for differentiation processes. This suggests a scenario for secondary metabolite production with low effects on plant fitness. Competition in resource-rich environments selects for a growth-dominated strategy, whereas stress or resource-poor environments select for a differentiation-dominated strategy (Herms and Mattson 1992). This theory predicts moderate concentrations of secondary metabolites in slow-growing plants. Intermediate resource availability leads to high concentrations of secondary metabolites but medium growth. High resource availability results in high growth rates and intermediate secondary metabolite levels. However, only a few studies have tested this hypothesis so far, too few to be conclusive (Stamp 2003).

The hypotheses on chemical defense introduced above assume that plants are well defended. But, are plants really that well defended? The low hit rates in pharmaceutical screenings suggest that biological activities are a rather rare phenomenon in nature (Jones and Firn 1991; Firn and Jones 1996, 2003). Firn and Jones formulated the *screening hypothesis*, which recognizes the fact that biological activity is a rare phenomenon in nature because there exists a fundamental physiochemical constraint: biological activity requires the active compound to bind to a receptor pocket of an enzyme or membrane protein. To do so, in most cases, the candidate molecule has also to travel to the target and pass several membranes on its way. Further, the screening hypothesis provides also an argument for why, from an evolutionary viewpoint, it makes sense for organisms to maintain compounds without evident benefit: inactive molecules provide essential precursors for active molecules, or conversely have been active in the past but the target organism is already extinct. Finally, the overall metabolic process acts as an ultimate constraint on the metabolic diversity of a secondary metabolite producing organism. In my opinion, this thesis explains many inconsistencies of the various defense hypotheses. However, the reception of the screening hypothesis has been (e.g., Berenbaum and Zangerl 1996; but see Stamp 2003) and still is (Pichersky et al. 2006, Firn and Jones 2006) controversial.

Reactive oxygen species (ROS), which include hydrogen peroxide, singlet oxygen, superoxide, and the hydroxyl radical, are byproducts of photosynthesis and respiration. Small-molecule antioxidants, such as carotinoids, tocopherols, ascorbic acid, and gluthathione, have been recognized as part of a regulation system of oxygen radical concentrations in the cell, and serve as important signal molecules to coordinate responses of the cell to various abiotic and biotic stresses (Desikan et al. 2005; Foyer and Noctor 2005). Maintaining the redox homeostasis is mandatory for cells to maintain the numerous metabolic processes requested for sustaining life. Many phenolic plant secondary metabolites, such as phenylpropanoids and flavonoids, which are induced by various abiotic and biotic stresses, serve primarily as antioxidants in the cell but, under certain conditions, may show also prooxidant properties (Grace 2005). Conversely, generation of oxygen radicals may cause toxicity, also in fungi; the perylquinone cercosporin, a fungal metabolite, induces the formation of superoxide anions both in the light and in the dark (Xing et al. 2003). Hypericin, the red pigment in the glands of *Hypericum perforatum*, is a perylquinone occurring in plants (Fig. 10.7)

We may assume that most secondary metabolites possess prooxidative and antioxidative properties, always depending on the redox state of the cell. So far, this

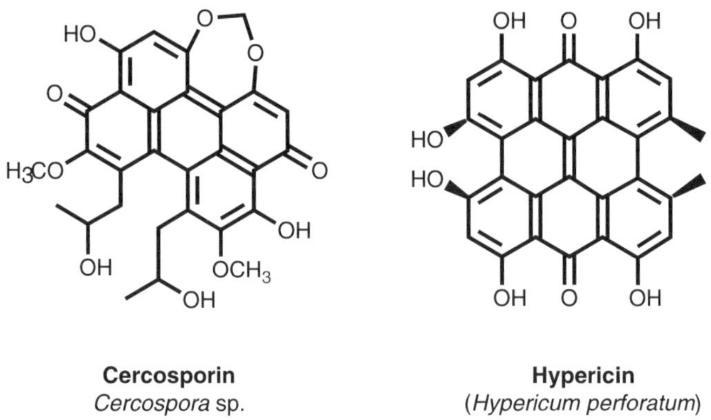

Fig. 10.7 Photosensitizing perylenequinone pigments from fungi and higher plants: cercosporin and hypericin

aspect has been explored only for a few selected secondary metabolites. As most of the known secondary metabolites contain oxygen functions, they will most likely interact with the redox homeostasis of cells, always depending on their ability to enter actively or passively.

Another criterion of activity is dose. Some inhibitory compounds stimulate in low concentrations; this effect is called hormesis and is considered to be widespread in biological activities (Calabrese and Baldwin 2003). In pharmacological screenings though, decisions of whether a tested compound is active or not depend largely on a comparison of the end point with that of positive controls of established drugs. The question is now: Can we relate this procedure also to the assessment of a specific plant secondary metabolite in chemical defense against some predator? Basically yes; a comparison of effects comprising a range of concentration is always practicable and reasonable. By this procedure we will at least obtain information on which compounds are more active and which less. And, if we increase the concentrations, some compounds, certainly not all, will also significantly decrease the performance of the organism tested. Plants accumulate their constitutive secondary metabolites in specific compartments in their tissues, and thus local concentrations may be extremely high. Destruction of the tissues, either by damage or by decomposition, then releases the secondary metabolites in a concentration-gradient-dependent fashion into the cytosol of the damaged cells from various compartments such as the vacuole or specialized cells or tissues.

The determination of the actual concentration of focused secondary metabolites in plant tissues remains problematic because all extraction methods available to us depend on destruction of the tissue and rarely allow discrimination between different tissues with good resolution. The lack of suitable methods poses a bottleneck to comprehensive studies of the dynamics of secondary metabolites on the cell level. Some secondary metabolites have high enough proportions of unsaturated bonds so that a monitoring of their cellular dynamics becomes possible by fluorescence microscopy; this works well for many aromatic compounds but usually not for terpenoids, the saponin avenacin A-1 (Fig. 10.3) being an exception (Buschmann et al. 2000; Osbourn et al. 2003). Immunoanalytical methods present a methodologically appealing alternative, but the laborious efforts to develop and obtain the appropriate antibodies have restricted the application of these methods to mycotoxin analyses in food stuffs so far (Ho and Durst 2003).

Considering all the aspects discussed, the following conclusions are plausible: (1) roots and other plant organs may contain some compounds that suppress the growth of some fungi; (2) not all extant plants contain antifungal compounds with pronounced antifungal activity in their organs but may rely on the accumulation effect as efficient defense; (3) not all fungi present in the rhizosphere are equally suppressed, some of them may be even tolerate specific secondary metabolites inhibitory to others; and (4) a better understanding of the mode of action of secondary metabolites is needed for an explanation of the observed phenomena. One study focusing on the sensitivity to avenacin A-1 by a range of root-colonizing fungi of oat and wheat provides a exemplary illustration of the spectrum of effects that are to be expected in investigations of antifungal defense mechanisms of plant secondary metabolites on a community level (Carter et al. 1999).

10.3 Secondary Metabolites Maintain Plant Diversity

10.3.1 Theory in Community Ecology

Plants are members of distinct communities. So far, in plant community ecology, *niche-based theories* (Tilman 1982; Tilman and Pacala 1993) and *neutral theories* (Hubbell 2001) are offered to predict mechanisms that lead to organismic biodiversity. Niche-based theories predict that patterns of plant diversity correlate with the quality of nutrient gradients (interspecific competition and resulting tradeoffs) and that habitat heterogeneity practically facilitates unlimited biodiversity. However, they do not predict limitations to diversity and do not provide any explanation for relative species abundances. In contrast, neutral theory assumes that species are ecologically equivalent in their responses to all constraints. Species are rare or abundant, not because of their traits and the traits of their competitors, but solely owing to stochastic drift in the density of the competitively identical species. This would provide us with an elegant explanation for relative species abundance patterns. However, numerous observations contradict the absence of any relation between the traits of species and their abundance in plant communities (Tilman 2004). As a result, Tilman (2004) suggested including stochastic processes into tradeoff theory (*stochastic niche theory*). This is founded on three observations: (1) success and failure of propagules from invaders determine community assembly; (2) successful invaders must efficiently utilize resources that remain unconsumed by established species; and (3) resource requirements closely related to those of the established species facilitate the success of the invader (note that "invader" is not understood in a geographical context).

Facilitation includes positive interactions that make a local environment more favorable to a cooccurring organism either by direct or by indirect effects (Callaway

1995). It differs from *mutualism* by its one-sided nature, and compared with competition has hitherto been less focused on in ecological theory development (Bruno et al. 2003). Recently, a scenario that also included negative interactions was added to the debate about what may be determining community structure. Bever (Bever 1994, 2003; Bever et al. 1997) suggested focusing on biotic interactions; his work is centered on mechanisms of how *positive feedbacks* from mutualists and *negative feedbacks* from antagonists and pathogens might affect plant community structure (Fig. 10.8). Surprisingly, this notion is not considered at all in the general debate

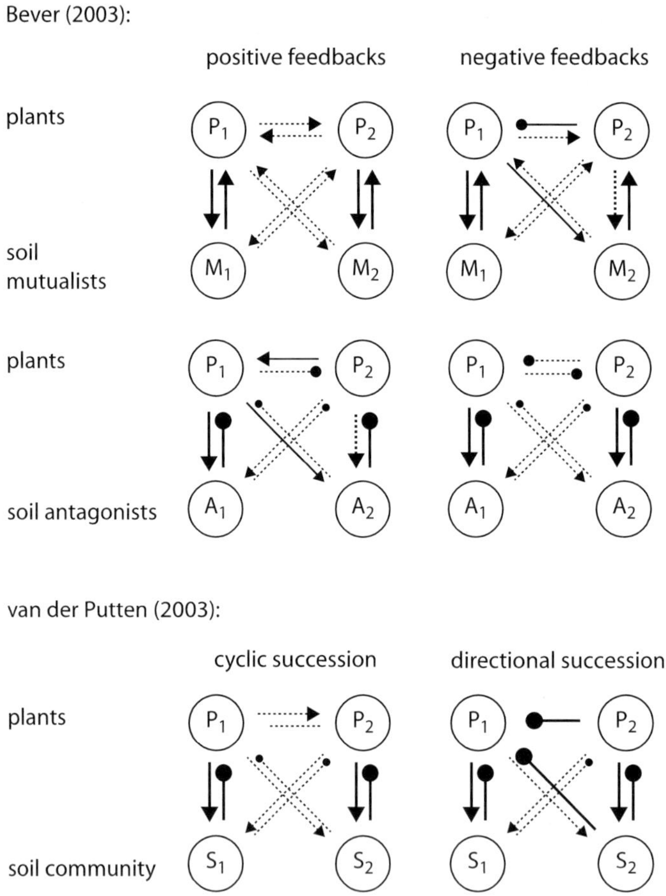

Fig. 10.8 Plant–soil community feedback patterns: the soil community comprises both mutualistic and antagonistic organisms. The net effect may be either positive (arrows) or negative (*lines with a circle*); the thickness of the arrows and lines with a circle (full or dashed lines) indicates the strength of the interactions. Asymmetry in the feedbacks may cause positive feedbacks to become negative, and negative feedbacks to become positive (Bever 2003), as well as directional succession (van der Putten 2003). (Redrawn from Bever et al. 1997; Bever 2003; van der Putten 2003)

about niche-based versus neutral theories; the concept of the latter even categorically excludes organismic interactions (Hubbell 2001). Bever (2003) developed a model of soil community feedbacks affecting plant community structure and an exploration suggested that negative feedbacks facilitate the coexistence of plant species much more than positive ones (Fig. 10.9). It is quite feasible that secondary metabolites from roots and also from leaf litter may significantly be a part of these negative feedbacks.

However, feedbacks affecting plant community structure do not solely include belowground interactions; linkages between aboveground and belowground factors should also be taken into consideration (Bardgett and Wardle 2003; Wardle et al. 2004). In fertile and productive ecosystems, fast-growing and short-lived plants with high litter quality prevail owing to reduced levels of phenolics, lignin, and other structural carbohydrates and high levels of nitrogen. Herbivores consume a high percentage of the net primary production of these plants and, in consequence, massively return labile fecal material to the soil. In soil food webs, this favors bacterial populations, and as a consequence, populations of earthworms increase compared with those of microathropods. Rapid decomposition and nutrient mineralization but low carbon sequestration leads to high nutrient availability. Conversely, in infertile and unproductive ecosystems, slow-growing and long-lived plants dominate. They are usually characterized by high carbon allocation to secondary metabolites and thus low forage quality. Herbivores consume a low percentage of the net primary production and thus return low amounts of fecal material to soil. This causes an accelerated succession to plants with low litter quality. The litter has low nitrogen levels but high contents of phenolics, lignin, and other structural carbohydrates.

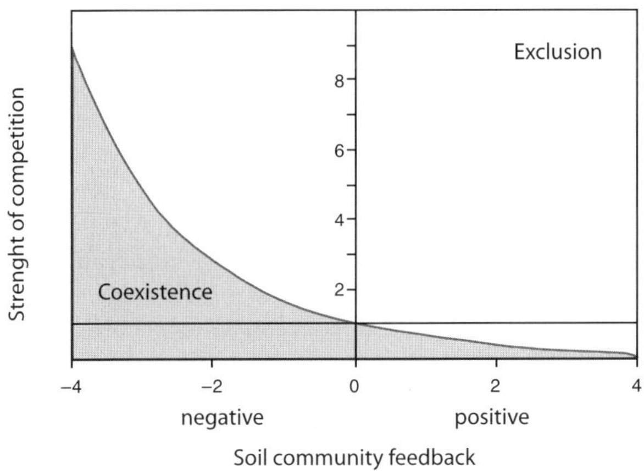

Fig. 10.9 Conditions of coexistence for competing plant species in the presence of the soil community feedback. The *shaded regions* represents the parameter values for which the coexistence of competing species is possible as a function of the strength of interspecific competition and soil community feedback. (Redrawn from Bever 2003)

This favors fungi as decomposers, and as a consequence, the development of populations of enchytraeid worms and macro- and microathropods. The mixing of the soil is low, and slow decomposition and nutrient mineralization as well as high carbon sequestration result in low nutrient supply rates. Further, we have to consider that aboveground herbivores may affect metabolic processes in and exudation of roots, and vice versa, belowground herbivores may induce quantitative as well as qualitative levels of secondary metabolites in leaves (Bezemer and van Dam 2005).

10.3.2 A Survey of Studies Supporting the Functions of Secondary Metabolites Maintaining Biodiversity

The notion that secondary metabolites may have functions in processes shaping biodiversity is especially evident from the literature exploring plant–herbivore interactions (Southwood 1985; Jones and Lawton 1991; Rosenthal and Berenbaum 1992; Foley and Moore 2005), but also to a lesser extent from plant–microbe interactions, e.g., the correlation of leaf endophytic fungi and leaf chemistry reported by Arnold and Herre (2003). However, concerning their function in maintaining this biodiversity, there exist even fewer studies by many orders of magnitude. I will review some which particularly caught my attention.

10.3.2.1 Chemical Warfare Promotes Microbial Diversity

Rhizodeposition constitutes a valuable carbon source for bacteria and fungi. Both are known to produce toxic metabolites, a phenomenon that is called antibiosis. The majority of studies that focus on this aspect are directed at identifying bacterial or fungal strains with potential for application in biocontrol (Whipps 2001). However, saprophytes competing for resources may also produce antibiotic metabolites. Interestingly, one of the most potent antifungal classes of secondary metabolites, the strobilurins, was isolated from the culture broth of a wood-decomposing basidiomycete (Sauter et al. 1999). This finding suggests that production of toxic secondary metabolites may not be uncommon among saprophytic fungi. Competition for carbon results in low availability of this resource and this leads to the activation of secondary metabolite biosynthesis genes via specific transcription factors (Demain 1996). This behavior clearly indicates that secondary metabolites are used by fungi for competitive chemical warfare. Another often-reported observation is that fresh isolates of fungi usually readily produce secondary metabolites and spores. However, during continued cultivation on artificial carbon-rich media this trait disappears.

If we do not follow the neutral theory and do not assume that the huge diversity of bacteria and fungi is accidentally caused by stochastic drift, we have to identify factors that define the numerous microniches required for the maintenance of this diversity. In soils, microbial diversity by far exceeds plant diversity, and even if variation in the quality of the carbon source is taken into consideration, it does not explain why such a huge diversity exists in such a comparatively homogenous environment. A possible answer to this question is provided by two papers. Czárán et al. (2001) propose that antibiosis contributes to the maintenance of microbial biodiversity by a mechanism that is based on a spatially explicit game theory model. This model agrees with the prediction of Bever's model (Bever 2003) that negative feedbacks with the soil microbial community facilitate the coexistence of plant species rather more than effects of positive feedbacks, the latter of which we might expect to be more efficient. Further, the findings of Kerr et al. (2002) suggest that the interactions and dispersal processes of microbes outlined have to be confined spatially to small scales in order to contribute to the maintenance of biodiversity. There are good reasons to view secondary metabolites produced by competing microbes as well as secondary metabolites from root exudations or from decomposing litter as necessary determinants of microniches that are required for microbial biodiversity in soils and, ultimately, also for plant biodiversity.

10.3.2.2 Chemodiversity Hypothesis

So far, the effects of secondary metabolites on maintaining organismic diversity has been noted for prokaryotes. Iason et al. (2005) provided the first study known to me that explicitly tests a *chemodiversity hypothesis* for higher plants. The main prediction of such a hypothesis is that diversity of secondary metabolites in lower trophic levels is required to maintain species diversity of the community. In their study, Iason et al. (2005) focus on individual Scots pine trees (*Pinus sylvestris*) and the associated ground vegetation. Needles of pine trees contain monoterpenes and the individual trees show genetically determined constitutive variation of this trait. The authors could show that diversity of terpenoids of the needles is significantly correlated with nonwoody vascular plant diversity beneath the respective trees (Fig. 10.10). The accumulated monoterpenes affect the litter quality of each tree; and as litter quality affects nutrient availability (Wardle 2002), the chemical variation in the litter diversifies niches to be occupied by different sets of ground vegetation. This study does not provide any insights into the mechanistic processes involved but its outcome highly suggests further and more intensified exploration of the chemodiversity hypothesis.

Another study also points in this direction: Ehlers and Thompson (2004) compared the performance of brome grass in soil from underneath different *Thymus vulgaris* chemotypes and noted facilitation of accessions from the same site, whereas material grown from seed collected from other sites of other chemotypes performed less vigorously. This study does not explicitly discuss the effects of constitutive secondary metabolites on maintaining biodiversity but provides evidence that constitutive secondary metabolites of a plant species may act as a constraint on cooccurring species and may select for adaptive traits in them. However, the mechanisms involved are still unclear and require further study.

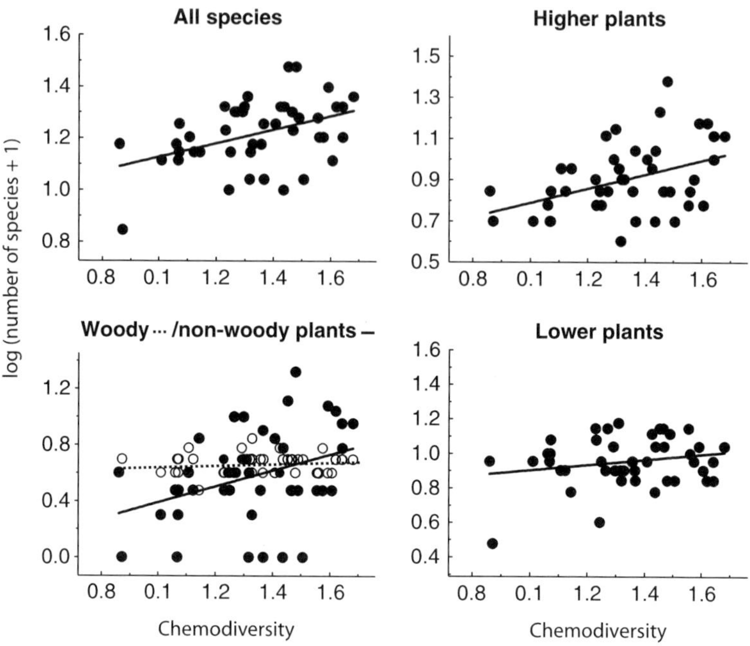

Fig. 10.10 The relationship between plant species diversity (natural logarithm of 1 plus the number of plant species) under Scots pine trees and the chemical diversity (mixture of monoterpens) of needles (Shannon index). Each *circle* represents a tree. (Redrawn from Iason et al. 2005)

To obtain deeper insights into the contributions of secondary metabolites to plant defense, and in a larger context to a chemodiversity hypothesis, we have to pay more attention to plant defense strategies (van der Putten 2003; Wardle et al. 2004). To date, nearly all studies have focused on a particular class of secondary metabolites rather than on the complete accumulation pattern (metabolic capacity *sensu* Firn and Jones 2003). This is partly due to methodological constraints—a complete assessment of a plant's secondary metabolites is usually hampered by selective sensitivity of the analysis methods, only combinations of diode-array detection and mass spectrometry detection usually yield the required quality of results (Fig. 10.11) —and it is also partly due to the fact that the focus is usually directed towards a specific derivative or class of secondary metabolites, not only for reasons of known

Fig. 10.11 a Metabolic profiling of secondary metabolites from a dominant (*Peucedanum cervaria*) and rare plant (*Peucedanum alsaticum*) of a species-rich grassland community near Vienna, Austria. Gas chromatography–electron impact mass spectrometry (GC-EIMS) (silylated chloroform phase of the methanol extract, quadrupole mass spectrometry, column 5% phenyl-methylsilicone, 20 m, 0.18-mm diameter, 0.18-μm film) and HPLC-DAD analyses (chloroform extract, for details of separation conditions see the legend to Fig. 10.12) illustrate the selective and reciprocal sensitivity of these analysis techniques. UV and mass spectrometry spectra provide

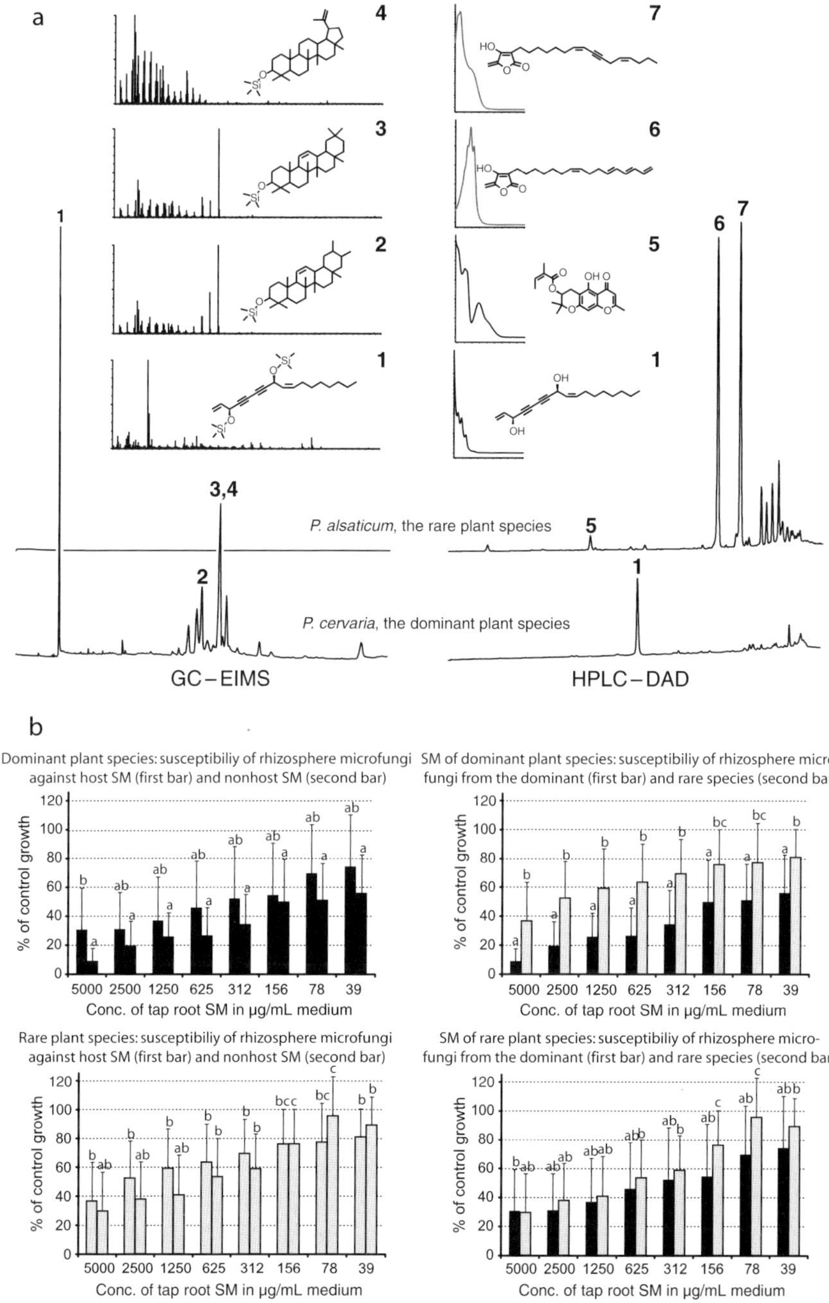

Fig. 10.11 (continued) some means of tentative structure assignment. **b** Susceptibilities of microfungi from the rhizosphere of both model plants. Twofold microdilution of secondary metabolites in aqueous solution suspended with water-soluble carbohydrates from the plant roots with Tween 80 as an emulsifier. Assessment of growth in percentage of control growth after 3 days of development at ambient temperature in the dark, evaluation by turbidity at 620 nm. *a*, *b*, and *c* denote differences at the 90% level of significance for all treatments (n = 10). *SM* secondary metabolites

biological activities but also for practicability. Hence, studies that focus on the whole defense strategy are rather rare. However, one existing study, though exploiting the coevolution hypothesis, explicitly supports the chemodiversity hypothesis: Farrell et al. (1991) presented a metaanalysis of the literature pointing out that resin and oil canal bearing plant lineages are more diversified than their sister groups lacking this trait. The study was performed exploring the chemical arms race hypothesis, but it also provides support for a chemodiversity hypothesis by suggesting that maintenance and accumulation of secondary metabolites results in diversification, certainly not only by the phenomenon alone, but also by the induced feedbacks with mutualists and antagonistic competitors.

10.3.2.3 Chemodiversity Belowground

Photosynthesis requires light and leaves represent the organs in which this process takes place. By contrast, roots are those organs that take up nutrients; this process is primarily located belowground. Indirect effects of the botanical composition of a plant community on carbon and nitrogen cycling occur (Hooper and Vitousek 1998; Schimel et al. 1998). The quality of leaf litter of a specific plant species may decisively affect nutrient uptake (Hättenschwiler and Vitousek 2002; Wardle 2002; Souto and Pellisier 2002). Similar effects are also to be expected for root litter but, to the knowledge of the author, no specific studies exist. This may be due to the fact that the input of root litter does not occur in such predictable circles as the annual input of leaf litter. Further, root litter is definitely much more difficult to locate than leaf litter. Thus, we may assume that (1) root litter, especially that of woody perennial roots, will take much longer to decompose than leaf litter; (2) during this process, secondary metabolites will be constantly released; and (3), when viewed over longer periods, root litter of dominant plant species will exert more pronounced feedbacks than rare species of the community, either directly on the soil microbial community, or indirectly on the coexisting plant species.

Sinkkonen (2003) introduced an interesting model characterizing the effects of secondary metabolites released from decaying plant litter. It was aimed at providing predictions regarding phytotoxicity in context with allelopathy. At the beginning of the decomposition process, stimulation occurs that is rapidly replaced by inhibition but reemerges in the final stage. Density dependence is another factor that may influence the actual outcome of an allelopathic interaction as high target plant density causes a dilution of the effect. Hence, this suggests allelopathy should function better if the target species occurs in low numbers and the producing species in high numbers. Thus, if density dependence is included, the model also includes scenarios in which observation of effects is not stringently predicted. Likewise, the question if dominant or rare plants of a community are involved is pointed out as an important criterion. As a consequence, we should expect to find any proof in favor of a chemodiversity hypothesis in dominant plant species rather than in rare plant species.

In the following I want to present some of my ongoing work. Plant communities are made up of dominant and rare plant species. If we locate a pair of plant species which belong to the same genus, we can turn them into an attractive model for exploring the chemodiversity hypothesis. If we want to identify effects of secondary metabolites that support the chemodiversity hypothesis, we have to consider the density dependency of such effects. Most likely, we will expect them to occur in dominant plants. I used two umbellifers from a dry grassland community, the broad-leaved spignel, *Peucedanum cervaria*, as a dominantly occurring plant, and *Peucedanum alsaticum* as a rare plant. Both umbellifers develop prominent tap roots that contain resin in the first instance and latex in the second instance. Polyactetylenes and triterpenes are prominent nonvolatile secondary metabolites in the resin of the former and alkylbutenolides and methylchromones occur in latex of the latter (Fig. 10.11). A survey of susceptibility against the two mixtures of secondary metabolites against a number of soil fungi from the rhizosphere of both species provided the following results: (1) the two rhizospheres supported different microfungi; (2) the rhizosphere fungi showed some adaptation to tolerate the secondary metabolites of their respective host; and (3) effects of secondary metabolites of the dominant plant species allowed discrimination of the source of the two groups of microfungi; the soil fungi from the rhizosphere of *Peucedanum alsaticum* distinctly showed more sensitivity when confronted with the *Peucedanum cervaria* metabolites than in the reversed test scenario (Fig. 10.11). Hence, these data suggest that plant secondary metabolites cause feedbacks on the composition of their associated rhizosphere fungal community and that a dominant plant species may show more pronounced effects in this aspect than a rare plant species. This can be regarded as evidence for a chemodiversity hypothesis as different dominant plants provide niches for different microfungal communities. Conversely, the different microfungal communities may cause feedbacks on the composition of plant species (Bever 2003). Direct observation of such processes will be hampered by the comparatively large time scales in which successions in species-rich communities take place, and it is most likely that these mechanisms occur in later rather than in earlier successional stages.

The effects of secondary metabolites on soil microbes are not confined to tolerance alone. Conversely, microbes may also utilize several secondary carbohydrates, especially phenolics, as a carbon source by expressing extremely versatile laccases (Cates 1996; Rahouti et al. 1999; Mayer and Staples 2002). The utilization of chlorogenic acid, a widely occurring plant phenolic, is illustrated by Fig. 10.12. We do not know much about other enzymatic capabilities of fungi to utilize secondary metabolites as a carbon source; however, their generally observed instability in soil (Schmidt and Ley 1999) implies that, besides tolerance, utilization capability may be another factor that affects chemodiversity-modulated processes.

To sum up, chemodiversity should not be regarded as the only factor that facilitates species diversity. However, if incorporated into existing hypothetical frameworks, it will contribute to improved predictions of the actual scenarios.

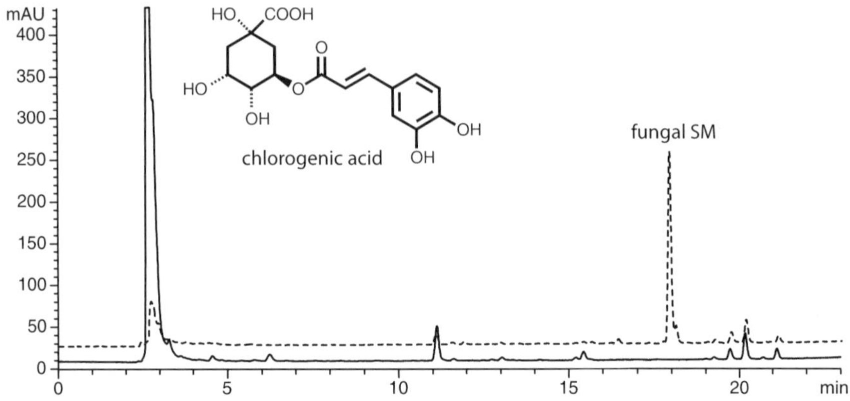

Figure 10.12 Chlorogenic acid, a major metabolite in tubers of the umbellifer *Cicuta virosa*, water hemlock, is metabolized by an endophytic *Fusarium avenacum* strain. Fungal metabolites were identified by cochromatography of a culture medium extract of the identical fungus. HPLC analysis of the methanolic extract (RP C18, Spherisorb C18, 5 μm, 250 mm × 4 mm, 60–100% methanol in aqueous buffer, pH 3, signal 230 nm)

10.3.3 Emerging Methodical Approaches for Insights into Secondary Metabolite Functions in Exploring Ecological Theory

The conceptual framework is presented appealing because of its to obtain better insights into biological functions of secondary metabolites and because it is not hampered by some idiosyncrasies as are the defense hypotheses. What is still definitely lacking are more mechanistic studies to obtain a better understanding of plant defense strategies. Here, the difficulties with the analysis of secondary metabolites may certainly deter many ecologists. The structural diversity of secondary metabolites in living organisms is huge and their inclusion as traits in ecological studies requires a tentative identification at least. Metabolomics may constitute a possible avenue of a future methodological approach (Bezemer and van Dam 2005). Today, this term is often mentioned in connection with omics technologies, such as genomics and proteomics. Another term, which was previously used but has been given up in favor of metabolomics, is metabolic profiling (Fiehn et al. 2000; Jenkins et al. 2004). There exist great and in my opinion also founded expectations that metabolomics will provide improved insights into proportions of primary and secondary metabolites in feedbacks between plants and their predators and pathogens (Bezemer and van Dam 2005). Metabolomics utilizes a systems biology approach by utilizing pattern analysis with chemometric methods. There also exists one inherent problem: various chromatographic and spectroscopic methods are available for the metabolite profiling. However, methods with high sensitivity, such as gas chromatography linked to mass-selective detection, discriminate against a number of analytes. In contrast, universal techniques, such as nuclear magnetic resonance spectroscopy are less sensitive in comparison.

10.4 Conclusions

This chapter has reviewed hypotheses which were introduced to explore the efficacy of secondary metabolites as chemical plant defense exemplified by antifungal root secondary metabolites. Some case studies unambiguously provide support for a chemical defense function. A few other studies suggest that secondary metabolites may have functions beyond just defining the outcome of a specific biotic interaction. Plant secondary metabolites may also affect nutrient cycling, either by leaching from leaf litter or by being released from decomposed roots. Consequently, we may ascribe various direct and indirect feedbacks to secondary metabolites that may affect other cooccurring living organisms. In this context, the chemodiversity hypothesis presents an attractive additional avenue to explore mechanisms leading to biodiversity and may serve as an interesting contribution to the ongoing debate.

However, a substantial methodological bottleneck exists in assigning structures to the secondary metabolites focused on. Expertise in secondary metabolites is mostly confined to pharmacologists, who attempt to tap biodiversity to find new drugs. Fundamental financial investments and competition in these screening procedures have resulted in low availability of public databases which otherwise might assist biologists in incorporating secondary metabolites explicitly in their studies. The acceptance and ultimately also the success of metabolomics as a tool to elucidate the complex functions of secondary metabolites in ecology will also depend on how soon informative databases will be available. Regarding this aspect, biologists should not hope for too much help from neighboring disciplines, such as pharmacology and chemistry, but should become active themselves.

References

Agrawal AA, Tuzun S, Bent E (1999) Induced plant defenses against pathogens and herbivores. APS, St Paul

Arnold AE, Herre AE (2003) Canopy cover and leaf age affect colonization by tropical fungal endophytes: ecological patterns and process in *Theobroma cacao*. Mycologia 95:388–398

Asiegbu FO (2000) Adhesion and development of the root rot fungus (*Heterobasidion annosum*) on conifer tissues: effects of spore and host surface constituents. FEMS Microbiol Ecol 33:101–110

Bardgett RD, Wardle DA (2003) Herbivore-mediated linkages between aboveground and belowground communities. Ecology 84:2258–2268

Barz W (1997) Phytoalexins. In: Hartleb H, Heitefuss R, Hoppe H-H (eds) Resistance of crop plants against fungi. Fischer, Jena, pp 183–201

Berenbaum MR (1983) Coumarins and caterpillars: a case for coevolution. Science 201:532–534

Berenbaum MR, Zangerl AR (1996) Phytochemical diversity: adaptation or random variation. In: Romeo JT, Saunders JA, Barbosa P (eds) Phytochemical diversity and redundancy in biological interactions. Plenum, New York, pp 1–24

Bever JD (1994) Feedback between plants and their soil communities in an old field community. Ecology 75:1965–1977

Bever JD (2003) Soil community feedback and the coexistence of competitors: conceptual frameworks and empirical tests. New Phytol 157:465–473

Bever JD, Westover KM, Antonovics J (1997) Incorporating the soil community into plant population dynamics: the utility of the feedback approach. J Ecol 85:561–573

Bever JD Schultz PA, Pringle A, Morton JB (2001) Arbuscular mycorrhizal fungi: more diverse than meets the eye, and the ecologicl tale of why. Bioscience 51:923–931

Bezemer TM, van Dam NM (2005) Linking aboveground and belowground interactions via induced plant defenses. Trends Ecol Evol 20:617–624

Bohlmann F, Burkhardt T, Zdero C (1973) Naturally occurring polyacetylenes. Academic, London

Bohlmann J, Gershenzon J, Aubourg S (2000) Biochemical, molecular genetic and evolutionary aspects of defense related terpenoids mechanisms in confers. In: Romeo JT, Ibrahim R, Varin L, De Luca V (eds) Evolution of metabolic pathways. Pergamon, Amsterdam, pp 109–150

Bones AM, Rossiter JT (1996) The myrosinase-glucosinolate system, its organization and biochemistry. Physiol Plant 94:194–208

Brader G, Vajrodaya S, Greger H, Bacher M, Kalchhauser H, Hofer O (1998) Bisamides; lignans, triterpenes, and insecticidal cyclopenta[*b*]benzofurans from *Aglaia* species. J Nat Prod 61:1482–1490

Brimecombe MJ, De Leij FA, Lynch JM (2001) The effect of root exudates on rhizosphere microbial communities. In: Pinton R, Varanini Z, Nannipieri P (eds) The rhizosphere. Biochemistry and organic substance at the soil–plant interface. Dekker, New York, pp 95–104

Bruno JF, Stachowicz JJ, Bertness MD (2003) Inclusion of facilitation into ecological theory. Trends Ecol Evol 18:119–125

Bryant JP, Chapin FS III, Klein DR (1983) Carbon/nutrient balance of boreal plants in relation to vertebrate herbivory. Oikos 40:357–368

Buschmann H, Rodriguez MX, Tohme J, Beeching JR (2000) Accumulation of hydroxycoumarins during post-harvest deterioration of tuberous roots of cassava (*Manihot esculenta* Crantz). Ann Bot 86:1153–1160

Calabrese EJ, Baldwin LA (2003) Toxicology rethinks its central belief. Nature 421:691–692

Callaway R (1995) Positive interactions among plants. Bot Rev 61:306–349

Carter JP, Spink J, Cannon PF, Daniels MJ, Osbourn AE (1999) Isolation, characterization, and avenacin sensitivity of a diverse collection of cereal-root-colonizing fungi. Appl Environ Microbiol 65:3364–3372

Cates RG (1996) The role of mixtures and variation in the produxtion of terpenoids in conifer–insect–pathogen interactions. In: Romei JT, Saunders JA, Barbosa P (eds) Phytochemical diversity and redundancy in ecological interactions. Plenum, New York, pp 179–216

Clay K, Schardl C (2002) Evolutionary origins and ecological consequences of endophyte symbiosis with grasses. Am Nat 160:S90–S127

Coley PD (1987) Effects of plant growth rate and leaf lifetime on the amount and type of antiherbivore defense. Oecologia 74:531–536

Coley PD, Bryant JP, Chapin FS III (1985) Resource availability and plant antiherbivore defense. Science 230:895–899

Czárán TL, Hoekstra RF, Pagie L (2001) Chemical warfare between microbes promotes biodiversity. Proc Natl Acad Sci USA 99:786–790

Deacon J (1997) Modern mycology. 3rd edn. Blackwell, Oxford

de Bruxelles GL, Roberts MR (2001) Signals regulating multiple responses to wounding and herbivores. Crit Rev Plant Sci 20:487–521

Demain AL (1996) Fungal secondary metabolism: regulation and function. In: Sutton B (ed) A century of mycology. Cambridge University Press, Cambridge, pp 233–254

Desikan R, Hancock J, Neill S (2005) Reactive oxygen species as signaling molecules. In: Smirnoff (ed) Antioxidants and reactive oxygen species in plants. Blackwell, Oxford, pp 169–196

Dethier VG (1954) Evolution of feeding preferences in phytophagous insects. Evolution 8:33–54

Duffy B, Schouten A, Raaijmakers (2003) Pathogen self-defense: mechanisms to counteract microbial antagonism. Annu Rev Phytopathol 41:501–538

Ehlers BK, Thompson J (2004) Do co-occurring plant species adapt to one another? The response of *Bromus erectus* to the presence of different *Thymus vulgaris* chemotypes. Oecologia 141:511–518

Ehrlich PR, Raven PH (1964) Butterflies and plants: a study in coevolution. Evolution 18:586–608

Eissenstat DM, Wells CE, Yanai RD, Whitbeck JL (2000) Building roots in a changing environment: implications for root longevity. New Phytol 147:33–42

Farrell BD, Mitter C (1993) Phylogenetic determinants of insect/plant community diversity. In: Ricklefs RE, Schluter D (eds) Species diversity in ecological communities. University of Chicago Press, Chicago, pp 253–266

Farrell BD, Dussourd DE, Mitter C (1991) Escalation of plant defense: do latex and resin canals spur plant diversification? Am Nat 138:881–900

Feeny P (1975) Biochemical coevolution between plants and their insect herbivores. In: Gilbert LE, Raven PH (eds) Coevolution of animals and plants. University of Texas Press, Austin, pp 3–19

Feeny P (1976) Plant apparency and chemical defense. In: Wallace JW, Mansell RL (eds) Biochemical interactions between plants and insects. Plenum, New York, pp 1–40

Felton GW, Korth KL, Bi JL, Wesley SV, Huhman DV, Mathews MC, Murphy JB, Lamb C, Dixon RA (1999) Inverse relationship between systemic resistance of plants to micro-organisms and to insect herbivory. Curr Biol 9:317–320

Fiehn O, Kopka J, Dörmann P, Altmann T, Trethewey RN, Willmitzer L (2000) Metabolite profiling for plant functional genomics. Nat Biotechnol 18:1157–1161

Firn RD, Jones CG (1996) An explanation of secondary metabolite "redundancy". In: Romeo TJ, Saundes JA, Barbosa P (eds) Phytochemical diversity and redundancy in biological interactions. Plenum, New York, pp 295–312

Firn RD, Jones CG (2003) Natural products—a simple model to explain chemical diversity. Nat Prod Rep 20:382–391

Firn RD, Jones CG (2006) Response to Pichersky et al.: Correcting a misconception about the screening hypothesis. Trends Plant Sci 11:422

Fliegmann J, Schröder G, Schanz S, Britsch L, Schröder J. (1992) Molecular analysis of chalcone and dihydropinosylvine synthase from scots pine, and differential regulation of these and related enzyme activities in stressed plants. Plant Mol Biol 18:489–503

Flor HH (1971) Current state of the gene-for-gene concept. Annu Rev Phytopathol 9:275–296

Foley WJ, Moore BD (2005) Plant secondary metabolites and vertebrate herbivores—from physiological regulation to ecosystem function. Curr Opin Plant Biol 8:430–435

Foyer CF, Noctor G (2005) Redox homeostasis and antioxidant signaling: a metabolic interface between stress perception and physiological responses. Plant Cell 17:1866–1875

Fraenkel GS (1959) The raison d'être of secondary plant substances. Science 129:1466–1470

Fridman E, Pichersky E (2005) Metabolomics, genomics, proteomics, and the identification of enzymes and their substrates and products. Curr Opin Plant Biol 8:242–248

Futuyma D (1976) Food plant specialization and environmental predictability in Lepidoptera. Am Nat 110:285–292

Futuyma D (2000) Some current approaches on the evolution of plant–herbivore interactions. Plant Species Biol 15:1–19

Gershenzon J (2002) Secondary metabolites and plant defense. In: Taiz L, Zeiger E (eds) Plant physiology, 3rd edn. Sinauer, Sunderland, pp 283–308

Govaerts R (2001) How many species of seed plants are there? Taxon 50:1085–1090

Guo H, Ecker JR (2004) The ethylene signalling pathway: new insights. Curr Opin Plant Biol 7:40–49

Grace SC (2005) Phenolics as antioxidants. In: Smirnoff (ed) Antioxidants and reactive oxygen species in plants. Blackwell, Oxford, pp 141–168

Gräfe U (1992) Biochemie der Antibiotika. Spektrum, Heidelberg

Grayer R, Harborne JB (1994) A survey of antifungal compounds from higher plants, 1992–1993. Phytochemistry 37:19–42

Hadacek F (2002) Secondary metabolites as plant traits: current assessment and future perspectives. Crit Rev Plant Sci 21:273–322

Hartmann T (1986) Diversity and variability of plant secondary metabolism: a mechanistic view. Entomol Exp Appl 80:177–188

Hättenschwiler S, Vitousek PM (2000) The role of polyphenols in terrestrial ecosystem nutrient cycling. Trends Ecol Evol 15:238–243

Herms DA, Mattson WJ (1992) The dilemma of plants: to grow or to defend. Q Rev Biol 67:283–335

Hilker M, Meiners T (2002) Induction of plant responses to oviposition and feeding by herbivorous arthropods: a comparison. Entomol Exp Appl 104:181–192

Hiltner L (1904) Über neue Erfahrungen und Probleme auf dem Gebiete der Bodenbakteriologie. Arb Dtsch Landwirtsch Ges 98:57–78

Ho JAA, Durst RA (2003) Detection of fumonisin B1: comparison of flow-injection liposome immunoanalysis with high-performance liquid chromatography. Anal Biochem 312:7–13

Holloway PJ (1982) The chemical constitution of plant cutins. In: Cutler DF, Alvin KL, Price CE (eds) The plant cuticle. Academic, London, pp 45–85

Holopainen JK (2004) Multiple functions of inducible plant volatiles. Trends Plant Sci 9:529-533

Homans AL, Fuchs A (1970) Direct bioautography on thin-layer chromatograms for detecting fungitoxic substances. J. Chromatogr 51:242–245

Hooper DU, Vitousek PM (1998) Effects of plant composition and diversity on nutrient cycling. Ecol Monogr 68:121–149

Hubbell SP (2001) The unified neutral theory of biodiversity and biogeography. Monographs in population biology 32. Princeton University Press, Princeton

Iason GR, Lennon JJ, Pakeman RJ, Thoss V, Beaton JK, Sim DA, Elston DA (2005) Does chemical composition in individual scots pine trees determine the biodiversity of their associated ground vegetation?. Ecol Lett 8:364–369

Ikediugwu FEO (1976) The interface in hyphal interference by *Peniophora gigantea* against *Heterobasidium annnosum*. Trans Br Mycol Soc 66:291–296

Ikediugwu FEO, Webster J (1970) Hyphal interference in a range of coprophilous fungi. Trans Br Mycol Soc 54:205–210

Jarvis BB (2000) The role of natural products in evolution. In: Romeo JT, Ibrahim R, Varin L, De Luca V (eds) Evolution of metabolic pathways. Pergamon, Amsterdam, pp 1–24

Jarvis BB, Miller JD (1996) Natural products, complexity and evolution. In: Romei JT, Saunders JA, Barbosa P (eds) Phytochemical diversity and redundancy in ecological interactions. Plenum, New York, pp 265–293

Jenkins H, Hardy N, Beckmann M, Draper J, Smith AR, Taylor J, Fiehn O, Goodacre R, Bino RJ, Hall R, Kopka J, Lane GA Lange BM, Liu JR, Mendes P, Nikolau BJ, Oliver SG, Paton NW, Rhee S, Roessner-Tunali U, Saito K, Smedsgaard J, Sumner LW, Wang T, Walsh S, Wurtele ES, Kell D (2004) A proposed framework for the description of plant metabolomics experiments and their results. Nat Biotechnol 22:1601–1606

Jones CG, Firn RD (1991) On the evolution of plant secondary chemical diversity. Philos Trans R Soc Lond Ser B 333:273–280

Jones CG, Lawton JH (1991) Plant chemistry and insect species richness of British umbellifers. J Anim Ecol 60:767–777

Jose S (2002) Black walnut allelopathy: current state of science. In: Inderjit, Mallik AU (eds) Chemical ecology of plants: allelopathy in aquatic and terrestrial ecosystems. Birkhäuser, Basel, pp 149–172

Karban R, Baldwin IT (1997) Induced responses to herbivory. University of Chicago Press, Chicago

Kemp MS (1978) Falcarindiol—antifungal polyacetylene from *Aedopodium podagraria*. Phytochemistry 17:1002

Kerner von Marilaun A (1890) Pflanzenleben, vol 1. Verlag des bibliographischen Instituts, Leipzig

Kerr B, Riley MA, Feldman MW, Bohannan JM (2002) Local dispersal promotes biodiversity in a real-life game of rock–paper–scissors. Nature 418:171–174

Kessler A, Halitschke R, Baldwin IT (2004) Silencing the jasmonate cascade: induced plant defenses and insect populations. Science 305:665–668

Laloi C, Apel K, Danon A (2004) Reactive oxygen signalling: the latest news. Curr Opin Plant Biol 7:323–328

Landeweert R, Hoffland E, Finlay RD, Kuyper TW, Breemen N (2001) Linking plants to rocks: ectomycorrhizal fungi mobilize nutrients from minerals. Trends Ecol Evol 16:248–254

Langenheim JH (1994) Higher plan terpenoids: a phytocentric overview of their ecological roles. J Chem Ecol 20:1223–1280

Langenheim JH (2003) Plant resins. Timber, Portland

Lindahl BO, Taylor AFS, Finlay RD (2002) Defining nutritional constrainst on carbon cycling in boreal forests—towards a less phytocentric perspective. Plant Soil 242:123–135

Loomis WE (1932) Growth-differentiation balance vs carbohydrate-nitrogen ratio. Proc Am Soc Hortic Sci 29:240–245

Loomis WE (1953) Growth and differentiation—an introduction and summary. In: Loomis WE (ed) Growth and differentiation in plants. Iowa State College Press, Ames, pp 1–17

Lutzoni F, Pagel M, Reeb V (2001) Major fungal lineages are derived from lichen symbiotic ancestors. Proc Natl Acad Sci USA 411:937–940

Maleck K, Dietrich RA (1999) Defense on multiple fronts: how do plants cope with animal diversity. Trends Plant Sci 4:215–219

Mayer AM, Staples RC (2002) Laccase: mew functions for an old enzyme. Phytochemistry 60:551–565

McKey D (1974) Adaptive patterns in alkaloid physiology. Am Nat 108:305–320

McKey D (1979) The distribution of secondary metabolites in plants. In: Rosenthal GA, Janzen DH (eds) Herbivores: their interactions with secondary plant metabolites. Academic, New York, pp 55–133

Mendelsohn R, Balick M (1995) The value of undiscovered pharmaceuticals in tropical forests. Econ Bot 49:223–228

Mendgen K, Deising H (1993) Infection structures of fungal plant pathogens. New Phytol. 124:193–213

Metting FB (1993) Soil microbial ecology. Applications in agricultural and environmental management. Dekker, New York

Naigre R, Kalck P, Rogues C, Roux I, Michel G (1996) Comparison of antimicrobial properties of monoterpenes and their carbonylated products. Planta Med 62:275–277

Neill SJ, Desikan R, Hancock JT (2003) Nitric oxide signalling in plants. New Phytol 159:11–35

Nelson EB (2004) Microbial dynamics and interactions in the spermosphere. Annu Rev Phytopathol 42:271–309

Nishida R (2002) Sequestration of defensive substances from plants by Lepidoptera. Annu Rev Entomol 47:57–92

Okubara PA, Paulitz TC (2005) Root defense responses to fungal pathogens. Plant Soil 274:215–226

Osbourn AE, Qi X, Townsend B, Qin B (2003) Dissecting plant secondary metabolism—constitutive chemical defenses in cereals. New Phytol 159:101–108

Pichersky E, Gang DR (2000) Genetics and biochemistry of secondary metabolites in plants an evolutionary perspective. Trends Plant Sci 5:439–445

Pichersky E, Sharkey TD, Gershenzon J (2006) Plant volatiles: a lack of function or a lack of knowledge? Trends Plant Sci 11:421

Rahouti M, Steiman R, Seigle-Murandi F, Christov LP (1999) Growth of 1044 strains and species of fungi on 7 phenolic lignin model compounds. Chemosphere 38:2549–2559
Rapp RA, Wendel JE (2005) Epigenetics and plant evolution. New Phytol 168:81–91
Remy W, Taylor TN, Hass H, Kerp H (1994) Four-hundred-million-year old vesicular arbuscular mycorrhizae. Proc Natl Acad Sci USA 91:11841–11843
Rhoades DF (1979) Evolution of plant chemical defense against herbivores. In: Rosenthal GA, Janzen DH (eds) Herbivores: their interactions with secondary plant metabolites. Academic, New York, pp 1–55
Rhoades DF, Cates RG (1976) Towards a general theory of plant antiherbivore chemistry. In: Wallace JW, Mansell RL (eds) Biochemical interactions between plants and insects. Plenum, New York, pp 168–213
Rodriguez RJ, Redman RS (1997) Fungal life-styles and ecosystem dynamics: biological aspects of plant pathogens, plant endophytes and saprophytes. In: Tommerup IC, Andrews JH (eds) Advances in botanical research, vol 24. Academic, Amsterdam, pp 169–193
Rosenthal GA, Berenbaum MR (eds) (1992) Herbivores: their interactions with plant secondary metabolites. Academic, San Diego
Rubinstein B (1992) Similarities between plants and animals for avoiding predation and disease. Physiol Zool 65:473–492
Saikkonen K, Wali P, Helander M, Faeth SH (2004) Evolution of endophyte–plant symbioses. Trends Plant Sci 9:275–280
Sanders WB (2001) Lichens: the interface between mycology and plant morphology. Bioscience 51:1025–1035
Sauter H, Steglich W, Anke T (1999) Strobilurins: evolution of a new class of active substances. Angew Chem Int Ed 38:1328–1349
Scheffer RP (1991) Role of toxins in evolution and ecology of plant pathogenic fungi. Experientia 47:804–811
Schimel JP, Cates RG, Ruess R (1998) The role of balsam poplar secondary chemicals in controlling soil nutrient dynamics through succession in the Alaskan taiga. Biogeochemistry 42:221–234
Schmidt SK, Ley RE (1999) Microbial competition and soil structure limit the expression of allelochemicals in nature. In: Inderjit, Dakshini KM, Foy CL (eds) Principles and practice in plant ecology. Allelochemical interactions. CRC, Boca Raton, pp 339–351
Schröder J (2000) The family of chalcone synthase-related proteins: functional diversity and evolution. In: Romeo JT, Ibrahim R, Varin L, De Luca V (eds) Evolution of metabolic pathways. Pergamon, Amsterdam, pp 55–89
Schwab W (2003) Metabolome diversity: too few genes, too many metabolites. Phytochemistry 62:837–849
Scott AC, Stephenson J, Chaloner WG (1992) Interaction and coevolution of plants and arthropods during Palaeozoic and Mesozoic. Philos Trans R Soc Lond Ser B 335:129–165
Sesma A, Osbourn AE (2004) The rice leaf blast pathogen undergoes developmental processes typical of a root-infecting fungus. Nature 431:582–586
Shah J (2003) The salicylic acid loop in plant defence. Curr Opin Plant Biol 6:365–371
Sinkkonen A (2003) A model describing chemical interference caused by decomposing residues at different densities of growing plants. Plant Soil 250:315–322.
Soledade M, Pedras C, Nycholat CM, Montaut S, Xu Y, Khan AQ (2002) Chemical defense of crucifers: elicitation and metabolism of phytoalexins and indol-3-acetonitrile in brown mustard and turnip. Phytochemistry 59:611–625
Somerville C, Somerville S (1999) Plant functional genomics. Science 285:380–383
Southwood TRE (1985) Interactions of plants with animals: patterns and processes. Oikos 44:5–11
Souto XC, Pellisier F (2002) Feedback mechanisms in the chemical ecology of plants: role of soil microorganisms. In: Inderjit, Mallik AU (eds) Chemical ecology of plants: allelopathy in aquatic and terrestrial ecosystems. Birkhäuser, Basel, pp 89–98
Stamp N (2003) Out of the quagmire of plant defense hypothesis. Q Rev Biol 78:23–55

Stratmann JW (2003) Long distance run in the wound response—jasmonic acid is pulling ahead. Trends Plant Sci. 8:247–250

Taga ME, Bassier BL (2003) Chemical communication among bacteria. Proc Natl Acad Sci USA 100:14549–14554

Talbot NJ, Kershaw MJ, Wakley GE, de Vries OMH, Wessels JGH, Hamer JE (1996) *MPG1* encodes a fungal hydrophobin involved in surface interactions during infection-related development of *Magnaporthe grisea*. Plant Cell 8:985–999

Thines E, Anke H, Weber RWS (2004) Fungal secondary metabolites as inhibitors of infection-related morphogenesis in phytopathogenic fungi. Mycol Res 108:14–25

Tilman D (1982) Resource competiton and community structure. Monographs in population biology 17. Princeton University Press, Princeton

Tilman D (2004) Niche tradeoffs, neutrality, and community structure: a stochastic theory of resource competion, invasion, and community assembly. Proc Natl Acad Sci USA 101:10854–10861

Tilman D, Pacala S (1993) The maintenance of species richness in plant communities. In: Ricklefs RE, Schluter D (eds) Species diversity in ecological communities. University of Chicago Press, Chicago, pp 13–25

Tuomi K, Niemelä P, Chapin FS III, Bryant JP, Sirén S (1988) Defensive responses of trees in relation to their carbon/nutrient balance. In: Mattson JW, Levieux J, Barnard-Dagan C (eds) Mechanisms of woody plant defenses against insects: search for pattern. Springer, New York, pp 57–52

Tuomi J, Fagerstrom T, Niemelä P (1991) Carbon allocation phenotypic plasticity, and induced defenses. In: Tallamy DW, Raupp MJ (eds) Phytochemical induction by herbivores. Wiley, New York, pp 85–104

Vajrodaya S, Bacher M, Greger H, Hofer O (1998) Organ-specific chemical differences in *Glycosmis trichanthera*. Phytochemistry 48:897–902

Van der Heijden MGA, Klironomos JN, Ursic M, Moutogolis P, Streitwolf-Engel R, Boller T, Wiemken A, Sanders IR (1998) Mycorrhizal fungal diversity determines plant biodiversity, ecosystem variability and productivity. Nature 396:69–72

van der Putten WH (2003) Plant defense belowground and spatiotemporal processes in natural vegetation. Ecology 84:2269–2280

Veluri R, Weir TL, Bais HP, Stermitz FR, Vivanco JM (2004) Phytotoxic and antimicrobial activities of catechin derivatives. J Agric Food Chem 52:1077–1082

Wardle DA (2002) Communities and ecosystems. Linking the aboveground and belowground components. Monographs in population biology 34. Princeton University Press, Princeton

Wardle DA, Bardgett RD, Klironomos JN, Setäla H, van der Putten WH, Wall DH (2004) Ecological linkages between aboveground and belowground biota. Science 304:1629–1633

Wendt KU, Schulz GE, Corey EJ, Liu DR (2000) Enzyme mechanisms for polycyclic triterpene formation. Angew Chem Int Ed 39:2812–2833

Wink M (2003) Evolution of secondary metabolites from an ecological and molecular phylogenetic perspective. Phytochemistry 64:3–19

Whipps JM (2001) Microbial interactions and biocontrol in the rhizosphere. J Exp Bot 52:487–511

Wittstock U, Gershenzon J (2002) Constitutive plant defense and their role in the defense against herbivores and pathogens. Curr Opin Plant Biol 5:300–307

Xing M-Z, Zhang X-Z, Sun Z-L, Zhang H-Y (2003) Perylenequinones act as broad-spectrum fungicides by generating reactive oxygen species both in the dark and light. J Agric Food Chem 51:7722–7724

Chapter 11
Root Exudates Modulate Plant–Microbe Interactions in the Rhizosphere

Harsh P. Bais, Corey D. Broeckling, and Jorge M. Vivanco(✉)

11.1 Introduction

The chemical, physical, and biological interactions between roots and the surrounding environment of the rhizosphere are some of the most complex experienced by land plants. Over the last few years the field of rhizosphere biology has recognized the biological importance of root exudates in mediating interactions with other plants and microbes (Bais et al. 2004a, b; Walker et al. 2003; Weir et al. 2004). Chiefly, root exudates comprise two different classes of compounds. Low molecular weight compounds include amino acids, organic acids, sugars, phenolics, and various secondary metabolites, whereas high molecular weight exudates primarily include mucilage (high molecular weight polysaccharides) and proteins. Root exudation clearly represents a significant carbon cost to the plant (Uren 2000); however, the molecular mechanisms regulating exudation are still poorly understood. The roots of some plants also release border cells into the rhizosphere but literature discussing this phenomenon will not be covered in this chapter (for information see Hawes et al. 2000; Vicre et al. 2005).

The rhizosphere comprises the area of soil immediately surrounding a plant root and represents a highly dynamic environment involving interactions with competing roots and pathogenic/nonpathogenic microbes and invertebrates (Hirsch et al. 2003). The focus of this chapter will be on root–microbe interactions that can be broadly divided into positive interactions including classic symbioses, association with bacterial biocontrol agents, epiphytes, and mycorrhizal fungi; and negative interactions including associations with parasitic plants, pathogenic bacteria, fungi, and invertebrate herbivores. Microbial colonization of the rhizosphere is important not only as the first step in pathogenesis of soil-borne

Jorge M. Vivanco
Department of Horticulture and Landscape Architecture; and Center for Rhizosphere Biology, Colorado State University, Fort Collins, CO 80523 1173, USA
e-mail: j.vivanco@colostate.edu

P. Karlovsky (ed.), *Secondary Metabolites in Soil Ecology. Soil Biology 14.*

microorganisms, but also is crucial in the application of microorganisms for beneficial purposes. It has long been assumed that many microorganisms are attracted by carbon-containing compounds exuded by plant roots. Indeed, this phenomenon was first described in 1904 by Hiltner (1904), who observed increased numbers and activity of microorganisms in the vicinity of plant roots. Evidence continues to accumulate which suggests a major role for root exudates in determining the fate of microbial species in the rhizosphere. This chapter attempts to draw together some of these studies, with an emphasis on the importance of root exudates in defining root–microbe interactions.

11.2 Positive Interactions Mediated by Root Exudates

Plant–microbe interactions can positively influence plant growth through a variety of mechanisms. Such mechanisms include transfer of fixed nitrogen to leguminous plants by *Alphaproteobacteria* and *Betaproteobacteria* members (Moulin et al. 2001), increased biotic and abiotic stress tolerance imparted by the presence of endophytic microbes (Schardl et al. 2004), and direct and indirect advantages imparted by plant-growth-promoting rhizobacteria (PGPR; Gray and Smith 2005). Several well-described interactions establish the importance of root exudates in positive plant–microbe interactions, where they act either directly or indirectly.

11.2.1 Nodulation of Legumes by Rhizobia

The nodule is a specialized organ found in legume roots which contains and protects nitrogen-fixing rhizobacteria. This structure allows access by the bacterium to plant carbohydrates and organic acids as well as direct access by the plant to nitrogen fixed by the bacterium. This relationship is initiated by bacterial perception of and attraction to root-exuded flavonoids (Peters et al. 1986; Redmond et al. 1986). These root exudates elicit expression of bacterial *nod* genes, which are required for initiation of the root nodule. Additionally, bacterial derived lipo-oligosaccharides have been demonstrated to induce flavonoid biosynthetic genes in legume roots. Thus, chemical communication between bacteria and plant is bidirectional—with each transcriptionally responding to diffusible signals from the other. Bacterial initiation of symbiotic nodules can be divided into several stages, including root hair curling, infection thread formation, and ultimately nodule development. Recent research has revealed many of the early molecular events which are required for formation of the nodule structure and recognition of bacterial lipo-oligosaccharides (Endre et al. 2002; Krusell et al. 2002; Limpens et al. 2003; Ana et al. 2004; Levy et al. 2004).

11.2.2 Mycorrhizal Associations

Fungal mycorrhizal and bacterial rhizobial associations are thought to derive from a common ancestral plant–microbe interaction likely of fungal origin. This position is supported by the fact that the activity of some host proteins regulates both mycorrhizal and rhizobial associations (Levy et al. 2004). As with rhizobial recognition of root exudates, mycorrhizal fungi are able to recognize the presence of a compatible host plant through detection of root exudates (Nagahashi and Douds 2000; Tamasloukht et al. 2003). Though root exudates have long been suspected to serve a communicative role in mycorrhizal associations, identification of specific molecular structures has remained elusive. Recently, a sesquiterpene which triggers hyphal branching in dormant mycorrhizal fungus was identified from plant root exudates (Akiyama et al. 2005), establishing a novel role for root exudates. This positive effect of root exudates on mycorrhizal growth can also indirectly protect a plant from pathogens, through suppression of growth of pathogenic species in the soil by mycorrhizae, thereby preventing infection (Chakravarty and Hwang 1991). In addition, root exudates from some species are able to inhibit hyphal growth of mycorrhizal fungi. An ecologically intriguing example of this is the inhibition of mycorrhizal fungi growth by three species of *Lupinus*, which are nodule-forming legumes (Oba et al. 2002). This mechanism may improve the competitive advantage of *Lupinus* by suppressing the formation of mycorrhizal associations in other plants, which would otherwise be expected to improve the fitness of competing mycorrhizal plant species.

11.2.3 Endophytic Associations

Plants often support internal nonpathogenic fungal and bacterial species, termed "endophytes," which are either beneficial to the plant or are nondetrimental. Knowledge of the mechanistic details of endophyte–plant mutualism is lacking in most cases, although the grass–clavicep fungi association is relatively well studied. Plants harboring endophytes demonstrate increased resistance to a variety of biotic and abiotic stresses. The presence of the endophyte species can affect root exudation of the host plant, altering the secretion of phenolics from roots and modulating regulation of rhizosphere pH by the plant roots, which contributes to increased tolerance of mineral deficiencies (Malinowski and Belesky 2000). Many endophytes are seed-transmitted, and therefore a chemotaxic response is not observed or necessary. However, other endophytic associations are opportunistic, and chemical communication in the soil is likely, though yet to be documented.

11.2.4 Plant-Growth-Promoting Bacteria

Soil microbial communities can also positively regulate plant growth and traits such as disease or stress resistance through more indirect mechanisms than those discussed above. PGPR have been found to positively influence plants through a broad variety of direct and indirect mechanisms (Gray and Smith 2005). Though bacteria are likely to locate plant roots through cues exuded from the root, little is known of the role of root exudates in establishment or regulation of PGPR populations and activities. However, the inverse communication is known to occur, where volatiles generated by PGPR are able to affect plant growth and resistance to pathogens (Ryu et al. 2003, 2004), suggesting that chemical communication between plants and PGPR is critical and relevant to plant health.

11.3 Antagonistic Interactions Mediated by Root Exudates (Antimicrobial, Biofilm Inhibitors, and Quorum-Sensing Mimics)

11.3.1 Antimicrobials

Plants have an almost limitless ability to synthesize secondary metabolites, many of which are phenols or their oxygen-substituted derivatives (Dixon 2001). At least 12,000 secondary metabolites have been isolated, less than 2% of which have been found in root exudates (Dixon 2001; D'Auria and Gershenzon 2005). In many cases, these exuded metabolites serve as plant defense mechanisms against predation by microorganisms, insects, and herbivores (Foley and Moore 2005). Root exudates also serve as nutrients for the microbial community, and a relationship between plant exudates and increased microbial activity in the rhizosphere has been recognized for years. Plants are known to use diverse chemical molecules for defense (for example, isoflavonoids in the Leguminosae, sesquiterpenes in the Solanaceae), although some chemical classes are used for defensive functions across taxa (for example, phenylpropanoid derivatives). Most antimicrobial plant natural products have relatively broad spectrum activity, and specificity is often determined by whether or not a pathogen has the enzymatic machinery to detoxify a particular host product (Bourab et al. 2002). Accumulation of inducible antimicrobial compounds is often orchestrated through signal-transduction pathways linked to perception of the pathogen by receptors encoded by host resistance genes. Most studies have focused on accumulation patterns of secondary metabolites in plant roots without considering the role of those same metabolites in the rhizosphere. Research performed in the last 5 years has clarified the antimicrobial properties of root secretions. For instance, Bais et al. (2002a) identified rosmarinic acid, a caffeic acid ester in the root exudates of hairy root cultures of sweet basil (*Ocimum basilicum*), as a secreted compound elicited by cell wall extracts from *Phytophthora cinnamoni*. Basil roots were also induced to exude rosamarinic acid by fungal in situ challenge with *Pythium ultimum*, and rosamarinic acid demonstrated potent antimicrobial activity against an array of soil-borne microorganisms, including the opportunist plant pathogen *Pseudomonas aeruginosa* (Bais et al. 2002a). Similar studies by Brigham et al. (1999) with *Lithospermum erythrorhizon* hairy roots reported cell-specific production of pigmented naphthoquinones upon elicitation, and other biological activity against soil-borne bacteria and fungi. Recently, it has been shown that *Centaurea maculosa* Lam. (spotted knapweed) roots exude a flavonoid, (+)-catechin, which inhibits an array of soil-borne bacteria and fungi (Bais et al. 2002b; Veluri et al. 2004). *C. maculosa* also secretes (–)-catechin, which is phytotoxic (Bais et al. 2002b; Veluri et al. 2004). Similarly, the constitutive production of antifungal compounds in the root exudates of *Gladiolus* spp. L. determines resistance against *Fusarium oxysporum* sp. *gladioli*. These results showed, using spore germination tests, that the resistant cultivar exudes substances having a negative influence on microconidial germination of the pathogen; root exudates from a susceptible cultivar proved to have no effect on the fungal spores (Taddei et al. 2002). The ability of resistant cultivars to inhibit conidial germination of *F. oxysporum gladioli* can be mainly related to the presence of a higher relative amount of aromatic-phenolic compounds. Taken together, these studies strongly suggest that root defense against pathogens begins in the rhizosphere, before pathogen contact with the root.

Plants produce both constitutive as well as inducible defense metabolites (Dixon 2001). The terms dividing constitutive (phytoanticipins) from inducible (phytoalexins) compounds are often confusing because of the overall in vivo antimicrobial activity in the plants. Inducible metabolites are often localized in the cells under attack from a pathogen. The exemplar study discussing the suite of inducible compounds upon pathogen attack involves phenylpropanoids (Dixon 2001). The intracellular antimicrobial composition in roots differs dramatically from the composition of the antimicrobials found in the root exudates (Bednarek et al. 2005; Tan et al. 2004, Walker et al. 2003). A recent report demonstrates the role of root exudates in defining pathogenicity in plant–microbe interactions (Bais et al. 2005). Arabidopsis roots treated with a nonhost pathogen (*Pseudomonas syringae* pv. *phaseolicola*) exuded compounds in large titers compared with treatment with the host pathogen (*P. syringae* pv. DC3000). The aforementioned studies clearly outline the direction of this research, which should lead to the discovery of novel antimicrobial compounds in the root exudates and to the unraveling of exciting rhizospheric microbial interactions.

11.3.2 Biofilm Inhibitors and Quorum-Sensing Mimics

11.3.2.1 Biofilm Inhibitors

Microscopy-based studies of bacterial colonization in the rhizosphere indicate that, as in the phyllosphere, bacteria generally form microcolonies or aggregates on root surfaces and that these colonies have a patchy, nonuniform distribution called biofilms (Costerton et al. 1999; Bais et al. 2004a). Numerous species of bacteria have been observed to form microcolonies or aggregates when colonizing root surfaces. These include strains of saprophytic fluorescent *Pseudomonas* spp. that have potential as biological control agents (Bianciotto et al. 2001) or as plant-growth-promoting bacteria (Somers et al. 2004), as well as other plant-growth-promoting bacteria such as *Enterobacter agglomerans* (Verma et al. 2001). Most studies of colonization patterns of roots corroborate the notion that bacteria on root surfaces are present primarily as microcolonies at sites of root exudation, as proposed by Rovira et al. (1974), although this concept was not experimentally tested until recently. It has been speculated that production of antimicrobials from plant roots affects the ability of microbes to quorum and form biofilms. To counter these antimicrobials, microbes synthesize exopolysaccharides or other exopolymeric material which may enhance bacterial survival and the potential for colonization of roots. It is possible that the antimicrobial products secreted from plant roots could easily be cloistered in rhizospheric biofilms, leading to loss of activity of these compounds on microbes. Alternatively, if these compounds diffuse in a controlled manner or have the capability to degrade the established biofilm matrix, then antimicrobial activity could be highly concentrated and potentially more effective. Very few studies have targeted such complex possibilities of this rhizospheric interaction. Recently, one study demonstrated the antimicrobial nature of rosamarinic acid from *O. baslicum* (sweet basil) against an opportunist pathogen *Pseudomonas aeruginosa* (Walker et al. 2004). Interestingly, this study showed that rosamarinic acid was potent against the planktonic form of *P. aeruginosa* but lacked the ability to penetrate the biofilm on the root surfaces. Incidentally, *P. aeruginosa* biofilm inhibitors have recently been identified in human mucous secretions and some plant extracts (Singh et al. 2002; Bjarnsholt et al. 2005); thus, it might be possible to adapt the nondestructive method developed from *O. basilicum–P. aeruginosa* interactions to screen root exudates for biofilm inhibitors. Such inhibitors might greatly enhance the effectiveness of known antibacterial agents used against *P. aeruginosa* infections in humans. While few plant pathogens are known to produce biofilms on root surfaces, bacterial aggregates or biofilm-like structures are commonly observed on the leaf surfaces with plant pathogens such as *Xanthomonas campestris*, which produces biofilms upon infection of its plant host (Dow et al. 2003). Strains defective in biofilm formation had smaller lesions owing to a reduced ability to spread through the leaf vasculature (Dow et al. 2003). A recent study demonstrates, via microscopic examination and in vitro adhesion assays, that *P. syringae* produces biofilm-like communities on Arabidopsis root surfaces (Bais et al. 2004a). Interestingly, others have shown that *P. syringae* alginate mutants were found to be significantly impaired in their ability to colonize leaves, form less severe lesions, and reach lower population densities than wild-type strains (Keith et al. 2003). The abovementioned lines of research are encouraging as this would help in screening root exudates for potential biofilm inhibitors against several plant pathogens and also lead to the discovery of some of the regulatory pathways of biofilm formation in plant pathogens.

11.3.2.2 Quorum-Sensing Mimics

Quorum sensing (QS) is another very important regulatory system that has been implicated in virulence in nearly every bacterial pathogen studied (Miller and Bassler 2001; Donabedian 2003). QS systems in Gram-negative bacteria typically consist of an autoinducer, which produces a free diffusible molecule, and a receptor/transcriptional activator protein, which monitors the concentration of the autoinducer. As the bacterial population grows, the level of autoinducer in the environment increases. Both Gram-negative and Gram-positive bacteria, including important plant pathogenic bacteria such as *Erwinia* spp., *Pseudomonas* spp., and *Agrobacterium* spp., possess QS systems (reviewed in Newton and Fray 2004). Recent studies demonstrate that both plants and red algae are able to mimic QS signals produced by several bacteria by secreting compounds that structurally mimic the bacterial QS molecules (Newton and Fray 2004; Bauer and Mathesius 2004). The most widely studied QS "mimic" compounds are halogenated furanones, which are produced by the marine red alga *Delisea pulchra*. Givskov et al. (1996) recognized that the *Delisea* furanones are similar in structure to *N*-acyl homoserine lactones (AHLs), the most common QS signals among most of the Gram-negative bacteria. They showed that the furanones specifically inhibit AHL-regulated behaviors in several bacteria. Interestingly, it is known that some of the higher plants, such as pea, tomato, *Medicago truncatula*, and rice, also rhizosecrete compounds that affect AHL QS regulation in bacteria (Teplitski et al. 2000). Additionally, several other QS synthetic signals have been chemically identified (Whitehead et al. 2001), though it is not yet known if plants synthesize or secrete these QS mimics. Additionally, recent studies have also shown that *M. truncatula, Chlamydomonas reinhardtii*, and *Chlorella* spp. secrete unidentified substances that stimulate or inhibit an AI-2-specific reporter (Teplitski et al. 2000). It might be that bacteria use the hormonal signals of a eukaryotic host as cues to trigger the QS-regulated machinery for infection of that host. When considering natural encounters between plants and bacteria, the disruption of QS regulation by other bacteria may be as important as disruption by QS mimics from the host plant. Various bacterial species that produce AHL QS signals have been found to activate gene expression in a *Pseudomonas* reporter strain in a native wheat rhizosphere (Pierson et al. 1998), suggesting that there is significant QS cross-talk between bacteria on a plant root. It appears that AHL-responding bacteria do not need to be particularly close to the AHL-producing cells on the host root surface (Steidle et al. 2002). Thus, the rhizosphere might be thought of as a region of overlapping, communicating populations of bacteria, each defined by mutual recognition of specific QS signals, and each affected in different ways by the secretion of QS mimics and other allelochemicals by the host plant. There is no evidence to date, however, to suggest that plants make or use AHL-degrading enzymes, and such enzymes might hinder beneficial bacteria. As mentioned above it is most likely that plants find it more useful to "listen" to bacterial QS signals and to mimic them than to destroy them. Thus, it is possible that roots may have developed defense strategies by secreting compounds into the rhizosphere that interfere with bacterial QS responses such as signal mimics, signal blockers, and/or signal-degrading enzymes, but future studies are required to isolate and characterize these compounds from root exudates of several species to unravel yet more interesting interactions.

11.4 Tritrophic Interactions (Plant, Microbe, and Nematode Interactions)

Previous sections of this chapter describe the involvement of root exudates in determining microbial status in the rhizosphere. Rhizospheric nematodes often eavesdrop on chemical communication between microbes and plants. Unlike plants and microbes, rhizospheric nematodes are highly mobile and may readily avoid or respond to any underground chemical signals. There is also a tempting possibility that the final outcome of plant–microbe interactions may also be detrimental to nematodes' survival in the rhizosphere. Until recently there was little work on the impact of these root secretions on rhizospheric interaction between plant roots, microbes, and nematodes. Yeates (1999) found that infection by a root-feeding nematode *Heterodera trifolii* on *Trifolium repens* leads to an increase of photosynthetically fixed carbon in the rhizosphere, partially through increased root exudation in the soil. This study suggests that infection by parasitic nematodes of plant roots may lead to extra carbon availability in the soil for a possible increased microbial turnover.

Tritrophic interactions are best described in the context of rhizobial species and vesicular arbuscular mycorrhizal fungi (Khan et al. 2000). The research outlined from such studies has shown that a tritrophic interaction in the rhizosphere occurs, in which nematodes and microorganisms act in synergistic associations to influence plant growth. A recent study also emerged which redefined the beneficial association of the tritrophic components of plant roots, microbes, and nematodes. This new study shows that the soil-dwelling nematodes like *Caenorhabditis elegans* could also mediate the interaction between roots and rhizobia in a positive way leading to nodulation (Horiuchi et al. 2005). The work demonstrated that *C. elegans* transfers the rhizobium species *Sinorhizobium meliloti* to the roots of the legume *M. truncatula* in response to plant-root-released volatiles that attract the nematode. This study reveals a new, biologically relevant, and largely unknown interaction in the rhizosphere that is multitrophic, may contribute to the initiation of the symbiosis, and is mediated by root-released volatile compounds. A similar study reported the attraction of entomopathogenic nematodes to insect-damaged corn roots in a field setting. This attraction was found to be mediated by β-caryophyllene, and caryophyllene-deficient plants did not attract nematodes (Rasmann et al. 2005). At present, relatively little is known of how plants affect the nematophagous and antagonistic microflora in their rhizospheres. Also, the influence of the nematodes on these relationships is of fundamental importance, and research on the tritrophic interactions between plants, nematodes, and their microbial natural enemies will contribute much to our understanding of signaling systems mediated by root exudates in the rhizosphere.

11.5 Conclusions

The exudation of organic compounds from roots is an important way by which plants can respond to and alter their immediate environment. By modifying the biochemical and physical properties of the rhizosphere, plants increase nutrient availability and buffer the effects of hostile surroundings. Over the last several years research has targeted the biological significance of root-exuded compounds in the rhizosphere. Efforts should now focus on the need to understand more and more hidden plant–microbe conversations in the complex rhizosphere. The current challenge is to clone the genes that encode pumps and channels involved in root exudation. Attempts to modify exudation from roots, by changing the activity of biosynthetic enzymes with gene manipulation, will be the research focus for the next several years. Progress will rely on researchers developing plants with a greater capacity for root exudation. Finally, it will be important to chemically characterize those components of root exudates that favor disease suppression and facilitate more beneficial relationships in the rhizosphere.

References

Akiyama K, Matsuzaki K-i, Hayashi H (2005) Plant sesquiterpenes induce hyphal branching in arbuscular mycorrhizal fungi. Nature 435:824

Ana JM, Kiss GB, Riely BK, Penmetsa RV, Oldroyd GED, Ayax C, Levy J, Debelle F, Baek JM, Kalo P, Rosenberg C, Roe BA, Long SR, Denarie J, Cook DR (2004) *Medicago truncatula* DMI1 required for bacterial and fungal symbioses in legumes. Science 303:1364

Bais HP, Walker TS, Schweizer HP, Vivanco JM (2002a) Root specific elicitation and antimicrobial activity of rosmarinic acid in hairy root cultures of sweet basil (*Ocimum basilicum* L.). Plant Physiol Biochem 40:983

Bais HP, Walker TS, Stermitz FR, Hufbauer RA, Vivanco JM (2002b) Enantiomeric-dependent phytotoxic and antimicrobial activity of (+/-)-catechin. A rhizosecreted racemic mixture from spotted knapweed. Plant Physiol 128:1173

Bais HP, Fall R, Vivanco JM (2004a) Biocontrol of *Bacillus subtilis* against infection of Arabidopsis roots by *Pseudomonas syringae* is facilitated by biofilm formation and surfactin production. Plant Physiol 134:307

Bais HP, Park SW, Weir TL, Callaway RM, Vivanco JM (2004b) How plants communicate using the underground information superhighway. Trends Plant Sci 9:26

Bais HP, Prithiviraj B, Jha AK, Ausubel FM, Vivanco JM (2005) Mediation of pathogen resistance by exudation of antimicrobials from roots. Nature 434:217

Bauer WD, Mathesius U (2004) Plant responses to bacterial quorum-sensing signals. Curr Opin Plant Biol 7:429

Bednarek P, Schneider B, Svatos A, Oldham NJ, Hahlbrock K (2005) Structural complexity, differential response to infection, and tissue specificity of indolic and phenylpropanoid secondary metabolism in Arabidopsis roots. Plant Physiol 138:1058

Bianciotto V, Andreotti S, Balestrini R, Bonfante P, Perotto S (2001) Mucoid mutants of the biocontrol strain *Pseudomonas fluorescens* CHA0 show increased ability in biofilm formation on mycorrhizal and nonmycorrhizal carrot roots. Mol Plant Microbe Interact 14:255

Bjarnsholt T, Jensen PO, Rasmussen TB, Christophersen L, Calum H, Hentzer M, Hougen HP, Rygaard J, Moser C, Eberl L, Hoiby N, Givskov M (2005) Garlic blocks quorum sensing and promotes rapid clearing of pulmonary *Pseudomonas aeruginosa* infections. Microbiology 151:3873

Bouarab K, Melton R, Peart J, Baulcombe D, Osbourn A (2002) A saponin-detoxifying enzyme mediates suppression of plant defences. Nature 418:889

Brigham LA, Michaels PJ, Flores HE (1999) Cell-specific production and antimicrobial activity of naphthoquinones in roots of *Lithospermum erythrorhizon*. Plant Physiol 119:417

Chakravarty P, Hwang SF (1991) Effect of an ectomycorrhizal fungus, *Laccaria-Laccata*, on Fusarium damping-off in Pinus-Banksiana seedlings. Eur J For Pathol 21:97

Costerton JW, Stewart PS, Greenberg EP (1999) Bacterial biofilms: a common cause of persistent infections. Science 284:1318

D'Auria JC, Gershenzon J (2005) The secondary metabolism of *Arabidopsis thaliana*: growing like a weed. Curr Opin Plant Biol 8:308

Dixon RA (2001) Natural products and plant disease resistance. Nature 411:843

Donabedian H (2003) Quorum sensing and its relevance to infectious diseases. J Infect 46:207–214

Dow JM, Crossman L, Findlay K, He YQ, Feng JX, Tang JL (2003) Biofilm dispersal in *Xanthomonas campestris* is controlled by cell-cell signaling and is required for full virulence to plants. Proc Natl Acad Sci USA 100:10995

Endre G, Kereszt A, Kevei Z, Mihacea S, Kalo P, Kiss GB (2002) A receptor kinase gene regulating symbiotic nodule development. Nature 417:962

Foley WJ, Moore BD (2005) Plant secondary metabolites and vertebrate herbivores—from physiological regulation to ecosystem function. Curr Opin Plant Biol 8:430

Givskov M, de Nys R, Manefield M, Gram L, Maximilien R, Eberl L, Molin S, Steinberg PD, Kjelleberg S (1996) Eukaryotic interference with homoserine lactone-mediated prokaryotic signaling. J Bacteriol 178:6618

Gray EJ, Smith DL (2005) Intracellular and extracellular PGPR: commonalities and distinctions in the plant-bacterium signaling processes. Soil Biol Biochem 37:395

Hawes MC, Gunawardena U, Miyasaka S, Zhao X (2000) The role of root border cells in plant defense. Trends Plant Sci 5:128

Hiltner L (1904) Über neue Erfahrungen und Probleme auf dem Gebeit der Bodenbackteriologie und unter besonderer Berucksichtigung der Grundung und Brache. Arb Dtsch Landwirsch Ges 98:59

Hirsch AM, Bauer WD, Bird DM, Cullimore J, Tyler B, Yoder JI (2003) Molecular signals and receptors: Controlling rhizosphere interactions between plants and other organisms. Ecology 84:858

Horiuchi J, Prithiviraj B, Bais HP, Kimball BA, Vivanco JM (2005) Soil nematodes mediate positive interactions between legume plants and rhizobium bacteria. Planta 222:848

Keith RC, Keith LM, Hernandez-Guzman G, Uppalapati SR, Bender CL (2003) Alginate gene expression by *Pseudomonas syringae* pv. tomato DC3000 in host and non-host plants. Microbiology 149:1127

Khan AG, Kuek C, Chaudhry TM, Khoo CS, Hayes WJ (2000) Role of plants, mycorrhizae and phytochelators in heavy metal contaminated land remediation. Chemosphere 41:197

Krusell L, Madsen LH, Sato S, Aubert G, Genua A, Szczyglowski K, Duc G, Kaneko T, Tabata S, de Bruijn F, Pajuelo E, Sandal N, Stougaard J (2002) Shoot control of root development and nodulation is mediated by a receptor-like kinase. Nature 420:422

Levy J, Bres C, Geurts R, Chalhoub B, Kulikova O, Duc G, Journet EP, Ane JM, Lauber E, Bisseling T, Denarie J, Rosenberg C, Debelle F (2004) A putative Ca^{2+} and calmodulin-dependent protein kinase required for bacterial and fungal symbioses. Science 303:1361

Limpens E, Franken C, Smit P, Willemse J, Bisseling T, Geurts R (2003) LysM domain receptor kinases regulating rhizobial Nod factor-induced infection. Science 302:630

Malinowski DP, Belesky DP (2000) Adaptations of endophyte-infected cool-season grasses to environmental stresses: mechanisms of drought and mineral stress tolerance. Crop Sci 40:923

Miller MB, Bassler BL (2001) Quorum sensing in bacteria. Annu Rev Microbiol 55:165

Moulin L, Munive A, Dreyfus B, Boivin-Masson C (2001) Nodulation of legumes by members of the beta-subclass of Proteobacteria. Nature 411:948

Nagahashi G, Douds DD (2000) Partial separation of root exudate components and their effects upon the growth of germinated spores of AM fungi. Mycol Res 104:1453

Newton JA, Fray RG (2004) Integration of environmental and host-derived signals with quorum sensing during plant-microbe interactions. Cell Microbiol 6:213

Oba H, Tawaraya K, Wagatsuma T (2002) Inhibition of pre-symbiotic hyphal growth of arbuscular mycorrhizal fungus Gigaspora margarita by root exudates of Lupinus spp. Soil Sci Plant Nutr 48:117

Peters NK, Frost JW, Long SR (1986) A plant flavone, luteolin, induces expression of Rhizobium meliloti nodulation genes. Science 233:977

Pierson LS 3rd, Wood DW, Pierson EA (1998) Homoserine lactone-mediated gene regulation in plant-associated bacteria. Annu Rev Phytopathol 36:207

Rasmann S, Kollner TG, Degenhardt J, Hiltpold I, Toepfer S, Kuhlmann U, Gershenzon J, Turlings TC (2005) Recruitment of entomopathogenic nematodes by insect-damaged maize roots. Nature 434:732

Redmond JW, Batley M, Djordjevic MA, Innes RW, Kuempel PL, Rolfe BG (1986) Flavones induce expression of nodulation genes in Rhizobium. Nature 323:632

Rovira AD, Newman EI, Bowen HJ, Campbell R (1974) Quantitative assessment of the rhizosphere microflora by direct microscopy. Soil Biol Biochem 6:211

Ryu CM, Murphy JF, Mysore KS, Kloepper JW (2004) Plant growth-promoting rhizobacteria systemically protect *Arabidopsis thaliana* against Cucumber mosaic virus by a salicylic acid and NPR1-independent and jasmonic acid-dependent signaling pathway. Plant J 39:381

Ryu CM, Farag MA, Hu CH, Reddy MS, Wei HX, Pare PW, Kloepper JW (2003) Bacterial volatiles promote growth in Arabidopsis. Proc Natl Acad Sci USA 100:4927

Schardl CL, Leuchtmann A, Spiering MJ (2004) Symbioses of grasses with seedborne fungal endophytes. Annu Rev Plant Biol 55:315

Singh PK, Parsek MR, Greenberg EP, Welsh MJ (2002) A component of innate immunity prevents bacterial biofilm development. Nature 417:552

Somers E, Vanderleyden J, Srinivasan M (2004) Rhizosphere bacterial signalling: a love parade beneath our feet. Crit Rev Microbiol 30:205

Steidle A, Allesen-Holm M, Riedel K, Berg G, Givskov M, Molin S, Eberl L (2002) Identification and characterization of an N-acylhomoserine lactone-dependent quorum-sensing system in *Pseudomonas putida* strain IsoF. Appl Environ Microbiol 68 6371

Taddei P, Tugnoli V, Bottura G, Dallavalle E, Zechini D'Aulerio A (2002) Vibrational, 1H-NMR spectroscopic, and thermal characterization of gladiolus root exudates in relation to *Fusarium oxysporum* f. sp. gladioli resistance. Biopolymers 67:428

Tamasloukht M, Sejalon-Delmas N, Kluever A, Jauneau A, Roux C, Becard G, Franken P (2003) Root factors induce mitochondrial-related gene expression and fungal respiration during the developmental switch from asymbiosis to presymbiosis in the arbuscular mycorrhizal fungus *Gigaspora rosea*. Plant Physiol 131:1468

Tan J, Bednarek P, Liu J, Schneider B, Svatos A, Hahlbrock K (2004) Universally occurring phenylpropanoid and species-specific indolic metabolites in infected and uninfected *Arabidopsis thaliana* roots and leaves. Phytochemistry 65:691

Teplitski M, Robinson JB, Bauer WD (2000) Plants secrete substances that mimic bacterial N-acyl homoserine lactone signal activities and affect population density dependent behaviors in associated bacteria. Mol. Plant Microbe Interact 13:637

Uren NC (2000) Types, amounts and possible functions of compounds released into the rhizosphere by soil grown plants. In: Pinton R, Varanini Z, Nannipieri P (eds) The rhizosphere: biochemistry and organic substances at the soil interface. Dekker, New York, pp 19–40

Veluri R, Weir TL, Bais HP, Stermitz FR, Vivanco JM (2004) Phytotoxic and antimicrobial activities of catechin derivatives. J Agric Food Chem 52:1077

Verma SC, Ladha JK, Tripathi AK (2001) Evaluation of plant growth promoting and colonization ability of endophytic diazotrophs from deep water rice. J Biotechnol 91:127

Vicre M, Santaella C, Blanchet S, Gateau A, Driouich A (2005) Root border-like cells of Arabidopsis. Microscopical characterization and role in the interaction with rhizobacteria. Plant Physiol 138:998

Walker TS, Bais HP, Grotewold E, Vivanco JM (2003) Root exudation and rhizosphere biology. Plant Physiol 1320:44

Walker TS, Bais HP, Deziel E, Schweizer HP, Rahme LG, Fall R, Vivanco JM (2004) *Pseudomonas aeruginosa*-plant root interactions. Pathogenicity, biofilm formation, and root exudation. Plant Physiol 134:320

Weir TL, Park SW, Vivanco JM (2004) Biochemical and physiological mechanisms mediated by allelochemicals. Curr Opin Plant Biol 7:472

Whitehead NA, Barnard AM, Slater H, Simpson NJ, Salmond GP (2001) Quorum-sensing in Gram-negative bacteria. FEMS Microbiol Rev 25:65

Yeates G (1999) Effects of plants on nematode community. Annu Rev Phytopathol 37:127

Part V
Functional Aspects

Chapter 12
The Impacts of Selected Natural Plant Chemicals on Terrestrial Invertebrates

Neal Sorokin(✉) and Jeanette Whitaker

12.1 Introduction

Plants produce thousands of chemicals that are not necessarily involved in their primary metabolism, but are likely to be involved in plant defence, communication and competition. These chemicals may be stored within plant tissues, e.g. to act as a defence from herbivorous predators, or may be actively released into the surrounding environment. Natural chemicals can enter the environment via a number of mechanisms, including volatilisation, exudation from roots, leaching from plant material and decomposition of plant residues, and also through direct transfer via root and shoot grafts, mycorrhizal fungi or haustorial connections of parasitic vascular plants (Rice 1984). Once released into the soil, these chemicals have the potential to positively or negatively affect the environment (soil structure, nutrient availability) and the organisms in an exposed area.

Other chapters have investigated the effects of natural chemicals on organisms such as plants, fungi and bacteria. In this chapter we concentrate on the potential effects of plant toxins on soil invertebrates. Invertebrates play an essential role in the functioning and conditioning of soils. Organisms such as earthworms act as primary decomposers breaking down and distributing organic matter, are important in soil aeration and drainage, have phytopathological importance by reducing plant disease and are an important component of the terrestrial food web (Lokke and Gestel 1998). Conversely, soil invertebrates can also have deleterious effects destroying crops or competing with beneficial organisms. By investigating the impact of plant toxins on soil invertebrates, we can better understand the interactions of plants and insects, gain knowledge of the influence of plants on soil community structure and in some instances identify chemicals which may be advantageous to man, e.g. pesticide development.

Neal Sorokin
Reckitt Benckiser Healthcare (UK) Ltd., Dansom Lane Hull, HU8 7DS
e-mail: neal.sorokin@reckittbenckiser.com

P. Karlovsky (ed.), *Secondary Metabolites in Soil Ecology. Soil Biology 14.*

The following text provides a review of the effects of selected plant toxins on soil invertebrates, summarising the exposure conditions and effects data available for a range of invertebrate species. In addition, the results of preliminary soil toxicity tests using standard procedures and organisms are presented for comparison. Given the vast array of plant toxins and the large amount of data available, a small number of model chemicals have been selected for review and testing.

12.2 The Effects of Natural Chemicals on Soil Invertebrates

Plant secondary metabolites have been studied in great detail in the literature, with detailed information available on the secondary metabolites present in a wide-range of plant species globally, their chemical forms, biological activities and interactions with flora and fauna aboveground (Harborne and Baxter 1996). However, information on the effects of secondary metabolites in soil ecosystems has not previously been collated or reviewed systematically. In order to determine interaction between plant secondary metabolites and soil flora and fauna, additional information is required. For example, do these compounds have a route into the soil, e.g. through leaching or decomposition, and once in the soil how persistent are they? Locating this type of information proved to be a complex task as there was no clear source of information and much of the data were anecdotal or unpublished, e.g. gardening Web sites.

For this review data on the sources, fate and effects of natural chemicals in the environment were compiled from various electronic and hard-copy data sources. A database of over 190 natural plant toxins, identified in over 500 plant species, was compiled. Information collated included chemical class and structure, mode of action, plant species, concentration in plant and plant organ, routes into soil and toxicity to target organisms, e.g. which chemicals have insecticidal, herbicidal, antimicrobial, fungicidal, nematacidal and molluscicidal activity. The compounds identified belong to a range of chemical classes including terpenes, phenolics, glucosinolates, polyacetylenes, alkaloids and glycosides. In addition, a literature database was compiled containing over 800 references relating to the compounds identified.

The information available was fragmented and patchy for the majority of compounds identified; however, 20 chemicals were identified which could potentially have effects in soil ecosystems owing to their chemical properties, i.e. there was evidence that they were naturally present in soils, showed some degree of persistence, and there was evidence of toxicity to soil organisms (Table 12.1). From the shortlist we selected three compounds to illustrate the types of data available on their sources, fate and effects in the environment, and the gaps in experimental data which need to be filled in order to investigate more fully the role of these compounds in soil ecosystems. The compounds described in the following text come from three different chemical groups and are present in different groups of plant species: they are juglone (naphthaquinone), α-pinene (monoterpene) and gramine (indole amine).

Table 12.1 Information available on properties of plant secondary metabolites in soil ecosystems—selected examples

Chemical	Chemical group	Presence in soil	Persistence and fate	Toxic
(-)-Lariciresinol	Lignin	✓	ND	?
α-Pinene	Monoterpene	✓	X	✓
Adipic acid	Dicarboxylic acid	✓	X	?
Ailanthone	Quassinoid triterpene	✓	X	✓
Apigenin-4-*O*-methyl	Flavanoid	✓	ND	?
DIMBOA/DIBOA	Hydroxamic acid	✓	✓	✓
Emodin	Anthraquinone	✓	✓	✓
Gramine	Indole alkaloid	✓	✓	✓
Grayanotoxin	Diterpene	✓	ND	✓
Juglone	Napthoquinone	✓	✓	✓
Kaempferol-3-*O*-methyl	Flavanoid	✓	ND	✓
Kaempferol-3,7-di-*O*-methyl	Flavanoid	✓	ND	✓
Monoterpenoids (α-pinene)	Mono-terpenoid	✓	✓	✓
Oleandrin	Glycoside	?	?	✓
Parthenin	Sequiterpenoid lactone	✓	?	✓
p-Hydroxybenzoic acid	Phenolic acid	✓	?	?
Physicon	Anthraquinone	✓	✓	✓
Scopoletin	Coumarin	✓	ND	✓
Sorgoleone	Dihydroquinone	✓	?	✓
Taxane	Diterpenoid	✓	?	?
Tricolorin A	Macrolactone oligosacharide	✓	ND	✓
Vanillic acid	Phenolic acid	✓	?	?

DIMBOA 2,4-dihydroxy-7-methoxy-2*H*-1,4-benzoxazin-3(4*H*)-one, *DIBOA* 2,4-dihydrozy-2*H*-1, 4-benzoxazin-3(4*H*)-one, ✓ adequate data set (fulfils criteria), *X* unlikely to have effects in soil (fails criteria), *ND* no data available,? data are inconclusive

12.2.1 Biological Activity of Juglone

The napthaquinone juglone (5-hydroxy-1,4-naphthoquinone) has been identified primarily in species of the Juglandaceae, specifically in the genus *Juglans* (Willis 2000), but has also been identified in the *Lomatia* (Proteaceae) (Moir and Thomposon 1973) and *Caesalpinia* (Caesalpiniaceae) (Nageshwar et al. 1984). Juglone is most abundant in walnuts as the 4-glucoside of the corresponding 1,4,5-trihydroxynaphthalene (α-hydrojuglone); however, the free aglycone form will also occur in the plant (Gottingen and Zimmermann 1979; Muller and Leistner 1978).

Juglone is released into the environment via three routes: (1) exudation from tree roots; (2) leaching from leaves in periods of rainfall; and (3) during the decomposition of plant material (Jose and Gillespie 1998). Once in the environment juglone is likely to be of moderate to low persistence. Some evidence suggests that juglone is extractable from soils for 45 days after addition under dry conditions and for 90 days in wet conditions (Fisher 1978); however, other studies have reported much lower recovery rates (50% recovery after 1 h) (DeScisciolo et al. 1990).

Juglone is one of the most widely studied plant secondary metabolites, in terms of its ecological effects in soil. Much of the literature data for this chemical relate to its allelopathic effects on plants, and its fate and degradation in soil. However, it has also been investigated as an antifeedant and deterrent to phytophagous insects, as a toxin for the eggs of pests and as a nematacide.

For example, Bernays and Cornelius (1992) investigated the effect of juglone on the feeding and reproduction of the alfalfa weevil (*Hypera brunneipennis*) by dosing alfalfa (*Medicago sativa*) leaves with 0.1% juglone, a concentration reported to be within that found naturally in walnut leaves (Bernays and Cornelius 1992). The 0.1% concentration significantly deterred feeding compared with feeding for the controls but appeared to have no effect on fecundity. Owing to the units used in the experiments, it is difficult to assess the severity of these effects and relate them to expected concentrations in walnut; suffice to say that an antifeeding effect was seen in the experiments.

Thiboldeaux et al. (1994) also investigated the effects of dietary juglone on saturnid moths. They exposed two species of moth, one (*Actias luna*) that was known to feed on walnut and likely to be tolerant to its effects and a second species (*Callosamia promethea*) that was not expected to feed on walnut. Larvae of the two species were exposed to juglone via birch leaves containing 0.05% (leaf wet weight) juglone and the survival, growth and development of the species were monitored (Thiboldeaux et al. 1994). As expected the *Actias luna* species showed no effect on development or mortality. On the other hand, *Callosamia promethea* showed a reduced growth rate and a threefold reduction in consumption rate (Thiboldeaux et al. 1994). However, the validity of these results may be in question as the majority of the *Callosamia promethea* moths, including the controls, died during the test as a result of disease.

Other studies have shown juglone to be toxic to the eggs of Cotton stainer (*Dysdercus koenigii*), with 100 mg l^{-1} reducing egg survival by approximately 28% (24 h) (Satyanarayana and Gujar 1995). In addition, nematodes have also been shown to be sensitive to juglone, with motility of *Haemonchus contortus* larvae reduced by 40% (24 h) with 10 mg l^{-1} (Fetterer and Fleming 1991).

These studies all demonstrate the insecticidal/antifeedant activity of juglone in organisms living aboveground; there have, however, been no studies on the effects of juglone on soil invertebrates.

12.2.2 Biological Activity of α-Pinene

α-Pinene, a monoterpene, is found in a wide array of plants and has been identified in, amongst others, the Coniferae, Cyperaceae, Poaceae, lamiaceae, Lauraceae, Liliaceae, Myrtaceae, Rutaceae and Umbelliferae (Rudolf 1975). α-Pinene, can often be the primary constituent of essential oils and volatile organic carbon emissions and can make up 10–30% of plant essential oils (Owen et al. 1997, 2001; Geron et al. 2000; Sabillon and Cremades 2001; Rinne et al. 2002; Tani et al. 2002). Owing to their physical and chemical properties, monoterpenes are primarily released from plants as volatile gases from the leaves, from roots (Hayward et al. 2001) and may also be released from decaying plant matter (Jordan et al. 1993; Kainulainen and Holopainen 2002). Once in the soil, α-pinene may adsorb to organic matter, degrade via autoxidation reactions or undergo aerobic or anaerobic biodegradation. It is only likely to be moderately persistent and may be degraded within days or weeks (Harder and Probian 1995; Misra et al. 1996; Misra and Pavlostathis 1997; Harder and Foss 1999; Kleinheinz et al. 1999).

The effects of α-pinene have been tested on various invertebrate species using different exposure pathways such as diet, aerial exposure and via direct contact. It has been shown not only to affect the survival of organisms, but also to act as a deterrent to phytophagus insects.

The antifeedant effects of α-pinene have been tested on the pine weevil (*Hylobis abietis*) and the cutworm (*Spodoptera litura*) (Klepzig and Schlyter 1999; Mukherjee 2003). Growth of the cutworms was significantly inhibited, whilst in contrast α-pinene appeared to have no effect on the feeding rate of weevils (Klepzig and Schlyter 1999). However, given the pine weevils preferred food type (coniferous trees) it may not be surprising to find that they are not deterred by α-pinene, which is often the primary constituent of conifer essential oils.

In contact tests where the organisms are exposed to filter paper containing α-pinene, a concentration of up to $0.1\,mg\,cm^{-2}$ had no effect on the survival of adults of the adzuki bean weevil (*Callosobruchus chinensis*) after a 24-h exposure (Park et al. 2003). Slightly higher concentrations ($0.18\,mg\,cm^{-2}$) had a more significant effect on the survival of the rice weevil (*Sitophilus oryzae*), with 6% mortality after 4 days. The highest concentration tested ($0.26\,mg\,cm^{-2}$) had more significant effects, with 18, 30 and 36% mortality after 2, 3 and 4 days, respectively (Park et al. 2003).

The fumigant activity of α-pinene has also been tested on various species, including *Acanthoscelides obtectus*, a bruchid pest of kidney beans, and the American cockroach (*Periplaneta americana*) (Regnault-Roger and Hamraoui 1995; Ngoh et al. 1998). Bruchids were exposed to the fumes of α-pinene emanating from treated filter paper in a fumigant test chamber. Effects on survival, reproduction and larval hatching were assessed. LC_{50} (concentration causing mortality in 50% of exposed organisms) values of 31.6 and $25.1\,mg\,l^{-1}$ were reported for 24 and 48 h, respectively. No significant effects were seen on reproduction at a concentration up to $0.8\,mg\,l^{-1}$, although a significant effect on larval ability to penetrate kidney beans was reported (Regnault-Roger and Hamraoui 1995).

Tests with the American cockroach (*Periplaneta americana*) indicate that this species is more tolerant to the fumigant effects of α-pinene. α-Pinene was added to filter paper and allowed to volatilise in a fumigant chamber. Cockroaches were then exposed to the vapours over 24 h. Adults appeared to be relatively tolerant of the chemical with EC_{50} (concentration that causes knockdown/immobilisation of 50% of test individuals) values greater than $0.7\,mg\,cm^{-2}$ (Ngoh et al. 1998). However, α-pinene did appear to have a reasonable repellent effect on this species, with an EC_{50} (repellency) of 0.059 mg cm^{-2}.

In addition to the effects on surface-dwelling terrestrial organisms, the effects of the isomers of α-pinene have been tested on the larvae of the mosquito (*Culex pipiens*) (Traboulsi et al. 2002). Larvae were exposed to each isomer in an aquatic system for 24 h. The reported LC_{50} values indicate similar toxicity of each isomer to this species, with values of 47 mg l^{-1} for the (1*R*)-(+)-α-pinene isomer and 49 mg l^{-1} for the (1*S*)-(−)-α-pinene isomer (Traboulsi et al. 2002).

12.2.3 Biological Activity of Gramine

Gramine (*N,N*-dimethyl-3-aminomethylindole), an indole amine, has been identified in various plant families, including the Acaraceae, Leguminosae and Gramineae (Smith 1977). However, the majority of the research on gramine has focussed on its role in the life cycle of barley (*Hordeum* Sp.) (Lovett and Hoult 1995). It is released into the environment in barley root exudates and washed from leaf surfaces (Liu and Lovett 1990; Yoshida et al. 1993).

The effects of gramine have been studied in a number of species. However, owing to its production by the barley plant much of the research has concentrated on its impacts on barley pests such as aphids. Nevertheless, gramine has been shown to affect not only the growth and survival of invertebrates, but also their reproduction.

The effect of dietary gramine has been tested on grasshoppers. Westcott et al. (1992) reported significant effects on the weight and survival of grasshopper nymphs (*Melanoplus sanguinipes*) fed on a diet containing more than 1% (10 g kg^{-1}) gramine (Westcott et al. 1992). However, these effects appear to occur at concentrations above that found naturally in higher plants (above 1 mg kg^{-1} fresh weight). Bernays (1991) also reported no effect on the feeding rate or growth rate of the grasshopper *Schistocerca americana* at environmentally relevant concentrations.

The most widely studied aspect of the role of gramine in higher plants relates to its antifeedant and deterrent properties. Salas and Corcuera (1991), in their studies on the effect of environmental factors on gramine production, investigated the susceptibility of the aphid *Schizaphis granium* to the associated changes in gramine concentration. A gramine-containing cultivar and a gramine-free cultivar of barley were infested with aphids and then exposed to a temperature range of 20–35 °C. Internal gramine concentrations of the plants and population growth rates of the aphids were then measured. Aphids infesting the gramine-free cultivar had a constant population growth rate over all temperatures; however, the population growth rates of aphids on the gramine-containing cultivar were significantly lower. As the temperature increased the internal gramine concentration also increased and as a consequence the aphid population growth rate decreased. At 21 °C the internal gramine concentration of the plant was approximately 500 mg kg^{-1} (fresh weight) and the population growth rate of the aphids was 0.51, but at 35 °C the internal gramine concentration was over 1,200 mg kg^{-1} (fresh weight) and the population growth rate was reduced to 0.12 (Salas and Corcuera 1991). A similar trend was found when the two cultivars were exposed to differing light regimes. A longer photoperiod increased the concentration of the gramine-containing cultivar and lowered the population growth rate. Aphids infesting the gramine-free plants showed no variation in their population growth rate regardless of photoperiod (Salas and Corcuera 1991).

Similar results have been shown with barley under normal conditions. Plants grown at 25 °C for 6 days were infested with *Schizaphis granium* and the gramine content of the plant and the population growth rate of the aphid were monitored over 20 days (Zuniga et al. 1985). A significant negative correlation was found between gramine content and population growth rate. Aphids were also reared on artificial diets and survival and reproduction were recorded after 24 and 72 h, respectively. Aphid survival was severely affected by gramine in the diet, with an LD_{50} (lethal dose causing effects in 50% of the test population) of 139 mg kg^{-1} (Zuniga et al. 1985). Reproduction was also significantly affected, with the reproductive index (average number of nymphs to the average number of adults) reducing from 3 in the controls to 1 at a concentration of 41.8 mg kg^{-1} of gramine in the diet (Zuniga et al. 1985).

Corcuera (1984) also reported LD_{50} values for *Schizaphis granium* and *Rhopalosiphum maidis* exposed to gramine-containing diets. The 48-h LD_{50} values were 122 and 505 mg gramine per kilogram of food for the two aphids, respectively. Interestingly, in exposure experiments of 10-h duration, higher gramine contents had no effect on survival. This was attributed to a deterrency effect of gramine, with aphids avoiding food with high gramine concentrations (Corcuera 1984).

Barley plants have also been shown to increase their gramine content in response to aphid attack. Velozo et al. (1999) reported that the internal gramine content of three cultivars of barley (Frontera, Libra and Acuario) increased when they were infested with aphids (*Schizaphis granium*), whereas the concentration in aphid-free plants remained unchanged. Gramine content in infested plants increased by 123% in the Frontera cultivar, 720% in the Acuario cultivar and 1,080% in the Libra cultivar (Velozo et al. 1999). Gramine concentration was strongly correlated with aphid density, with higher densities resulting in higher gramine accumulation. However, after 4 days plant vigour was seriously affected in the density experiments and the correlations no longer held true (Velozo et al. 1999).

Gramine may also have sublethal effects on organisms. Gramine is a well-known inhibitor of octopamine (Orr et al. 1985; Ismail et al. 1993). Octopamine plays an important role in regulating the nervous system of insects and acts as a neurotransmitter neurohormone and neuromodulator (Hirashima et al. 1999b). In the silkworm (*Bombyx mori*) a concentration of 30 mg kg^{-1} in food inhibited octopamine concentrations, resulting in delayed pupation; however, metamorphosis was unaffected even at concentrations as high as 1 g kg^{-1} (Hirashima et al. 1999b).

The literature data for the compounds described here suggest that they are toxic to a wide range of organisms and that they can have effects on a range of end points from growth to survival and reproduction. These chemicals can also exert their effects through a number of exposure pathways, including dietary exposure, aerial exposure and via direct contact. However, some of the data are of limited relevance to true soil-dwelling organisms (e.g. earthworms) and few express effects as a concentration in soil (e.g. milligrams per kilogram of soil), a more relevant measure of exposure in the rhizospere. To date no measure of the actual concentration of gramine in the rhizosphere has been carried out. Nor has the fate (biodegradation, persistence, etc.) of gramine in soil ecosystems been studied in any depth. Data are required that will better explain the effects of these chemicals through soil exposure.

12.3 Invertebrate Toxicity Assays

In order to understand the effects of natural chemicals in the terrestrial environment the experimental data that are generated should reflect, where possible, the exposure pathways observed in the soil environment. The literature data for the chemicals described in the previous sections are most relevant to the antifeedant and deterrent properties of the ecotoxins and provide only limited insight into the effects associated with soil exposures. In order to investigate this exposure pathway, we tested the effects of the three natural chemicals on the survival of earthworms and collembolans using standard test procedures and a uniform standardised soil, to demonstrate the types of data which would be needed in order to more fully understand the interactions between plant secondary metabolites and soil invertebrates.

Earthworms and collembolans were chosen as representatives of soil invertebrates as both organisms are well studied and standardised procedures have been developed for their testing. Although it would seem counterintuitive for plant-derived chemicals to have negative effects on beneficial organisms such as earthworms, data on the effects on these organisms can provide indications of the potential effects in other soil-dwelling fauna.

Soil toxicity screening tests were carried out using standard guidelines for earthworms (OECD 2004) and collembolans (ISO 2002) and were carried out in standard OECD soil (10% finely ground sphagnum peat 20% kaoline clay and 70% fine quartz sand adjusted to 35% of the water-holding capacity and pH 6 ± 0.5 using calcium carbonate).

Owing to the limited water solubility of the test chemicals, stock solutions were prepared for each chemical in acetone. A concentration series of 0.01, 0.1, 1, 10, 50, 100, 500 and 1,000 mg kg^{-1} (dry weight) was prepared by mixing the appropriate amount of stock solution into 10 g of sand. The sand/stock solution mixture was then left so the solvent completely evaporated and was then mixed into to the appropriate volume of test soil. In addition to the test soils, a control and solvent control were also prepared.

Collembolans (*Folsomia candida*) were taken from laboratory cultures. Synchronised cohorts for testing were obtained by transferring several hundred adults to clean culture vessels and allowing them to lay eggs over a 2–3-day period. Adults were then removed and the eggs monitored for hatching. Experiments were carried out using 10–12-day-old juveniles. The collembolan survival tests were carried out in

100-ml (screw-top) glass jars filled with 30 g (wet weight) of test soil. Ten, 10–12-day-old collembolans were placed into each vessel with 2 mg of baker's yeast added. Test vessels were maintained in a temperature-controlled room at 20 ± 1°C in a 16 h/8 h light/dark cycle. After 14 days the test was terminated. The jars were flooded with deionised water and the number of surviving adults counted.

Earthworms (*Eisenia fetida*) were purchased from Blades Biologicals UK. Worms were held for 3 weeks prior to testing in commercial culture soil and fed dried ground rabbit manure as required. The earthworm tests were carried out in 1-l flat-bottomed glass jars. The jars were filled with 1 kg (wet weight) of test soil and ten adult worms were placed into each vessel. The worms were fed dried and ground rabbit manure during the test. The test vessels were maintained in a temperature-controlled room at 20±1°C in a 16 h/8 h light/dark cycle. After 28 days, adult earthworms were dry sieved from the test vessels and the number of surviving adults recorded.

12.4 Inverterbate Toxicity Data for Selected Plant Secondary Metabolites

In soil exposures gramine and α-pinene had no significant effect on the survival of either earthworms or collembolans at concentrations up to 1,000 mg kg^{-1}, suggesting low toxicity in soils (Figs. 12.1, 12.2). It is difficult to compare these data directly with those in the literature as the exposure pathways are different, i.e. in the literature most organisms are reported as having been exposed via their food or

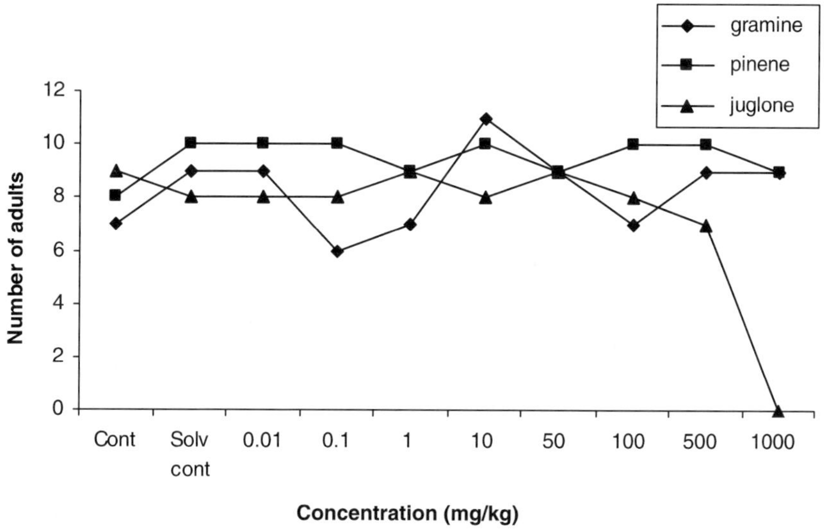

Fig. 12.1 Effects of the natural toxins gramine, α-pinene and juglone on the survival of adult earthworms (*Eisenia fetida*)

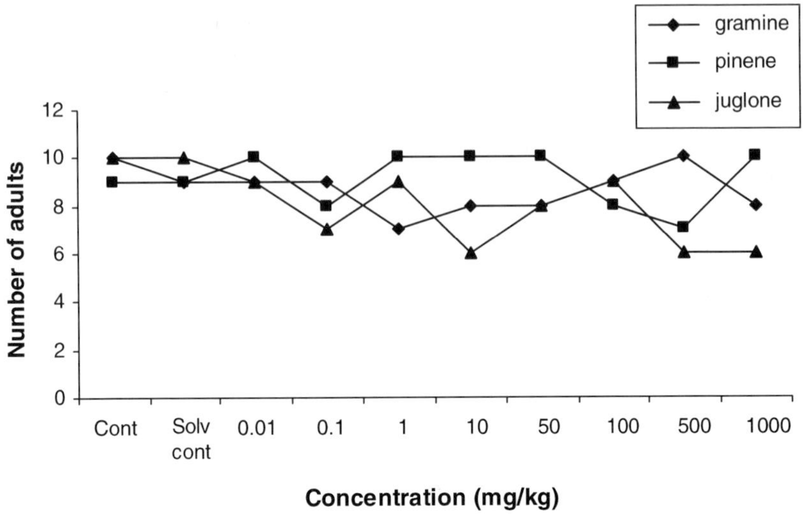

Fig. 12.2 Effects of the natural toxins gramine, α-pinene and juglone on the survival of collembola (*Folsomia candida*)

in the case of α-pinene by aerial exposures. Comparison of the soil data with dietary exposures to gramine indicate similarly low sensitivities, with population growth rates of aphids (*Schizaphis granium*) significantly affected at concentrations greater than 1,000 mg kg^{-1} food (Salas and Corcuera 1991) and grasshopper survival affected at concentrations of more than 10,000 mg kg^{-1} food (Westcott et al. 1992). However, some organisms appear to be more sensitive, with 50% effects on survival of aphids (*Schizaphis granium*) reported to occur at 130 mg kg^{-1} food (Corcuera 1984; Zuniga et al. 1985). The data for α-pinene are more difficult to compare as most of the effects are expressed as fumigant exposures or as surface-area applications in contact tests. Nevertheless, the data for soil exposures indicate low toxicity to worms and collembolans.

Comparison of the toxicity data with measured soil concentrations in the field suggests that there may be insufficient levels of the chemicals, in isolation, to have effects on invertebrate survival in soils. Liu and Lovett (1990) reported that barley seedlings released gramine in a bioassay system, with peak concentrations of gramine reaching 22 mg l^{-1} from washing of filter paper 4 days after germination. Yoshida et al. (1993) reported that gramine was washed from barley leaves when they were exposed to simulated normal rain (pH 6) and acid rain (pH 4). Measurable concentrations of gramine were found in the rainwater (1 mg kg^{-1} fresh weight of plant tissue), although no difference was found between the concentrations in the normal and the acid rain.

Environmental concentrations of α-pinene are slightly higher than those of gramine, but would still be insufficient, on the basis of the soil exposure data, to cause effects on the survival of soil invertebrates. Kanda et al. (2007) measured the concentration of α-pinene in soils associated with pure stands of Scots pine (*Pinus sylvestris*), Sitka spruce (*Picea sitchensis*) and Norway spruce (*Picea abies*). Soils were taken at various depths and distances away from the trees. The concentrations of α-pinene were greatest in surface soils directly under the trees, where the levels were reported to be 70, 114 and 16 mg kg^{-1} (dry weight) for Sitka spruce, Scots pine and Norway spruce, respectively (Kanda et al. 2007).

Juglone appears to be more toxic than either gramine or α-pinene, at least to earthworms. Exposure to a concentration of 1,000 mg kg^{-1} resulted in 100% mortality in *Eisenia fetida* (Fig. 12.1). Collembolans, on the other hand, were less sensitive, with no significant effect on survival up to 1,000 mg kg^{-1} (Fig. 12.2). The difference in the sensitivity of these organisms to juglone may be a result of species-dependent routes of uptake, where collembolans are mainly exposed via pore water, while earthworms are also exposed through dietary exposure to soil. Again it is difficult to compare the soil exposures with the literature data. The literature data relate primarily to dietary exposures but express the effects as percentages of juglone in the foodstuff. Without detailed information on the quality of the foodstuff it is difficult to determine an exact exposure dose, but by crudely comparing the proportions of juglone to the exposure medium some comparisons can be made. Exposure to leaves dosed with 0.1% (approximately1,000 mg kg^{-1}) and 0.05% juglone (approximately 500 mg kg^{-1}) resulted in significant effects on feeding of weevils (*Hypera brunneipennis*) and the survival and growth of saturnid moths (*Callosamia promethea*), respectively (Bernays and Cornelius 1992; Thiboldeaux et al. 1994). These data suggest very similar sensitivities to those found in earthworms exposed via soil, where significant effects occurred at 0.1% juglone (1,000 mg kg^{-1}). However, the relevancy of the comparisons is limited owing to the differences in the test systems and the expression of results.

As with gramine and α-pinene, comparison of the soil toxicity results with data for field concentrations of juglone indicate that the type of effects seen in the laboratory soil exposures are unlikely in the natural environment. Soil samples taken from the area around six *Juglans nigra* trees contained juglone concentrations ranging from 1.5 to 3.25 mg kg^{-1} (dry weight) (DeScisciolo et al. 1990). Ponder and Tadros (1985) report similar concentrations of juglone (3.65–3.95 mg kg^{-1} dry weight) in soils of a 14-year-old walnut plantation and a coplanted walnut and black alder stand (Ponder and Tadros 1985).

12.5 Conclusions

Data from the literature and from laboratory soil exposures indicate that the plant toxins gramine, juglone and α-pinene can have toxic effects on soil invertebrates. They appear to affect a wide range of species and can impact a range of end points from growth to reproduction. However, the literature and laboratory data suggest that reasonably high concentrations of these chemicals (above 100 mg kg^{-1}) would be required to have effects on the survival of invertebrates in the environment. Analyses of the data on environmental concentrations of these chemicals in soil indicate that such levels are unlikely to occur. On the other hand, such levels are frequently found internally within plants, suggesting that these chemicals act primarily as antifeedants and deterrents to phytophagus insects.

The lack of effects in the laboratory, at environmentally relevant soil concentrations, does not necessarily mean that these chemicals would have no effect in the environment. The laboratory toxicity data are surrogates for soil invertebrates in general and can only provide an indication of the potential effects. It is quite possible that more sensitive soil organisms are present in the terrestrial environment. These may well be specific pests of the plant species on which the secondary metabolites are specifically acting. In addition, the laboratory tests measure effects on survival, a rather extreme end point to assess the effects of these natural chemicals. It is possible that they have more subtle effects on parameters of the life cycle of invertebrates e.g. the effects of gramine on octopamine in silkworms (Hirashima 1999a, b), or they may act only as deterrents.

Another important factor is that the laboratory data with earthworms and collembolans only highlight the fact that, in isolation, these chemicals are not highly toxic to soil invertebrates. Natural chemicals are not released from plants in isolation and organisms can be exposed to a cocktail of chemicals, of which these model ecotoxins may be only one e.g. barley plants contain, in addition to gramine, vanillic, ferulic and *p*-hydroxybenzoic acids and hordenine (Borner 1960; Overland 1966). Mixtures of chemicals can often have more potent effects than the single chemicals (Inoue et al. 1992). The laboratory data suggest that gramine, juglone and α-pinene are not highly toxic in isolation. Consequently, if they were have an effect on the survival of invertebrates in soils surrounding the plant they would possibly have to be released in combination with other chemicals. However, further work is required to confirm this hypothesis.

References

Bernays EA (1991) Relationship between deterrence and toxicity of plant secondary compounds for the grasshopper Schistocerca americana. J Chem Ecol 17(12):2519–2526

Bernays EA, Cornelius M (1992) Relationship between deterrence and toxicity of plant secondary compounds for the alfalfa weevil *Hypera brunneipennis*. Entomol Exp Appl 64(3):289–292

Borner H (1960) Liberation of organic substances from higher plants and their role in the soil sickness problem. Bot Rev 26:393–424

Corcuera LJ (1984) Effects of indole alkaloids from Gramineae on aphids. Phytochemistry 23(3):539–541

DeScisciolo B, Leopold DJ, Walton DC (1990) Seasonal patterns of juglone in soil beneath Juglans nigra (black-walnut) and influence of J. nigra on understory vegetation. J Chem Ecol 16(4):1111–1130

Fetterer RH, Fleming MW (1991) Effect of plumbagin on development of the parasitic nematodes *Haemonchus contortus* and *Ascaris suum*. Comp Biochem Physiol C 100(3):539–542

Fisher RF (1978) Juglone inhibits pine growth under certain moisture regimes. Soil Sci Soc Am J 42:801–803

Geron C, Rasmussen R, Arnts R, Guenther A (2000) A review and synthesis of monoterpene speciation from forests in the United States. Atmos Environ 34(11):1761–1781

Gottingen AP, Zimmermann MH (1979) Encyclopedia of plant physiology, new series, vol. 8. Secondary plant products. Springer, Berlin

Harborne JB, Baxter H (1996) Dictionary of plant toxins. London, Wiley

Harder J, Foss S (1999) Anaerobic formation of the aromatic hydrocarbon p-cymene from monoterpenes by methanogenic enrichment cultures. Geomicrobiol J 16(4):295–305

Harder J, Probian C (1995) Microbial degradation of monoterpenes in the absence of molecular oxygen. Appl Environ Microbiol 61(11):3804–3808

Hayward S, Muncey RJ, James AE, Halsall CJ, Hewitt CN (2001) Monoterpene emissions from soil in a Sitka spruce forest. Atmos Environ 35(24):4081–4087

Hirashima A, Hirokado S, Ohta H, Suetsugu E, Sakaguchi M, Kuwano E, Taniguchi E, Eto M (1999a) Titres of biogenic amines and ecdysteroids: effect of octopamine on the production of ecdysteroids in the silkworm *Bombyx mori*. J Insect Physiol 45:843–851

Hirashima A, Suetsugu E, Hirokado S, Kuwano E, Taniguchi E, Eto M (1999b) Effect of octopamine on the activity of juvenile-hormone esterase in the silkworm *Bombyx mori* and the red flour beetle *Tribolium freemani*. Gen Comp Endocrinol 116:373–381

Inoue M, Nishimura H, Li HH, Mizutani J (1992) Allelochemicals from *Polygonum sachalinense* Fr Schm (Polygonaceae). J Chem Ecol 18(10):1833–1840

Ismail SMM, Dekeyser MA, Downer RGH (1993) Effect of dihydrooxadiazines on the octopamine-sensitive adenylate cyclase complex of two-spotted spider mite, *Tetranychus urticae* Koch. Pestic Biochem Physiol 47(1):1–7

ISO (2002) Soil quality—inhibition of reproduction of collembola (*Folsomia candida*) by soil pollutants. Draft international standard ISO/DIS 11267. International Organization for Standardization, Geneva

Jordan ED, Hsieh TCY, Fischer NH (1993) Volatiles from litter and soil associated with *Ceratiola ericoides*. Phytochemistry 33(2):299–302

Jose S, Gillespie AR (1998) Allelopathy in black walnut (Juglans nigra L.) alley cropping. I. Spatio-temporal variation in soil juglone in a black walnut corn (Zea mays L.) alley cropping system in the midwestern USA. Plant Soil 203(2):191–197

Kainulainen P, Holopainen JK (2002) Concentrations of secondary compounds in Scots pine needles at different stages of decomposition. Soil Biol Biochem 34(1):37–42

Kanda R, Sorokin N, Hartley-Whitaker J, Weeks JM, Black HIJ (2007) Determination of the natural ecotoxin α-pinene and other monoterpenes in the terrestrial environment using accelerated solvent extraction and gas chromatography mass spectrometry. Ann Chem (in press)

Kleinheinz GT, Bagley ST, John WPS, Rughani JR, McGinnis GD (1999) Characterization of alpha-pinene-degrading microorganisms and application to a bench-scale biofiltration system for VOC degradation. Arch Environ Contam Toxicol 37(2):151–157

Klepzig KD, Schlyter F (1999) Laboratory evaluation of plant-derived antifeedants against the pine weevil Hylobius abietis (Coleoptera: Curculionidae). J Econ Entomol 92(3):644–650

Liu DL, Lovett JV (1990) Allelopathy in barley: potential for biological suppression of weeds. In: Alternatives to the chemical control of weeds. Proceedings of an international conference. Ministry of Forestry, Rotorua

Lokke H, Gestel CV (1998) Handbook of soil invertebrate toxicity tests. Wiley, Chichester

Lovett JV, Hoult AHC (1995) Allelopathy and self-defense in barley. In: Inderjit, Dakshini, KMM, Einhellig FA (eds) Allelopathy: organisms, processes, and applications. ACS symposium series 582. American Chemical Society, Washington, pp 170–183

Misra G, Pavlostathis SG (1997) Biodegradation kinetics of monoterpenes in liquid and soil-slurry systems. Appl Microbiol Biotechnol 47(5):572–577

Misra G, Pavlostathis SG, Perdue EM, Araujo R (1996) Aerobic biodegradation of selected monoterpenes. Appl Microbiol Biotechnol 45(6):831–838

Moir M, Thomposon RH (1973) Naphthoquinones in *lomatia* species. Phytochemistry 12:1351–1353

Mukherjee S (2003) Influence of plant allelochemicals on growth rate, nutritional physiology and mid-gut esterase activity in fifth instar larvae of Spodoptera litura (F.) (Lepidoptera: Noctuidae). Invertebr Reprod Dev 43(2):125–132

Muller W-U, Leistner E (1978) Metabolic relation between naphthalene derivatives in *Juglans*. Phytochemistry 17:1735–1738

Nageshwar G, Radhakrishnaiah M, Narayana LL (1984) Chemotaxonomy of *Caesalpinia*. Curr Sci 53:813–814

Ngoh SP, Choo LEW, Pang FY, Huang Y, Kini MR, Ho SH (1998). Insecticidal and repellent properties of nine volatile constituents of essential oils against the American cockroach, Periplaneta americana (L.). Pestic Sci 54(3):261–268

OECD (2004) Earthworm reproduction test (*Eisenia fetida/andrei*). OECD guideline 222 for the testing of chemicals OECD, Paris

Orr GL, Gole JWD, Downer RGH (1985). Charecterisation of an octopamine-sensitive adenylate cyclase in haemocyte membrane fragments of the American cockroach *Periplaneta americana* L. Insect Biochem 15(6):695–701

Overland L (1966) The role of allelpathic substances in the "smother crop" barley. Am J Bot 53(5):423–432

Owen S, Boissard C, Street RA, Duckham SC, Csiky O, Hewitt CN (1997). Screening of 18 Mediterranean plant species for volatile organic compound emissions. Atmos Environ 31(Suppl 1):101–117

Owen SM, Boissard C, Hewitt CN (2001) Volatile organic compounds (VOCs) emitted from 40 Mediterranean plant species: VOC speciation and extrapolation to habitat scale. Atmos Environ 35(32):5393–5409

Park I-K, Lee S-G, Choi D-H, Park J-D, Ahn Y-J (2003) Insecticidal activities of constituents identified in the essential oil from leaves of *Chamaecyparis obtusa* against *Callosobruchus chinensis* (L.) and *Sitophilus oryzae* (L.). J Stored Prod Res 39:375–384

Ponder F, Tadros SH (1985) Juglone concentration in soil beneath black walnut interplanted with nitrogen-fixing species. J Chem Ecol 11(7):937–942

Regnault-Roger C, Hamraoui A (1995) Fumigant toxic activity and reproductive inhibition induced by monoterpenes on *Acanthoscelides obtectus* (Say) (Coleoptera) a bruchid of kidney bean (*Phaseolus vulgaris* L.). J Stored Prod Res 31(4):291–299

Rice EL (1984). Allelopathy, 2nd edn. Academic, New York

Rinne HJI, Guenther AB, Greenberg JP, Harley PC (2002) Isoprene and monoterpene fluxes measured above Amazonian rainforest and their dependence on light and temperature. Atmos Environ 36(14):2421–2426

Rudolf EV (1975) Volatile leaf oil analysis in chemosystematic studies of North American conifers. Biochem Syst Ecol 2:131–167

Sabillon D, Cremades LV (2001). Diurnal and seasonal variation of monoterpene emission rates for two typical Mediterranean species (*Pinus pinea* and *Quercus ilex*) from field measurements—relationship with temperature and PAR. Atmos Environ 35(26):4419–4431

Salas ML, Corcuera LJ (1991). Effect of environment on gramine content in barley leaves and suseptability to the aphid *Schizaphis granium*. Phytochemistry 30(10):3237–3240

Satyanarayana RV, Gujar GT (1995) Toxicity of plumbaign and juglone to the eggs of the cotton strainer *Dysdercus koenigii*. Entomol Exp Appl 77:189–192

Smith TA (1977) Tryptamine and related compounds in plants. Phytochemistry 16:171–175

Tani A, Nozoe S, Aoki M, Hewitt CN (2002) Monoterpene fluxes measured above a Japanese red pine forest at Oshiba plateau, Japan. Atmos Environ 36(21):3391–3402

Thiboldeaux RL, Lindroth RL, Tracy JW (1994) Differential toxicity of juglone (5-hydroxy-1,4-naphthoquinone) and related naphthoquinones to saturniid moths. J Chem Ecol 20(7):1631–1641

Traboulsi AF, Taoubi K, El-Haj S, Bessiere JM, Rammal S (2002) Insecticidal properties of essential plant oils against the mosquito *Culex pipiens* molestus (Diptera: Culicidae). Pest Manag Sci 58(5):491–495

Velozo JA, Alvarez RI, Wachter GA, Timmermann N, Corcuera LJ (1999) Increase in gramine content in barley infested by the aphid Schizaphis graminium R. Phytochemistry 52:1059–1061

Westcott ND, Hinks CF, Olfert O (1992) Dietary effects of secondary plant compounds on nymphs of *Melanoplus sanguinipes* (Orthoptera, Acrididae). Ann Entomol Soc Am 85(3):304–309

Willis RJ (2000) Juglans spp, juglone and allelopathy. Allelopathy J 7(1):1–55

Yoshida H, Tsumuki H, Kanehisa K, Corcuera LJ (1993) Release of gramine from the surface of barley leaves. Phytochemistry 34(4):1011–1013

Zuniga GE, Salgado MS, Corcuera LJ (1985) Role of an indole alkaloid in the resistance of barley seedlings to aphids. Phytochemistry 24(5):945–947

Chapter 13
The Role of Soil Microbial Volatile Products in Community Functional Interactions

Ron E. Wheatley

13.1 Introduction

13.1.1 Soil Ecosystems

Soil ecosystems are a complex of biotic and abiotic components.

There are a great many different types of soils with a wide range of different textures. Type and texture are determined by the relative proportions of organic material, from living and dead organisms, and mineral material, including sand, silt and clay. Texture and structure affect the amount and size of pore spaces in the soil, which contribute between 5 and 35% of the soil volume. These pore spaces are occupied by air and water, and may be connected to the troposphere and contain a similar atmosphere.

Soils therefore have a porous three-dimensional structure, with varying degrees of continuity and interconnection, and so have a large surface area. Soil organisms live, and function, inside these pores and on the surfaces provided by the soil particles. Under temperate grasslands there can be a total biota of more than 45 t ha^{-1} fresh weight, which is similar to the aboveground biomass, and equivalent to a stocking rate of several hundred sheep per hectare. In each gram of soil there are billions of bacteria, tens of kilometres of fungal hyphae, many thousands of protozoa, thousands of nematodes, hundreds of insects, adults and larvae, arachnids and worms, plant roots, and the occasional visiting mammal! Similarly these numbers are equalled in scale by the biodiversity found within such communities. Many molecular studies of environmental DNA have shown high levels of biodiversity, both within the general eubacterial populations (Torsvik et al. 1990; McCaig et al. 1999; Curtis et al. 2002) and with time (Wheatley et al. 2003), and also within specific functional groups (Mitchell et al. 2001).

Ron E. Wheatley
Environment–Plant Interactions Programme, Scottish Crop Research Institute, Invergowrie, Dundee, DD2 5DA, UK
e-mail: ron.wheatley@scri.ac.uk

P. Karlovsky (ed.), *Secondary Metabolites in Soil Ecology. Soil Biology 14.*

As well as being extremely complex and very biodiverse these soil microbial communities are dynamic, continually changing over both time and space, and perform many functions and transformations, with many taxonomic types contributing to a functional attribute. They play a vital part in sustaining biosphere functioning. Driven by the products of primary production, soil microorganisms are responsible for continued plant growth, through residue decomposition and nutrient cycling and mobilisation. For example, the nitrogen cycle is a series of transformations linked simply by substrate availability, rather than a cohesive sequential event. There are many competitors for any available substrate functioning in the same environmental and physiological niche. Microorganisms in soils are opportunists that are very responsive to inputs. Although in bulk soil they are generally energy-restricted, they respond rapidly when a substrate becomes available and will use energy simply because it is there. Examples of such inputs include root material, as senescing structures or root exudates during growth, and incorporated aerial parts, such as the leaves and stems that are taken into the soil and distributed through it by worms and larvae, etc. Such inputs tend to be unevenly distributed through the soil and temporally very variable. As a consequence of these variations in substrate availability soil microorganisms have to be capable of maintaining themselves minimally, during periods of low input, but be able to react quickly to any opportunity that arises when potential substrates are introduced into the soil.

Large spatial and temporal variations in functional rates, with rapid changes in rates of expression, have been reported many times in soil systems. Frequently the rates of change in such microbial functional dynamics in soils are so great and rapid that they cannot be explained by corresponding changes in inputs or environmental conditions, or indeed changes in population size and structure (Wheatley et al. 2003). One explanation for this could be that only some members of a functional group may be active at a particular time, then in response to changes in inputs other members of the group join in, or indeed some of the original contributors cease to be active, with resulting rapid changes in functional rates. Such variations are also an indication of the complexity of interactions between components of the soil microbial community, with inhibition, stimulation and feedback responses, possibly involving infochemicals, all involved in the competition for resources in the soil ecosystem.

Community members that have a means of sensing that others are active, perhaps exploiting a substrate that has become available, such as deposited plant debris or invading plant roots, or producing a specific substrate such as ammonium nitrogen, will have a great advantage over others. Advantageous responses could include growth towards the new source, by hyphal extension for example, when signals from rhizosphere organisms may indicate the appearance of a suitable host for a pathogen, or preparation for the availability of a substrate, such as ammonium nitrogen by nitrifying bacteria. Although the fungi could explore speculatively, via random hyphal extension, it would be much more efficient if such expeditions were directed by sensing through the soil atmosphere. The fungal colony could then find and establish itself in a 'new' more nutrient status favourable environment, most efficiently. Although many bacteria are mobile they have a strong tendency to clump in colonies on surfaces, so any advance notice of a potential substrate availability in the soil solution could again be very advantageous.

13.1.2 Interactions Between Microorganisms

Soil ecosystem functioning appears to result from a collection of random events, the dynamics of which are limited by conditions imposed by both soil and community structure at microsites, so local functional rates result from the response of only a very limited portion of the total microbial diversity present through the bulk soil. The importance of the relationship between these determining factors and the frequency of microsites highlights the potential importance of interactions between species and functional groups in regulating microbial process dynamics (Wheatley et al. 2001).

A means of sensing activities and opportunities in this complex environment would make functioning within soil ecosystems much more efficient. Compounds produced during microbial activity, both specifically as antimicrobial agents or coincidentally as secondary metabolites, can have profound effects on microbial functional dynamics and community structure in soil ecosystems. Such effects will have influenced the population composition that has developed in the soil over time, as well as affecting contemporary community dynamics.

13.2 Compounds

13.2.1 Antibiotic and Antimicrobial Product Production

Although soil-borne fungal pathogens can cause great damage to crop plants and significantly affect yields, and subsequent storage life, the greater majority of the fungi found in soils are not pathogenic. Indeed many of these non-pathogenic fungi contribute to soil-ecosystem functioning in a positive way, e.g. the decomposition of organic matter. They may even be beneficial to plants, by occupying ecological niches to the exclusion of pathogens, and also actively participating in the biocontrol of such pathogens. Biological control by ‘niche occupation’ with a non-pathogenic member of the same genus, e.g. *Fusarium oxysporum*, has been reported (Toyota et al. 1996). Microorganisms involved directly in biocontrol include *Streptomyces melanosporofaciens* EF-76, which produces geldanamycin, a polyketide with antibiotic/antimicrobial activity (De Boer et al. 1970), which gives effective protection for crops against several fungal diseases, and also against some Gram-positive bacteria (Rothrock and Gottlieb 1984; Toussaint et al. 1997).

Similarly, although many soil bacteria are pathogens most are not. Indeed many have been reported to be interactive in microbial community dynamics and have been used successfully in biocontrol (Fiddaman and Rossall 1993). In this type of interaction antibiotic/antimicrobial components are produced. In this way some bacteria that are frequently associated with plants, e.g. *Pseudomonas* spp., have been used to suppress crop diseases. Products of *Pseudomonas fluorescens* CHA0 can protect wheat plants against infection by *Gaeumannomyces graminis* var. *tritici* in greenhouse and field experiments, and are also effective against a range of other root diseases (Defago et al. 1990; Shaukat and Siddiqui 2003). Active antimicrobial products, including 2,4-diacetylphloroglucinol and pyoluteorin, have been identified (Duffy and Defago 1999) and these have also been reported as protecting cucumbers from several fungal pathogens, including *Pythium* spp. (Girlanda et al. 2001) when *Pseudomonas fluorescens* CHA0-Rif was introduced into the soil.

13.2.2 Other Interactive Microbial Compounds

There are many reports of chemically mediated interactions between individuals and groups, at the species and functional group level, in the biosphere, for example between insects, plants and mammals. Soil atmospheres contain a lot of different volatile organic compounds (VOCs). Many of these are secondary metabolites produced during microbial activity (Stotzky and Schnenck 1976; Stahl and Parkin 1999). Since microbial activity, in both function and degree, is determined by environmental factors, the type and amounts of these VOCs will be indicative of the functional dynamics occurring at that time in the soil ecosystem. These volatile compounds will be able to diffuse over distances ranging from micrometres to metres through the tortuous porous interconnected structure of soils, and, as most are soluble, will be capable of passing through any potential barriers in the soil caused by water-filled pores. These abilities, coupled with their production being a consistent response to functional activity in soils, supports the concept that VOCs make ideal candidates for consideration as infochemicals in soil ecosystems. Microbially produced VOCs will facilitate communication and interactions between members of the microbial population, which although large is distributed over a large surface area with a resultant considerable spatial separation. The consequences of such VOC interactions will have a strong influence on the development and evolution of populations in soil ecosystems.

Microorganisms produce a consistent VOC profile that is determined both by the organisms involved (Table 13.1) and by the conditions under which they are cultured (Table 13.2). The profile of VOCs produced by a microbial species is consistent when environmental parameters are constant, but responds to changes, such as in nutrient availability or temperature (Tronsmo and Dennis 1978; Zechman and Labows 1985; Giudici et al. 1990; Fiddaman and Rossall 1994; Wheatley et al. 1997; Bruce et al. 2000), to another consistent pattern. Many VOCs found in soil atmospheres have been identified, and some have been related to soil function (Table 13.3) (van Cleemput et al. 1983; Wheatley et al. 1996; see also Chap. 8 by Splivallo).

Table 13.1 Volatile organic compounds (VOCs) identified in the headspace samples of *Trichoderma* spp. (percentage of total output); on different media. (Reprinted from Wheatley et al. 1997)

		Trichoderma viride						*Trichoderma pseudokoningii*					
		Low medium			Malt medium			Low medium			Malt medium		
		A			B			C			D		
		Replicate			Replicate			Replicate			Replicate		
No.	Compound	1	2	3	1	2	3	1	2	3	1	2	3
1	Ethanol	40.7	29.9	30.7	24.9	29.8	23.1	38.3	27.5	33.6	58.3	57.1	43.4
2	2-Methylpentane	0.0	0.0	0.0	0.0	0.0	0.0	0.0	0.0	0.0	4.0	4.4	4.4
3	Propanal	0.0	0.6	0.0	1.3	1.5	1.1	0.0	0.0	0.0	0.0	0.0	0.0
4	2-Propanone	10.6	10.5	5.9	22.2	24.9	25.7	7.5	4.5	5.0	27.3	26.1	37.3
5	2-Propanol	0.0	0.0	0.0	0.0	8.9	0.0	0.0	0.0	0.0	0.0	0.0	0.0
6	4-Penten-2-ol	0.0	2.9	0.0	5.0	0.0	0.0	0.0	0.0	0.0	0.0	0.0	0.0
7	Hexane	0.0	7.1	0.0	25.5	15.0	15.7	0.0	1.2	1.1	2.5	4.0	3.2
8	5-Methyl-5-hexem-3-ol	1.0	0.0	1.3	0.0	0.0	0.0	0.0	0.0	0.0	0.0	0.0	0.0
9	Acetonitrile	4.2	4.7	2.8	8.2	10.7	11.1	0.3	0.4	0.7	2.5	3.0	3.2
10	3-Methyl-propanol	0.0	0.0	0.0	0.0	0.0	0.0	0.0	0.7	2.4	0.0	0.0	0.0
11	1-Propanol	3.1	3.3	2.2	0.0	0.0	0.0	21.7	24.7	17.4	0.0	0.0	0.0
12	Ethyl ester acetic acid	0.0	0.0	0.0	0.0	0.0	0.0	1.2	0.7	0.0	0.0	0.0	0.0
13	2,4,6-Trimethyl1 nonene	0.5	0.7	1.4	0.0	2.9	3.0	0.0	0.0	0.0	0.0	0.0	0.0
14	3-Methylhexane	0.0	0.0	0.0	1.9	0.0	0.0	0.0	0.0	0.0	0.0	0.0	0.0
15	Butanal	0.0	0.0	0.0	1.4	0.0	1.4	0.0	0.5	0.0	0.0	0.0	0.0
16	2-Butanone	4.2	2.6	4.4	1.0	0.0	1.4	1.5	1.2	3.9	0.0	0.8	0.0
17	Heptane	3.2	1.6	1.6	5.5	3.3	9.1	0.6	0.0	0.1	0.0	0.0	0.0
18	2-Methyl-l-propanol	29.2	30.1	38.7	0.0	0.0	0.0	15.3	16.6	18.0	5.2	4.6	8.4
19	Methylcyclohexane	2.3	1.3	1.4	3.1	2.9	7.5	0.4	0.0	0.0	0.0	0.0	0.0
20	1-Butanol	1.1	4.8	9.6	0.0	0.0	0.0	13.3	22.0	16.8	0.0	0.0	0.0
21	Octane	0.0	0.8	0.0	2.1	0.0	0.0	0.0	0.0	0.0	0.0	0.0	0.0
22	2-Propenylidene cyclobutene	0.0	0.0	0.0	0.0	0.0	0.0	0.0	0.0	0.0	3.7	6.1	4.6

(continued)

Table 13.1 (continued)

		Trichoderma viride						*Trichoderma pseudokoningii*					
		Low medium			Malt medium			Low medium			Malt medium		
		A			B			C			D		
		Replicate			Replicate			Replicate			Replicate		
No.	Compound	1	2	3	1	2	3	1	2	3	1	2	3
23	3-Methyl-1-butanol	11.9	6.5	7.3	0.0	0.0	0.0	43.8	30.8	38.6	6.2	6.1	3.6
24	2-Methyl-1 butanol	3.6	1.9	2.0	9.4	17.2	11.5	4.4	6.2	5.8	13.4	10.9	9.6
25	1-Pentanol	0.0	0.0	0.0	0.0	0.0	0.0	0.0	2.9	2.4	0.0	0.0	0.0
26	Hexanal	0.9	0.0	0.0	5.5	5.7	7.0	0.0	0.0	1.5	0.0	0.0	0.0
27	*p*-Xylene	0.0	0.8	0.0	0.0	0.0	0.0	0.0	0.0	0.0	0.0	3.8	2.5
28	1-Hexanol	0.0	0.0	0.0	0.0	0.0	0.0	0.0	3.7	3.5	0.0	0.0	0.0
29	2-Heptanone	0.7	0.0	0.0	0.0	0.0	0.0	0.0	0.0	0.0	0.0	0.0	0.0
30	Heptanal	0.8	0.0	0.0	3.6	4.5	4.7	0.0	0.0	1.4	0.0	4.5	3.7
31	2,2,4,6,6-Pentamethyl-3-heptene	0.0	0.0	0.0	0.0	0.0	0.0	20.7	25.7	17.7	37.5	19.3	38.1
32	Limonene	0.0	0.0	0.0	0.0	0.0	0.0	4.7	2.3	4.5	0.0	0.0	0.0
33	Formic acid heptyl ester	0.0	0.0	0.0	0.0	0.0	0.0	0.0	3.4	3.5	0.0	0.0	0.0
34	2-Octen-l-ol	0.0	0.0	0.0	0.0	0.0	0.0	0.0	2.5	0.0	0.0	0.0	0.0
35	6-Methyl-5-hepten-2-one	0.0	0.0	1.2	2.1	2.5	2.4	0.0	0.0	0.0	0.0	0.0	0.0
36	2-Octanone	0.8	0.0	0.0	0.0	0.0	0.0	1.9	0.0	0.0	0.0	0.0	0.0
37	Octanal	3.1	2.4	3.1	10.9	10.1	11.8	3.0	4.4	4.6	10.0	14.1	12.1
38	Benzaldehyde	0.8	0.0	0.0	2.4	3.0	2.9	0.0	0.0	0.0	0.0	0.0	0.0
39	2-Propyl-l-pentanol	64.7	78.7	72.4	15.6	16.4	13.4	11.6	10.4	8.0	18.7	22.8	12.7
40	Nonanal	6.4	3.7	5.1	18.4	19.1	20.0	0.0	0.0	0.0	0.0	0.0	0.0
41	Acetic acid, 2-ethyl ester	0.0	1.7	0.8	0.0	0.0	0.0	0.0	0.0	0.0	2.2	0.0	0.0
42	Decanal	5.0	2.4	4.5	13.5	12.0	13.2	9.7	7.6	8.4	8.2	12.5	13.1
43	Banzothiazole	1.4	1.1	1.8	0.0	4.2	0.0	0.0	0.0	0.0	0.0	0.0	0.0
44	Caryophyllene	0.0	0.0	0.0	16.4	5.2	13.1	0.0	0.0	0.0	0.0	0.0	0.0
45	Pentadecane	0.0	0.0	1.8	0.0	0.0	0.0	0.0	0.0	0.0	0.0	0.0	0.0

Table 13.2 VOCs produced by *Trichoderma aureoviride* when grown on low-nutrient media containing phenylalanine (*LNB-B*), arginine (*LNB-C*), glutamine (*LNB-D*) or all three amino acids (*LNB-A*). (Reprinted from Bruce et al. 2000)

VOC	LNM-A	LNM-B	LNM-C	LNM-D
Acetaldehyde	***		*	**
Ethanol	***	***	*	**
Propanal	**		*	
Acetone	***	*	*	
4-Methyl-3-penenoic acid			*	
Isobutane		*		**
5-Methyl-5-hexen-3-ol	***			**
1-Propanol		***		***
Heptane	***	***		
Chloroform	**		***	
1,3-Hexandien-5-yne		**	*	
2-Methyl-1-propanol	***			
2,4-Dimethyl heptane		**		
Heptanone	***	***	***	***
Nonane		**		
3-Methyl-2-hexanol	**			*
Heptanal		***	**	
7-Octen-4-ol	***			
Octanone		***	***	*
2-Octanone	***	***		*
Undecanone			**	
2,4-Dimethyl-1-decane				**
Decane		***		***
2-Propyl-1-pentanol		**		
2-Ethyl-4-methyl-1-pentanol	**	*	*	
2-Nonanone				**
Nonanone	***	***	***	
Nonanal		**	***	**
Benzyl alcohol		**		
Decanal		*	***	

* one replicate, ** two replicates, *** three replicates

Table 13.3 VOCs detected in the headspace of aerobically and anaerobically incubated soils. (Reprinted from Wheatley et al. 1996)

Alcohols	Ketones
Ethanol	Propane-2-one
Propan-1-ol	Butan-2-one
Butan-1-ol	Pentan-2-one
Butan-2-ol	Pentan-3-one
2-Methylpropan-1-ol	4-Methylpentan-2-one
2-Methylbutan-1-ol	5-Methylheptan-2-one
	3-Hydroxybutan-2-one

(continued)

Table 13.3 (continued)

Alcohols	Ketones
Aldehydes	
2-Methylbutan-1-al	Aromatics
3-Methylbutan-1-al	Benzene
	Ethyl benzene
Sulphides	Dimethyl benzene
Dimethyl sulphide	Trimethyl benzene
Dimethyl disulphide	Benzaldehyde
Dimethyl trisulphide	
2-Methylpropyl sulphide	Ethyl esters
	Acetic acid
Methyl esters	Butanoic acid
2-Methyl butanoic acid	2-Methylpropanoic acid
3-Methyl-butanoic acid	2-Methylbutanoic acid
	3-Methylbutanoic acid
Butyl esters	
Acetic acid	

13.3 Organisms

13.3.1 Organisms/Groups Involved in Interactions and Possible Biocontrol

In a series of interaction experiments, in which the only connection between the organisms was atmospheric, all of a range of soil bacterial isolates either significantly stimulated or inhibited the growth rate of at least one of four fungi. The fungi had been selected to represent a range of habitats, and were *Trichoderma viride*, a common soil saprophyte; *Phanaerochaete magnoliae*, a pathogen of beech trees; *Phytophthora cryptogea* a plant pathogen with a wide host range; and *Gaeumannomyces graminis* var. *tritici*, a specific pathogen of wheat. Fungal growth rates were inhibited, by up to 60%, in some couplets, and stimulated, by up to 35% ($P < 0.05$), in others (Fig. 13.1). No bacterial isolate was effective against all of the fungi, but the majority, 54%, inhibited the growth rate of some fungi and also stimulated others. A large proportion, 42%, only inhibited growth, but no individuals were solely stimulatory ($P < 0.05$) (Mackie and Wheatley 1999). Many of the fungi that had been inhibited only resumed growth after transfer onto fresh medium. Also cores from the growing margins of control cultures did not grow ($P < 0.05$) if transferred to medium that had previously been exposed to the bacterial cultures. This series of interactions showed that responses between the bacteria and fungi were species-specific, with each fungus responding uniquely to the products of each of the bacterial cultures. These products must have been volatile to diffuse through the atmospheric connection between the cultures, and then dissolved into the growth media.

When a random selection of 250 bacterial isolates from a range of soil associations were screened for suitability as biological control agents, on the basis of their action through the VOCs produced, against 14 biotypes of a pathogen of sports turf,

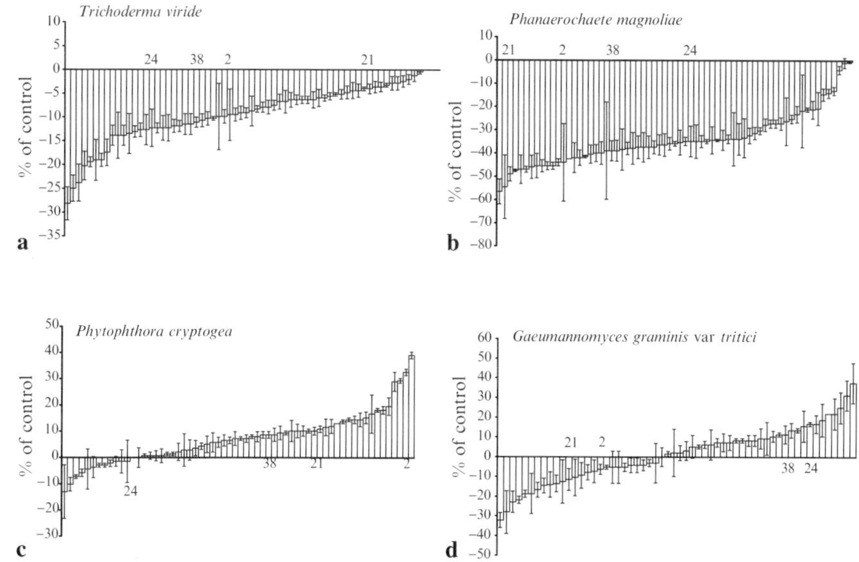

Fig. 13.1 The influence of volatile organic compounds (VOCs) produced by randomly selected soil bacterial isolates on mean radial growth of **a** *Trichoderma viride*, **b** *Phanaerochaete magnoliae*, **c** *Phytophthora cryptogea* and **d** *Gaeumannomyces graminis* var. *tritici*, expressed as the percentage of the mean growth rate of unexposed cultures. (Reprinted from Mackie and Wheatley 1999)

Microdochium nivale, a range of interactions, from inhibition to stimulation of the pathogen, of up to 60 and 40%, respectively, was observed (Fig. 13.2). Two bacterial isolates, one identified as *Citrobacter freundii*, and the other a strain of *Pseudomonas fluorescens*, caused the greatest inhibition. The activity of these bacterial isolates was not related to the soil type from which they were isolated or the cultivation techniques applied to the soils (Wheatley, unpublished results).

All of 21 strains of soil bacteria isolated from oil seed rape roots suppressed the pathogen *Verticillium dahliae* in both direct and indirect ways (Alstrom 2001). Nine of these were *Enterobacteriaceae*, one was identified as *Serratia proteamaculan*, and three were pseudomonads: *Pseudomonas putida, Pseudomonas acidovorans* and *Pseudomonas chlororaphis. Stenotrophomonas* sp. and *Alcaligenes* sp. were also identified. Some of the bacteria prevented symptom development in field rape plants. In a study of interactions between mycorrhizal fungi and soil organisms, Fitter and Garbaye (1994) reported that bacteria have an important role in promoting mycorrhizal formation. Similarly Azcon-Aguiler et al. (1986) state that both germination and hyphal growth of the arbuscular mycorrhizal fungus *Glomus mosseae* were improved when rhizosphere bacteria were present, and postulated that the organic products of soil bacteria may be responsible for these interactions (Azcon-Aguiler and Barea 1985).

VOCs produced by three bacteria, one unknown and two *Serratia* spp. and one yeast, *Saccharomyces cerevisiae*, that had previously been shown to be inhibitory to the sapstain fungi *Ophiostoma piliferum, Ophiostoma piceae, Sclerophoma*

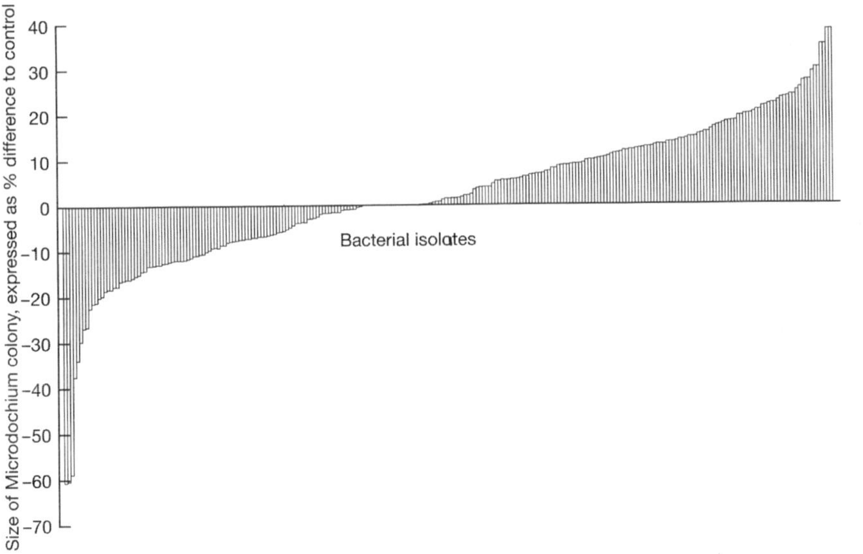

Fig. 13.2 Effects of 250 bacterial isolates on the growth of a biotype of *Microdochium nivale*, ranked according to effect

pithyophila, Aurebasidium pullulans and *Botrydiplodium theobromae* were identified (Bruce et al. 2004). Each sapstain fungus was inhibited by the VOCs produced from at least one of the antagonists. Couplet responses varied greatly, with both the microbial components involved and with the media on which they were growing. Both *Serratia* spp. and one of the unknown bacterial isolates completely inhibited the sapstain fungi when grown on tryptone soya agar (TSA). The other unknown bacteria inhibited *Sclerophoma pithyophila* by 90%, but had no significant effect on *Aurebasidium pullulans* and the two *Ophiostoma* spp. Inhibition caused by the yeasts was very varied, ranging between 20 and 100%, dependent on the target. Significant growth inhibition was only occasionally seen when the antagonists were grown on either minimal or malt extract agars.

The phenomenon that levels of inhibition were dependent on the substrate was exploited, using principal component biplot analyses, to nominate several candidate compounds that could be responsible for the inhibition. These included dimethyl disulphide, dimethyl trisulphide, 2-nonanone, 2 undecanone and undec-2-enylester acetic acid. Not all were produced by all of the isolates on all media. Contrastingly, Bruce et al. (2003) reported that when *Ophiostoma piliferum* was exposed to VOCs produced by *Saccharomyces cerevisiae*, grown on TSA, pigmentation was increased, with a changed morphology.

In a possibly more realistic input scenario, specific *Trichoderma* spp. strains were capable of the biocontrol of a range of wood-rot basidiomycetes when grown on a low-nutrient medium (Srinivasan et al. 1992). This medium was designed to be representative of fresh softwood, with a C/N ratio of 410:1 and an amino acid and glucose composition analogous to that found in the sap of growing coniferous trees.

Somewhat differently, Howell et al. (1988) reported that volatile products of the bacterium *Enterobacter cloacae* inhibited the growth of *Pythium ultimum*, responsible for the damping off of seedlings, and *Rhizoctonia solani*. Analysis of the volatiles identified ammonia as the candidate, and when ammonia production was suppressed, by adding sugars to the growth medium, biocontrol activity ceased.

Bacteria isolated from canola and soybean plants produced VOCs that inhibited sclerotia and ascospore germination and mycelial growth of *Sclerotinia sclerotiorum* (Fernando et al. 2005). Six inhibitory compounds were identified: benzothiazole, cyclohexanol, *n*-decanal, dimethyl trisulphide, 2-thyl-1-hexanol and nonanal. Ten of the 14 bacterial strains, 12 isolated from soil, were identified as three species of pseudomonad.

Payne et al. (2000) reported that the VOCs produced by the yeast *Debaryomyces hansenii* reduced staining of wood by blue stain fungi. In some cases fungal mycelial growth was coincidentally unaffected, but was reduced by two isolates. Similarly five bacterial isolates also effectively inhibited these stain fungi, and in both cases VOCs alone were responsible. They suggested that these yeasts and bacteria might be used as biocontrol agents at sawmills to prevent wood spoilage.

So both positive, i.e. stimulation, and negative, i.e. inhibition, interactions occur. These can involve pathogenicity, enzyme production, growth rates and form, etc., and are dependent on both the microbial couplets, as individuals or groups, and the environmental inputs involved.

The effects resulting from these interactions are usually transient, in that they are frequently reversible. So when the interactive scenario relating to that couplet is changed, such as substrate depletion, addition of a different substrate or removal of one of the protagonists, the inhibition or stimulation generally ceases.

13.3.2 Input-Specific VOC Production

Adjustments to growth conditions, such as relatively minor differences in inputs, result in changes in both the types and the amounts of VOCs produced. Providing one of three different amino-acids, L-phenylalanine or L-arginine or glutamine, in the growth medium but maintaining the same C/N ratio (280:1) and other cultural conditions resulted in significant, reproducible changes in VOC output by *Trichoderma* spp. The responses of the basidiomycete target cultures, *Neolentinus lepidus, Gloeophyllum trabeum* and *Coriolus versicolor* varied with both the *Trichoderma* spp. being used as the antagonist and the growth medium. Growth of the target was inhibited by between 20 and 60%, depending on the microbial couplet, both antagonist and target, and the amino acid in the antagonist's medium. (Bruce et al. 2000). The most suppressive VOCs were produced when cultures of *Trichoderma* spp. antagonists were grown on the L-arginine inclusive medium, and the least when grown on the L-phenylalanine medium. Principal component analyses

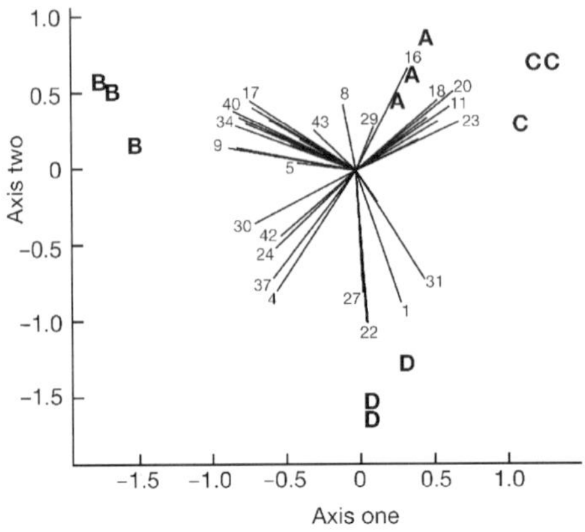

Fig. 13.3 Principal component analysis biplot separating VOCs produced by *Trichoderma viride* and *Trichoderma pseudokoningii* grown on malt extract and minimal media: replicates of *Trichoderma viride* growing on minimal media (*A*) and on malt extract (*B*); replicates of *Trichoderma pseudokoningii* growing on minimal media (*C*) and on malt extract (*D*). VOCs not associated with *A, B, C* or *D* are considered 'candidate' infochemicals, i.e. numbers *4, 24, 30, 37* and *42*. (Reprinted from Wheatley et al. 1997)

suggested that both aldehyde and ketone volatile products were associated with the greatest inhibition of these basidiomycetes (Fig. 13.3). The suite of VOCs produced by the *Trichoderma* spp. isolates differed between each amino acid, but was consistent for each individual amino acid medium (Table 13.2). Using a combination of all three amino acids together in the same substrate produced a different catalogue of VOCs from that of the sum of each when used individually.

When *Trichoderma pseudokoningii* and *Trichoderma viride* were cultured on more complex media similar differences in VOC outputs were also reported (Wheatley et al. 1997; Table 13.1). Biplot analyses of the VOCs produced by each of the isolates, on the different media, demonstrated a species-specific consistency in the VOCs produced and also suggested five possible 'candidate' chemicals, 2-propanone, 2-methyl-1-butanol, heptanal, octanal and decanal (numbers 4, 24, 30, 37 and 42 in Fig. 13.3), that might affect the growth rate of the basidiomycetes (Wheatley et al. 1997). Similar phenomena have been reported for the production of non-volatile antimicrobial compounds by *Pseudomonas fluorescens*. Rates of production of the antimicrobials 2,4-diacetylphloroglucinol and pyoluteorin were shown to be input-dependent, both being affected by the presence or absence of Zn^{2+} and glucose in the growth medium.

13.3.3 VOC Candidate Infochemicals

The effects of four VOCs, 2-propanone, 2-methyl-1-butanol, heptanal and octanal, that had been suggested as possible interactive compounds produced by *Trichoderma* spp. (Wheatley et al. 1997) were studied, over a range of concentrations, on the growth and respiration rates of *Neolentinus lepidus, Gloeophyllum trabeum, Postia placenta* and *Trametes versicolor* (Humphris et al. 2001). Growth of all four fungi was affected by at least one of the compounds. In most cases fungal growth was inhibited, but on occasions it was stimulated. There were significant effects on biomass development in 14 of the 16 fungal/chemical combinations, and on respiration rates in 15 of the combinations ($P < 0.05$). All four decay fungi were inhibited, to some degree, in atmospheric concentrations of 25 $\mu g\ ml^{-1}$ of both heptanal and octanal, and were totally inhibited at a concentration of 250 $\mu g\ ml^{-1}$. None were inhibited in atmospheres containing 2-propanone, but some were stimulated, and 2-methyl-1-butanol was only effective at the highest concentration of 2,500 $\mu g\ ml^{-1}$ (Humphris et al. 2000, 2001).

13.4 Mechanisms

13.4.1 Interaction Mechanisms

Many interactive mechanisms have been reported. For example, bacterial VOCs have been shown to affect both fungal mycelial growth and enzyme activity. When exposed to VOCs produced by a series of bacterial isolates, laccase activity in *Phanaerochaete magnoliae* ceased completely, and was significantly reduced in *Trichoderma viride*. Somewhat differently, tyrosinase activity in *Phanaerochaete magnoliae* increased, decreased or remained the same, depending on the bacterium to which it was exposed. However tyrosinase activity in *Trichoderma viride* was not affected by any of the bacterial isolates used (Mackie and Wheatley 1999).

Upregulation and downregulation of gene expression by messenger RNA and changes in protein synthesis have also been reported. Biotypes of *Serpula lacrymans* showed a change in its patterns of protein synthesis when exposed to the VOCs produced by *Trichoderma aureoviride* and *Trichoderma viride*, but not when exposed to those produced by *Trichoderma pseudokoningii* (Fig. 13.4) (Humphris et al. 2002). Rates of mycelial growth were affected in a similar way. VOCs produced by *Trichoderma aureoviride* completely inhibited the synthesis of two proteins, at 22.4 and 29.1 kDa (Fig. 13.4, lanes 11–13) in *Serpula lacrymans* (H28) and reduced mycelial growth by over 70%. Both synthesis of these proteins and mycelial growth resumed when the antagonist was removed. There were interesting differences between the responses of the biotypes to the same antagonist and also to the different antagonists. For example, VOCs produced by *Trichoderma viride* completely inhibited the synthesis of the 32.2-kDa protein by *Serpula lacrymans* (H28) but had no effect on the synthesis of this protein by *Serpula lacrymans* (Forfar). Consideration of these idiosyncrasies, and the similarity in differences of effects on mycelial growth (Table 13.4), highlights the specificity and detail involved in such VOC-mediated interactions. Not only do these results show that VOC-mediated microbial interactions occur at a functional level, but they also

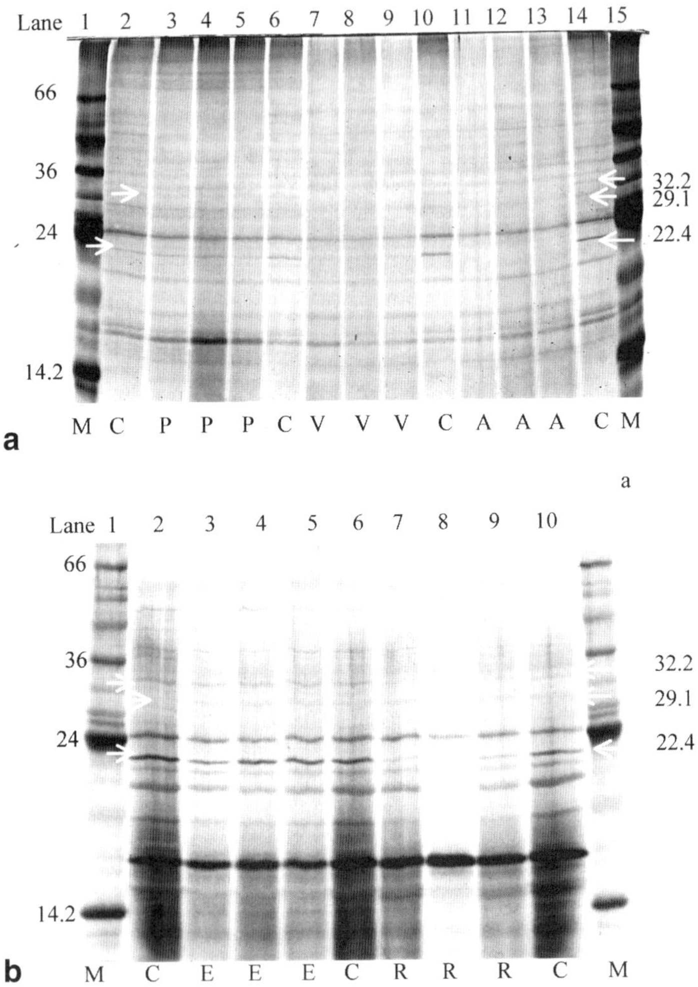

Fig. 13.4 **a** Protein profile of *Serpula lacrymans*. *Lanes 2, 6, 10* and *14* are from a 7-day culture with no antagonist; *lanes 3–5* are from a 7-day culture grown in the presence of *Trichoderma pseudokoningii; lanes 7–9* are from a 7-day culture grown in the presence of *Trichoderma viride; lanes 11–13* are from a 7-day culture grown in the presence of *Trichoderma aureoviride. Lanes 1* and *15* are molecular mass markers (kilodaltons). **b** Protein profile of *Serpula lacrymans* after removal of the antagonist *Trichoderma viride; lanes 3–5*, new growing edge of the culture; *lanes 7–9*, the previously exposed mycelium; *lanes 2, 6* and *10*, 12-day control culture with no antagonist. *E* newly growing mycelium, *R* mycelium recovering from exposure to the VOCs. (Reprinted from Humphris et al. 2002)

show that such interactions are also indicative of the vast number of possible combinations, and consequences of, such phenomena.

Reductions in mycelial growth rates are frequently seen. The factors causing this, and changes in the growth form such as hyphal branching patterns, are not

Table 13.4 Inhibition of growth (percent) of *Serpula lacrymans* (H28) and *S. lacrymans* (Forfar) by VOCs produced by *T. aureoviride, T. viride* and *T. pseudokoningii*. (Reprinted from Humphris et al. 2002)

Target	*S. lacrymans* (Forfar)	*S. lacrymans* (H28)
T. aureoviride	91.56 ± 1.6	74.12 ± 3.69
T. viride	54.13 ± 10.44	30.17 ± 7.6
T. pseudokoningii	8.18 ± 1.46	1.46 ± 0.18

known. Other obvious changes in the target's physiological expression include a reduction in melanin production, and so staining in the wood, by sapstain fungi such as *Ophiostoma piliferum* and *Sclerophoma pithyophila* (Bruce et al. 2003). In some cases, depending on the antagonist and inputs used, the fungus may still grow through the wood, but staining does not occur. Conversely, with some couplings melanin production and hence staining are increased.

13.4.2 Communication

The ability to be effective over a wide range of scales makes VOCs ideal infochemicals. The influence of such volatile microbial products will range from proximal interactions, possibly due to aqueous diffusion, to much greater distances via 'atmospheric' diffusion through the soil pores, even into the open troposphere, so relaying activity on the rhizoplane to the bulk soil. These compounds will also be moved rapidly around the soil by the diurnal patterns, and mass flow, of water movement.

Communication between microorganisms will frequently be of advantage to at least some of those involved. Opportunist organisms will gain great competitive advantage by evolving to a state where they can simply switch on in response to a signal from others in the microbial community that substrate is being produced, rather that having to continually drain their energy resources by maintaining a constant state of readiness. Rapid responses to the production of a suitable substrate will not only result in rapid acquisition and use, but will also prevent competitors from using these substrates and so occupying any desired environmental niches. The substrate-dependent variation in VOC production will result in variations in microbial dynamics, and consequently system response. The effectiveness of more active exploratory organisms such as pathogenic fungi will be increased, as the organism will be able to follow a chemical gradient to a potential host rather than simply randomly spreading, in opportunistic hope, through the bulk soil.

Others may gain competitive advantage when their VOCs affect the functioning of other competitor organisms, causing them to function at a slight disadvantage. The degree of such inhibition, of growth or enzyme production, required to cause a disadvantage is not known, but could presumably be quite small in the competitive, and frequently substrate limited, soil environment.

13.5 Conclusions

The hypothesis that microbial volatile products are responsible for a wide range of functional interactions among large numbers of microorganisms in soil ecosystems is well supported by many reports. Hora and Baker (1972) reported that fungal spore germination was inhibited by volatiles produced from sterilised soils after recolonisation with actinomycetes, but not when they were uninoculated. Contrastingly Griebel and Owens (1972) reported stimulation of microbial respiration and growth in soil by low molecular weight alcohols and aldehydes released from undecomposed plant residues, although in further experimentation these compounds were used as substrates. Fungistasis by soil microbial products has been reported several times (Griffin 1962; Lockwood and Lingappa 1963), and the efficacy of this phenomenon has been reduced by the further addition of compounds such as vancomycin or chloramphenicol, or by reducing microbial numbers (Epstein and Lockwood 1984).

Indeed as there are reports of interactions between all members of the microbial community under some of the possible combinations, of inputs or organisms, it is probable that all microorganisms can, and at some time do, have an effect on other individuals or functional groups within the population. Such interactions may have significant positive or negative outcomes, or be mutual, and take many forms. These include changes in growth rates which may be increased or decreased, and changes in shape, depending on the circumstances. Similarly enzyme activity levels may be modified and the organism's physiological functioning changed, for example in protein synthesis and melanin production. These influences may occur at a fundamental level, where gene expression may be either upregulated or downregulated, rather than being the result of toxic action.

VOC production and the functional interactions that they mediate are also species-specific, consistent and responsive to environment and inputs, all essential requirements for a signalling system. These interactions are subtle, not necessarily fatal nor necessarily inhibitory and frequently reversible. The resulting functional responses of the organisms involved may give selective advantage to some community members, by stimulation or enabled awareness. Other members of the microbial community may be actively disadvantaged, by growth inhibition or physiological disturbance, or simply coincidentally disadvantaged as a consequence of other stimulation to greater levels of activity. The extent of change required, sometimes a growth inhibition or stimulation of up to 60% has been reported, to cause an individual significant advantage or disadvantage is not known. But in the competitive habitat of the soil ecosystem possibly only a very small change, of a few percent, is all that is required.

So it is possible that microbially produced VOCs have played a significant role during the evolution of microorganisms, in the context of their interactions, and community, population and functional dynamics. Heilman-Clausen and Boddy (2005) support the concept that the compounds responsible for these interactions should be viewed as functional infochemicals with fundamental importance for the spatial organisation and activity in microbial ecosystems (Wheatley 2002).

Examples that support this include the interactions between individual target fungi and soil bacteria. In experimental culture designed to mimic bacterial activity, and the consequent VOC production, in the rhizosphere, growth rates of soil-inhabiting fungal plant pathogens such as *Phytophthora cryptogea* and *Gaeumannomyces graminis* were stimulated. So in a soil system these pathogens will be advantaged, as they will only grow when they have received a VOC 'signal' from the bacteria growing on the root, and so will exploit the opportunities presented by the presence of a host plant only when investment in growth is liable to be profitable. Conversely, in a significant number of interactions soil bacteria significantly inhibit the development of potential pathogens. Although it may appear that it is the plant that benefits from such an association, the bacteria will also benefit as the plant remains vigorous and grows more roots. Similarly there are also reports of interactions between saprophytes. In soil the fungus *Trichoderma viride* and soil bacteria will be in direct competition. In soils resource scarcity is a severe limiting factor, so reducing the effectiveness of any potential competitors to reach any resources will give a distinct advantage. The effectiveness of a range of soil bacteria in effectively limiting the growth of this fungus has been reported (Mackie and Wheatley 1999). Although many of the examples of functional group and species interactions that have been described previously here only involve two partners in an interactive couplet, in soils the situation will be much more complex. Multiple microbial interactions will occur, with the dominant participant changing as circumstances change. These could be changes in the amounts and types of substrate available, which will of course change as the microorganisms exploit them. Changes in physical parameters such as moisture content, and hence aeration, temperature and pH, will also alter the balance between the antagonists.

The rapid changes in functional dynamics, such as nitrification and denitrification, reported in many soil systems could be a result, at least in part, of such interactive population dynamics. In such events population size does not change rapidly; indeed many of the changes in functional rates are much too fast to be attributable to population growth, but are rather due to the 'switching on' of the indigenous population, from a sessile to an active state. Such reactions will be very much an advantage for microbial populations that have to survive for long periods in a nutrient-deficient habitat, but that have to react quickly to any spasmodic improvement, if they are to compete successfully.

The volatile products of soil microorganisms are intrinsically involved in the regulation of functional interactions. These VOCs are secondary metabolites, and although they are not specifically produced for the purpose, the resulting effects on the metabolisms of the participants have a very definitive effect on the functional development of soil systems. In such associations some will benefit by using competition and detection to their advantage, and consequently others will be disadvantaged.

So it is plausible that soil microorganisms can be manipulated towards a greater suppression of soil-borne diseases as a result of suitable soil management regimes. The incorporation of various amendments into the soil could result in disease-suppressive soils, in which crop losses will be less than might otherwise have been expected.

References

Alstrom S (2001) Characteristics of bacteria from oilseed rape in relation to their biocontrol of activity against *Verticillium dahliae*. J Phytopathol 149:57–64

Azcon-Aguiler C, Barea J-M (1985) Effect of soil micro-organisms on formation of vesicular-arbuscular mycorrhizas. Trans Br Mycol Soc 84:536–537

Azcon-Aguiler C, Diaz-Rodriguez CM, Barea J-M (1986) Effect of soil micro-organisms on spore germination and growth of the vesicular-arbuscular mycorrhizal fungus *Glomus mossae*. Trans Br Mycol Soci 86:337–340

Bruce A, Wheatley RE, Humphris SN, Hackett CA, Florence M (2000) Production of volatile organic compounds by *Trichoderma* spp. in media containing different amino acids and their effect on selected wood decay fungi. Hölzforschung 54:481–486

Bruce A, Stewart D, Verrall S, Wheatley RE (2003) Effect of volatiles from bacteria and yeast on the growth and pigmentation of sap-stain fungi. Int Biodeterior Biodegradation 51:101–108

Bruce A, Verrall S, Hackett CA, Wheatley RE (2004) Identification of volatile organic compounds (VOCs) from bacteria and yeast causing growth inhibition of sapstain fungi. Hölzforschung 58:193–198

Curtis TP, Sloan WT, Scannell JW (2002) Estimating prokaryotic diversity and its limits. Proc Natl Acad Sci USA 99:10494–10499

De Boer C, Meulman PA, Wnuk RJ, Peterson DH (1970) Geldanamycin, a new antibiotic. J Antibiot 23:442–447

Defago G, Berling C-H, Burger U, Haas RJ, Kahr G, Keel C, Voisard C, Wirthner P (1990) Suppression of black root rot of tobacco and other diseases by strain *Pseudomonas fluorescens*: potential applications and mechanisms. In: Hornby D, Cook RJ, Henis Y, KO WH, Rovira AD, Schippers B, Scott PR (eds) Biological control of soil-borne plant pathogens. CAB International, Wallingford, pp 93–108

Duffy BK, Defago G (1999) Environmental factors modulating antibiotic and siderophore biosynthesis by *Pseudomonas fluorescens* biocontrol strains. Appl Environ Microbiol 65:2429–2438

Epstein L, Lockwood JL (1984) Effect of soil microbiota on germination of *Bipolaris victoriae* conidia. Trans Br Mycol Soc 82:63–69

Fernando WGD, Ramarathnam R, Krishnamoorthy AS, Savchuk SC (2005) Identification and use of potential bacterial organic antifungal volatiles in biocontrol. Soil Biol Biochem 37:955–964

Fiddaman DJ, Rossall S (1993) The production of antifungal volatiles by *Bacillus subtilis*. J Appl Bacteriol 74:119–126

Fiddaman DJ, Rossall S (1994) Effect of substrate on the production of antifungal volatiles from *Bacillus subtilis*. J Appl Bacteriol 76:396–405

Fitter AH, Garbaye J (1994) Interactions between mycorrhizal fungi and other soil organisms. Plant Soil 159:123–132

Girlanda M, Perotto S, Moenne-Loccoz Y, Bergero R, Lazzari A, Defago G, Bonfante P, Luppi AM (2001) Impact of biocontrol *Pseudomonas fluorescens* CHA0 and a genetically modified derivative on the diversity of culturable fungi in the cucumber rhizosphere. Appl Environ Microbiol 67:1851–1864

Giudici P, Romano P, Zambonelli C (1990) A biometric study of higher alcohol production in *Saccharomyces cerevisiae*. Can J Microbiol 36:61–64

Griebel GE, Owens LD (1972) Nature of the transient activation of soil microorganisms by ethanol or acetaldehyde. Soil Biol Biochem 4:1–8

Griffin GJ (1962) Production of a fungistatic effect by soil microflora in autoclaved soil. Phytopathology 52:90–91

Heilman-Clausen J, Boddy L (2005) Inhibition and stimulation effects in communities of wood decay fungi: exudates from colonized wood influence growth by other species. Microb Ecol 49:399–406

Hora TS, Baker R (1972) Soil fungistasis: microflora producing a volatile inhibitor. Trans Br Mycol Soc 59:491–500

Howell CR, Beier RC, Stipanovic RD (1988) Production of ammonia by *Enterobacter cloacae* and its possible role in the biological control of Pythium pre-emergence damping-off by the bacterium. Phytopathology 78:1075–1078

Humphris SN, Bruce A, Wheatley RE (2000) Volatile inhibition of *Serpula lacrymans* by *Trichoderma* spp. In: Proceedings 31st annual meeting of the international research group on wood preservation, Hawaii, USA

Humphris N, Wheatley RE, Bruce A (2001) The effects of specific volatile organic compounds produced by *Trichoderma* spp. on the growth of wood decay fungi. Hölzforschung 55:233–237

Humphris SN, Bruce A, Buultjens E, Wheatley RE (2002) The effects of volatile microbial secondary metabolites on protein synthesis in *Serpula lacrymans*. FEMS Microbiol Lett 210:215–219

Lockwood JL, Lingappa BT (1963) Fungistatic activity of autoclaved soil inoculated with soil micro-organisms. Phytopathology 53:917–920

Mackie A, Wheatley RE (1999) Effects and incidence of volatile organic compound interactions between soil bacterial and fungal isolates. Soil Biol Biochem 31:375–385

McCaig AE, Glover EA, Prosser JI (1999) Molecular analysis of bacterial community structure in unimproved and improved grass pastures. Appl Environ Microbiol 65:1721–1730

Mitchell SM, Davidson J, Wheatley RE, McNichol J, Daniell TJ, (2004) Relationship between soil nitrification activity and population dynamics. In: Proceedings of 10th international symposium on microbial ecology, Cancun, Mexico, August 2004, p 307

Payne C, Bruce A, Staines HY (2000) Yeast and bacteria as biological control agents against fungal discolouration of *Pinus sylvestris* blocks in laboratory-based tests and the role of antifungal volatiles. Holzforschung 54:563–569

Rothrock CS, Gottleib D (1984) Role of antibiosis in antagonism of *Streptomyces hygroscopicus* var. *geldanus* to *Rhizoctonia solani* in soil. Can J Microbiol 30:1440–1477

Shaukat SS, Siddiqui IA (2003) Impact of biocontrol agents *Pseudomonas fluorescens* CHA0 and its genetically modified derivatives on the diversity of culturable fungi in the rhizosphere of mungbean. J Appl Microbiol 95:1039–1048

Srinivasan U, Staines HJ, Bruce A (1992) Influence of media type on antagonistic modes of *Trichoderma* spp. against wood decay basidiomycetes. Mater Org 27:301–321

Stahl PD, Parkin TB (1999) Microbial production of volatile organic compounds in soil microcosms. Soil Sci Soc Am J 60:821–828

Stotzky G, Schenck S (1976) Volatile organic compounds and microorganisms CRC Crit Rev 4:333–381

Torsvik VL, Goksoyer J, Daae FL (1990) High diversity in DNA of soil bacteria. Appl Environ Microbiol 56:782–787

Toussaint V, Valois D, Dodier M, Faucher E, Dery C, Brzezinski R, Ruest L, Beaulieu C (1997) Characterisation of actinomycetes antagonistic to *Phytophthora fragariae* var. *rubi*, the causal agent of raspberry root rot. Phytoprotection 78:43–51

Toyota K, Ritz K, Young IM (1996) Microbiological factors affecting the colonisation of soil aggregates by *Fusarium oxysporum* f. sp. *raphani*. Soil Biol Biochem 28:1513–1521

Tronsmo A, Dennis C (1978) Effect of temperature on antagonistic properties of Trichoderma species. Trans Br Mycol Soc 71:469–474

van Cleemput O, El-Sebay AS, Baert L (1983) Evolution of gaseous hydrocarbons from soil: effect of moisture content and nitrate level. Soil Biol Biochem 15:519–524

Wheatley RE (2002) The consequences of volatile organic mediated bacterial and fungal interactions. Antonie Van Leeuwenhoek 81:357–364

Wheatley RE, Millar SE, Griffiths DW (1996) The production of volatile organic compounds during nitrogen transformations in soils. Plant Soil 181:161–163

Wheatley RE, Hackett C, Bruce A, Kundzewicz A (1997) Effect of substrate composition on production and inhibitory activity against wood decay fungi of volatile organic compounds from *Trichoderma* spp. Int Biodeterior Biodegradation 39:199–205

Wheatley RE, Ritz K, Crabb D, Caul S (2001) Temporal variations in potential nitrification dynamics related to differences in rates and types of carbon inputs. Soil Biol Biochem 33:2135–2144

Wheatley RE, Caul S, Ritz K, Daniell T, Crabb D, Griffiths BS (2003) Microbial population dynamics related to temporal variations in nitrification function in a field soil. Eur J Soil Sci 54:707–714

Zechman JM, Labows JN (1985) Volatiles of *Pseudomonas aeuroginosa* and related species by automated headspace concentration-gas chromatography. Can J Microbiol 31:232–237

Index

Printing: Krips bv, Meppel, The Netherlands
Binding: Stürtz, Würzburg, Germany